24-Hour Cities

24-Hour Cities is the first full-length book about America's cities that never sleep. Over the last 50 years, the nation's top live–work–play cities have proven themselves more than just vibrant urban environments for the elite. They are attracting a cross-section of the population from across the US and are preferred destinations for immigrants of all income strata. This is creating a virtuous circle wherein economic growth enhances property values, stronger real estate markets sustain more reliable tax bases, and solid municipal revenues pay for better services that further attract businesses and talented individuals.

Yet, just a generation ago, cities such as New York, Boston, Washington, San Francisco, and Miami were broke (financially and physically), scarred by violence, and prime examples of urban dysfunction. How did the turnaround happen? And why are other cities still stuck with the hollow downtowns and sprawling suburbs that make for a 9-to-5 urban configuration? Hugh F. Kelly's cross-disciplinary research identifies the ingredients of success, and the recipe that puts them together.

Hugh F. Kelly is a Clinical Professor of Real Estate at the NYU Schack Institute of Real Estate. He served as the 2014 Chair of the Board for the Counselors of Real Estate organization and heads his own consultancy practice. He frequently speaks at international academic and industry conferences.

'Hugh Kelly's *24-Hour Cities* is a must-read for today's real estate investors. It overflows with insights derived from the author's long career as a leading commercial real estate researcher gifted with a keen sense of the wider social sciences. Hugh Kelly is an astute analyst of the forces that make cities vibrant; he artfully reminds us that successful cities are where people want to live, work, and play.' *Martha S. Peyton, Managing Director, TIAA-CREF*

'Hugh Kelly proves that a well written, insightful economics book is not an oxymoron. In *24-Hour Cities* Kelly also validates an assertion made two decades ago and he connects today's discussion on the future and shape of cities to its historical evolution. This is a must-read for urbanists and those who wish to be ahead of the curve.' *Raymond G. Torto, Lecturer, Harvard Graduate School of Design*

'Five-hundred years ago, in 1516, Sir Thomas More's *Utopia* described the ideal city. We have not created it yet, and it is unlikely we will do so soon. In his formidable study *24-Hour Cities*, Hugh Kelly analyzes the evolution of American cities since World War II and discusses the current leading models: the 24-hour city, the 9-to-5 city, the large market city. There can be no one model, no one ideal city, since socioeconomic factors and public preferences differ. The fourteen cities Kelly examines are not problem-free. He shows how the best excel at solving problems and meeting challenges – of business opportunities, culture, livability and innovation. They optimize human capital, financial capital, physical capital and social capital. *24-Hour Cities* thoughtfully reviews the plus and minus factors that attract or repel various city occupants – the young, creative, entrepreneurial; the older, established, conventional; those seeking lowest cost vs. those willing to address higher costs for the benefits received. It will be a leading urban affairs text of 2016.' *Daniel Rose, Chairman, Rose Associates, Inc.*

'Hugh Kelly's book *24-Hour Cities* is a remarkable achievement. It puts empirical flesh on the bones of the 24-hour-city hypothesis. Of the fourteen largest commercial real estate metro areas, he names and analyzes the seven that are 24-hour cities and the seven that are not. There is nothing fuzzy or vague about why some are and some are not. This is an important book for real estate analysts as well as urban scholars.' *Matthew Drennan, Visiting Professor of Urban Planning, UCLA*

'When RERC and Equitable Real Estate coined the term "24-hour cities" in the 1995 issue of Emerging Trends in Real Estate, we envisioned these markets – the cities where Americans could best live, work and play – as the places that would lead the fledgling commercial real estate recovery at that time. This concept was originated through interviews for Emerging Trends in Real Estate, which I was part of as president of RERC. As president of Situs RERC today, I am pleased to state that with this book, Hugh Kelly has truly embraced the concept of 24-hour cities and has taken it to another level by investigating what a 24-hour city currently means (versus 9-to-5 marketplaces). Based on his case studies and qualitative research, Kelly disputes and refutes old claims, examines socioeconomic and demographic trends, and artfully describes how 24-hour cities have evolved while competitively attracting physical, financial, and human capital. I congratulate him on this outstanding portrayal of what we only began to conceive of 20-some years ago.' *Ken Riggs, President, Situs RERC*

'Cities are an increasingly hot topic. Unfortunately, much of the commentary in recent books on cities is anecdotal or agenda driven. Therefore, Kelly's original research and focus on 24-hour cities is most timely. Kelly is uniquely qualified to have written this seminal work, being actively engaged in both the private sector and academia. He is respected internationally. *24-Hour Cities* is topical, fact driven and highly readable. It will set a new standard in our understanding of the modern city state.' *William McCarthy, CRE*

24-Hour Cities

Real investment performance, not just promises

Hugh F. Kelly

First published 2016
by Routledge
2 Park Square, Milton Park, Abingdon, Oxon OX14 4RN

and by Routledge
711 Third Avenue, New York, NY 10017

Routledge is an imprint of the Taylor & Francis Group, an informa business

British Library Cataloguing-in-Publication Data
A catalogue record for this book is available from the British Library

Library of Congress Cataloging-in-Publication Data
A catalog record for this book has been requested

ISBN: 978-1-138-65317-7 (hbk)
ISBN: 978-1-138-80511-8 (pbk)
ISBN: 978-1-315-75249-5 (ebk)

Typeset in Galliard
by Florence Production Ltd, Stoodleigh, Devon, UK

Printed and bound in Great Britain by
TJ International Ltd, Padstow, Cornwall

To my family: my wife Betty Duggan, and children Beth, Neil, Joanna, and Luke Kelly, who have enriched my life in the midst of the live–work–play environment of Brooklyn, New York.

Contents

Foreword

Richard Florida

"Deep in the night, I am almost unaware how many people are on the street unless something calls them together, like the bagpipe," the late great Jane Jacobs wrote, in *The Death and Life of Great American Cities.* "Who the piper was and why he favored our street I have no idea. The bagpipe just skirled out in the February night," she continued. "Swiftly, quietly, almost magically a little crowd was there . . . when he finished and vanished, the dancers and watchers applauded, and applause came from the galleries too, half a dozen of the hundred windows on Hudson Street."

The very words "24-hour city" call up the paintings of Edward Hopper, the neon-drenched cinematography of James Wong Howe, or maybe a sad line from a blues song, such as Jimmy Reed's "bright lights, a big city, they went to my baby's head." The images are of sad men in suits and snap brim hats hunched over coffee counters; sloe-eyed cigarette girls in smoke-filled nightclubs; a stack of early-morning papers being tossed off a truck onto a rain-slicked sidewalk while the L thunders overhead. Sexy or sad, dreamy or dangerous, in almost every case, they're tinged with nostalgia – they're windows into an urban nightscape that no longer exists. But in fact it still does.

When most people call New York the city that doesn't sleep, what they mean is that it is a great city to party and have fun in. But the truth is, its "24-hour-ness" (and the 24-hour-ness of many other great cities) is correlated, not just with drinking and crime and the licit and illicit liaisons that go along with them, but also with robust economic growth, high levels of innovation, the ability to attract talent, and even the increased safety that comes from having more people on the street at all times of day and night.

Almost all of our understanding of cities and of urbanism and urban economic development is based on the city of the day – the city of work and industry and commuters and all of that. But in truth, you can't understand cities and urbanism in the modern talent-driven creative economy without understanding what I like to call the city at night. The city at night is a place where creative innovators mix and mingle and network to build their companies and careers – not just to blow off their proverbial heads of steam, although they do that too. The nocturnal activities of great cities are important large-scale contributors to their economic output and their lights are signal indicators of economic activity and progress. I have used satellite images of them to identify the globe's great economic centers and to show that, instead of the world being flat, as Tom Friedman alleges, it is concentrated, clustered, and spiky.

Hugh F. Kelly has done the research and provides the textbook on this new and more productive kind of 24-hour city. He has dug into not just New York but what he calls "24-hour-ness" in Boston, Chicago, Miami, Washington, DC, and San Francisco. And, he's developed a new set of empirical metrics to better understand it, based on patterns of electricity consumption, population density, the volume of vehicular traffic between 9:00 p.m. and 5:00 a.m., the share of commuters who use public transportation, the number of 24-hour drugstores within ten miles of the city center, the ratings of restaurants and ethnic restaurants, nightlife, and culture, and more.

The bottom line? 24-hour places are not only statistically different than sprawling 9–5 cities but have statistically superior economic performance.

24-Hour Cities does much more than prove a point that no longer needs any special pleading – that the pendulum is swinging back from car-dependent sprawl and single uses to mixed uses and density. Kelly gives us the required historical perspective, and brings the economic data vividly to life. Looking through one end of the glass, he traces the macroscopic processes that hollowed out so many great cities in the post-war years, and then brought some (but far from all of them) back to vibrant life in the 21st century. Looking through the other, he captures the texture of their ever-changing streets and neighborhoods in microscopic detail, with an ethnographer's precision and a novelist's flair for the telling phrase.

Kelly's portrait of his own 24-hour neighborhood of Kensington, Brooklyn, shows us how the urban transportation grid (the F-train that accounts for some 3 million trips in and out of the neighborhood every year; the taxis, gypsy cabs, and ubers that travel the nearby Brooklyn Queens Expressway to Manhattan and the airports; the trucks that drive through the neighborhood making local deliveries or on their way to Long Island) is so much more than a conveyor belt carrying people to and from their jobs. Rather, it is a way to disperse the energy of the downtown business district throughout the wider region. More interesting still, he shows how the diversity of Kensington's population – Orthodox Jews, Irish Catholics, Sunni Muslims, South Asians, Eastern Europeans, African Americans, and whites, who are old and young, straight and gay, married and single – is drawn together and held together by the city's and its own cultural, economic, and social vitality.

24-Hour Cities is a must-read for real estate professionals, economic and community developers, city leaders, and urbanites of all stripes. More than that, it's an engrossing read for anyone who is captivated by the speed, energy, and extravagant adaptability of the new age of 24-hour urbanity.

Richard Florida is the Director of the Martin Prosperity Institute at the University of Toronto's Rotman School of Management, Global Research Professor at New York University, and the co-founder and editor-at-large of *The Atlantic*'s CityLab. He is the author of the best-selling *The Rise of the Creative Class.*

Acknowledgments

It all begins with family, doesn't it? My grandparents, John and Catherine Hayes, made room for my parents and the first three of my generation in their own second-floor apartment on Fenimore Street, Brooklyn. Grandpa would rock me in the living room, reading stories or spontaneously rhyming verses, creating from the start a love of words and of learning. Later, the tables would be reversed, as they came to living in our basement in East Flatbush. The concept of a multigenerational household came into my life well before the term "boomerang child" entered our social vocabulary.

My parents taught me the work ethic and an ethic of openness and service. My father, John Kelly, was a bus driver, a loader of trucks, a bartender, a blue-collar guy who took pride in putting bread on the table. My mother, Margaret Kelly, gave living testimony to the virtues of love, neighborliness, and tolerance in a mixed-ethnic community. Many were the scraped knees and bloody elbows she bandaged from the motley group of friends her four sons hung out with.

I have been blessed by many great teachers, from kindergarten to graduate school. None opened my horizons more than Father James McMahon and Father Robert Lauder at Cathedral College, Douglaston, Queens, and Hannah Arendt and Reiner Schurmann at the New School for Social Research.

Before embarking on my real estate career, I worked intensively with many great priests, nuns, and laity in Brooklyn's inner-city parishes: Fathers Ray Gruhn, Jack Waldren, and Mike Breslin stand out, as do Sisters Elizabeth Folles, Eileen Payne, Terry Agliardi, and Ginny Hall. Add to these my close friends and colleagues Gene Tully, Andy Jordan, and Neil Griffin, and Michael Gecan of the Industrial Areas Foundation, so instrumental in organizing the 50 churches of East Brooklyn Congregations and developing the Nehemiah Program that has built more than 3,500 home in the poorest areas of Brownsville, New Lots, and the environs since the mid-1980s.

John R. White and Edgar B. Madsen gave me the opportunity to learn about real estate, cities, and the built environment, by taking a chance on me at Landauer Associates in 1978. Ken Patton and Rosemary Scanlon, my Deans at New York University's Schack Real Estate Institute, encouraged me to develop as a real estate economist in all ways and were particularly supportive as I plunged into research on 24-hour cities after 1999.

Dr. Raymond Torto generously allowed me access to the CBRE Econometrics' database for historical rents, vacancies absorption, construction, and other data for metropolitan markets and submarkets. Likewise, Robert White provided me with access to the Real Capital Analytics database on commercial property transactions, prices, capitalization rates, and aggregate sales volumes. Through an academic membership in NCREIF, I was able to use its Custom Query Facility to generate model portfolios for sets of cities and metropolitan areas, enabling the comparison of cumulative total returns on investment presented in this book. All real estate researchers benefit from the enormous work involved in developing and maintaining these databases and, I am sure, share in my gratitude that they provide a foundation for analysis and further investigation.

Stephen Roulac facilitated my introduction to the University of Ulster, in Northern Ireland. Stephen, Alastair Adair, and Stanley McGreal were my dissertation advisors as I put together the data and analysis that became the thesis upon which this book is based. They encouraged me as I completed the PhD degree at the age of 63 – making me the epitome of the "slow learner."

My daughters, Beth and Joanna, assisted in a couple of essential (and tedious) tasks. Beth compiled and arrayed the diurnal traffic counts, city by city, upon which the analysis in Chapter 8 is based. And Joanna took upon herself the task of searching the text for city references to be included in the Index. They spared me hours in taking this work upon themselves.

Coming full circle, my wife Betty Duggan, my grown kids and their significant others were there to watch me "walk" as I received my PhD at the University of Ulster, in Belfast, Northern Ireland, on July 4, 2012. I have been blessed, and I am grateful.

A primary question

What is a 24-hour city?

This book examines an assertion articulated at least two decades ago. In the 1995 edition of *Emerging Trends in Real Estate*, it was stated:

> For the future, we believe the premier investment opportunities will be available in the nation's "24-hour cities." These markets, urban or suburban, are places where people can comfortably and securely live, work, and shop. In contrast, "9-to-5" markets – those with weak residential fundamentals – have poor investment prospects.

As a working descriptive definition, the 24-hour city was considered to have several recognizable attributes: "attractive residential neighborhoods proximate to or integrated with the central commercial district; convenient shopping opportunities close to the workplace; a safe and secure environment; excellent mass transportation; and recreational, cultural, and environmental amenities."

Obviously, these definitions are qualitative and not a little subjective. But, as there was a claim that investment performance – a measurable indicator – would follow from a city's 24-hour status (or lack thereof), there has been a pressing need to develop similarly measurable criteria for judging a place's 24-hour status. Moreover, cities have proven very dynamic over time, under the influence of many forms of change. It is important to see how some cities have evolved their 24-hour character during the course of decades, while other cities have adopted a different urban form.

This book takes on that task, as a first step in examining the built environment in its interactions with social trends and investment objectives.

Introduction

The 24-hour-city hypothesis

The awning at my neighborhood greengrocer reads: "Open 24 hours. Golden Farms International Grocery. Russian, Ukrainian, Polish, Israeli, Turkish. Kosher. Organic. Gourmet Foods." The owner is Chinese. The cashiers and stock clerks are Mexican. My neighborhood is called Kensington, and it is in Brooklyn, New York City. Many have the impression that 24-hour cities are all about bright-lights downtowns, office skyscrapers, and high-end condominiums – a city for the elite. But, nestled within the vibrant 24-hour cities are small communities like mine: connected to the core, but amazingly vital in their own right.

Across the street from Golden Farms, along Church Avenue, are two sit-down restaurants, one Chinese and the other Japanese. On the far corner is a Rite-Aid drugstore. Along the block you can find Korean-staffed nail salons, a Palestinian-owned convenience store most residents call a "bodega," and a "sell everything store" that has a merchandise line running from air conditioners to cosmetics. A Russian gentleman operates a shoe-repair storefront, also retailing footwear. Further along are a pizzeria, a florist, a Carvel soft ice cream franchise, a 99-cent store, and a Halal butcher.

Nearby are a Bangladeshi grocery and an accountant/notary public office that also does insurance brokerage, run by a low-key but competent Southeast Asian immigrant. A T-Mobile retailer occupies a double-wide storefront, next to another 99-cent store. Another small greengrocer, named "New McDonald's Farm," faces the phone store. Answering the presence of Rite Aid, Walgreens Pharmacy recently planted its flag by purchasing a venerable neighborhood drugstore, Silverod's. A deli that is also a bagel bakery shares the street frontage. There is a candy store that sells lottery tickets, magazines, newspapers in several languages, and convenience items, and that is next to a branch of the Astoria Federal Savings Bank. Opposite the bank is the neighborhood gin mill, Denny's, which has been around as long as anyone can remember – the last of the Irish bars in this community.

Did I mention the 24-hour vegetarian Halal food cart near the subway entrance?

In the heart of darkest Brooklyn . . . but incredibly connected

On all four corners of Church and McDonald Avenues stairs descend to the F and G lines of the New York subway system. The F train works its way through the more upscale neighborhoods of Park Slope and Carroll Gardens before traversing downtown Brooklyn on its way to Manhattan's Greenwich Village and up Manhattan's Sixth Avenue. The G branches off – the only New York City subway line that never enters Manhattan – into the newly trendy areas of Fort Greene, Clinton Hill, and Williamsburg, and on into Queens.

Over the more than 30 years my family has lived in Kensington, the names on the storefronts and the businesses within have changed repeatedly. But the retail space has rarely sat empty for long. This is a community of great variety and unrelenting movement. It is alive, dynamically so.

The Church Avenue F and G station had a turnstile count of 3,201,585 in 2014. That makes it fairly modest in the New York City subway system, ranking only 167th among the 468 stations along the MTA's 660 miles of passenger track. Still, the tide of people averages more 10,000 passengers at Church Avenue each working day – a lot of foot traffic for the Kensington retailers.

The subway is not the only means of transportation in the neighborhood. The B35, B67, and B69 buses also service the area round the clock, intersecting at that same Church and McDonald Avenue crossing. The B35 ranks as New York's eighth busiest bus route, serving 10,735,351 passengers in 2014.[1]

Several 24-hour car services are near this intersection as well, licensed for pre-arranged trips locally and to the New York area airports. Though scores of New York City "medallion" yellow taxicab drivers live in Kensington, it is virtually impossible to hail the iconic New York taxi anywhere in the neighborhood.[2] The cabbies take advantage of the easy access to the Prospect Expressway Airport (about two blocks east of Golden Farms) leading to Manhattan and LaGuardia. Or, heading southbound on Ocean Parkway, they can be at JFK Airport in about 30 minutes, outside of rush hours.

The B35 runs from the Brooklyn waterfront at Sunset Park variously through that largely Hispanic community and skirting the orthodox Jewish neighborhood of Borough Park before entering Kensington. Then it proceeds up Church Avenue to the now-Caribbean neighborhood of East Flatbush (where I grew up in the 1950s and 1960s) and terminates at Mother Gaston Boulevard in impoverished Brownsville/East New York. Along the way, this bus line passes close by two hospital complexes – the Kings County Hospital and State University Downstate Medical Center, and the Brookdale Medical Center. Besides being teaching hospitals, these are among New York City's major trauma centers for accident victims and for gunshot wounds. Their round-the-clock operations fill the seats of the B35 through the nighttime hours.

The B67 and B69 share much of their route through the still largely Irish/ Italian Windsor Terrace area before passing through the thoroughly gentrified

"Brownstone Brooklyn" neighborhood of Park Slope, with its boutiques, restaurants, and landmark Methodist Hospital. Along 7th Avenue, in "the Slope," you can see a procession of steeples and spires that once earned Brooklyn the nickname "City of Churches." The buses pass by the brand new Barclays Center at Atlantic Yards, home of the NBA Brooklyn Nets, at the Flatbush Avenue terminus of the Long Island Railroad. They terminate in DUMBO, after running through Brooklyn's downtown judicial and business district. DUMBO stands for the District Under the Manhattan Bridge Overpass (actually it is between the Brooklyn Bridge and the Manhattan Bridge) and is a former warehousing district, now converted to upscale housing with spectacular views of the Lower Manhattan skyline.

Who's who in the 'hood?

Let's get beyond the ethnic labels and look at the people at little more closely.

My next-door neighbors to the north are Palestinian Arabs. There are four families that share a large house, and over the years the residents have frequently changed. It was once explained to me that the word "Arab" means "nomad," and in this case it has proved true that occupancy here has been transient. When we moved to this block in the mid-1980s, a gentleman named Benyamin was the familial leader, and there were slews of grandchildren. Many of the boys went away to the State University at Buffalo and studied engineering. When girls were married from the house, the women of the families would gather in preparation, and into the evening you could hear the ululating of Arabic chants and the rhythms of Middle Eastern music.

Many of the customs of the Levant were carried over, in a particularly Brooklyn sort of way. In the evenings, once the season turned warm enough, the families would gather in the little concrete front yard, arrayed in a circle, and have coffee and pastries until midnight or so. The men were shopkeepers, for the most part, and for a time one owned the Church Avenue bodega I've already mentioned. As a neighborly gesture, he would round down the prices for me when I dropped by for milk, or bread, or a bag of ice. Others owned furniture stores on nearby Coney Island Avenue.

In typical immigrant fashion, Benyamin and that first group of families eventually moved on to more suburban homes in Long Island or Staten Island. But the informal network of Palestinian families quickly found others to move into the vacated apartments, and the customs continued. The bodega is now owned by another Palestinian, from elsewhere in the neighborhood, and he has renamed it "Mother's Grocery." Mother, however, is never seen to work there.

After the 9/11 tragedy, I asked these neighbors what the impact on them had been. Sadly, they said that the children in school were suffering some ostracism. One of the men remarked to me, "This is what we came to America to get away from." Curious, I asked where exactly they had come from in Palestine. He answered, "Jerusalem."

Directly across the street from our house, to the east, are the Yonas. The Yonas are strictly observant Orthodox Sephardic Jews. I am their "Shabbos goy," a venerable Brooklyn role. It happens that one of the trash collection days on our block is Saturday, and the Yonas perform no work on the Sabbath, from sundown Friday until sundown Saturday. So I bring in their trash barrels from the curb into their yard, technically an extension of their house by Jewish law (marked by a wire called an *eruv*). This little service takes no more than a minute a week, but has been the basis of a very friendly connection with the Yonas.

Like our Palestinian neighbors, Levi and Ahlai Yona have many children. Ten, in fact. The Jewish kids are unfailingly polite and even sunny. Their dispositions and smiles brighten the austerity of their somber garb. On Purim and Hannukah, the kids bring over sweets to our house. We can't reciprocate though, as the Yona family is absolutely and strictly kosher. So our presents tend to be flowering plants and the like.

When the Yonas' oldest son got married, my wife Betty and I had the extraordinary honor of being invited to the wedding – the only non-Jews there. It was, by the way, an arranged marriage. Mrs. Yona went out of her way to explain that, though it would seem strange to those outside the orthodox community, the marriage customs – including the prenuptial period and the early months of marriage – have evolved to strengthen the marriage bond. Divorce, she noted, is far less frequent among the Sephardic families of her sect than in the general American population.

At the wedding I naturally was required to remain with the men – including the groom – who spent an hour in "davening" in Hebrew scripture reading and prayer on one floor of the wedding hall. Eventually, the men all accompanied the groom to meet the bride and all the women, and everyone witnessed the signing of the *katubbah*, or marriage contract. The men all sat on the left side of the aisle for the wedding ceremony itself, as the women sat on the right. Thought I, "We Catholics do call one side the groom's side and the other the bride's side in our church weddings, so this is just another logical step." But it was still a bit of a shock to realize that the "circle dancing" after the ceremony also followed the rule, "Men with men; women with women."

Both Mr. and Mrs. Yona are teachers, and it was a particular delight to see Mr. Yona's students of elementary school age enjoying the wedding festivities. I'd never thought of him as a rabbi, but of course that's what he is professionally – a teacher, more than a leader of religious ceremony.

Our next-door neighbors to the south are Tom Murphy and Lisa Frankel. Tom teaches literature at St. John's University in Queens, and Lisa is a camera operator for a cable television network in Manhattan. Both have children from prior marriages, and so we found ourselves with college-aged kids next door as well. With Tom, there are now at least three PhDs on the block (Don and Debby Marshall also moved in about the same time our family did. Don teaches English at New York University (NYU), where I teach in the Schack Real Estate Institute's masters degree program. Debby is a librarian at Hunter College of the City University of New York.) Betty and I share interests in gardening and backyard barbequing with many of these neighbors.

It is an interesting mix of ethnicity, occupation, religion, and interests for such a small sample, a cluster of five houses. But it is not atypical of the block or of the neighborhood as a whole.

The Kensington ZIP code (11218) has approximately 77,000 residents, living in an area of just 1.43 square miles, or about 53,700 persons per square mile.[3] That's exceptional urban density – the average population density in the US is 88 persons per square mile, and the 50 largest cities average about 4,250 persons per square mile. Kensington residents have more than 12 times that density. In fact, 11218 ranks 89th out of about 44,000 ZIP codes in total population, and 49th in population density.

The transportation infrastructure is a defining characteristic of Kensington as well. Most (62.3 percent) Kensington residents use public transportation to commute to work. An additional 9.3 percent walk to work, and 3.8 percent work from home. This is a far cry from the US norm,[4] which sees about 88 percent of workers using car, truck, or van to get to work, mostly in vehicles driven alone. The national average for public transit commutation is 4.7 percent, with 2.9 percent walking to work and 3.3 percent working from home.

The median age of the Kensington population is 32.4 years, younger than typical for the nation as a whole, which has a median age of 36.8 years. About 44.3 percent of Kensington residents are in the prime earning years of 25–54 years of age, above the national norm of 41.2 percent in that age range. Median household income is $50,938, 1.9 percent lower than the US standard. But the density of the population concentrates that income, and annual earnings in the ZIP code exceed $1.8 billion. That goes a long way to explaining the vitality of the Church Avenue shopping district.

More than 30,000 Kensington residents were born outside the United States, and 52.2 percent of them have become US citizens. The 2010 Census counted 27 percent of the population as foreign-born and non-citizen. History indicates that most of those non-citizens will take the citizen oath in the decade ahead. Many will take advantage of the two branches of the Brooklyn Public Library in ZIP code 11218.[5]

Of the foreign born, 9,419 have come from Eastern Europe, the largest single group of immigrants. South-of-the-border immigrants are nearly as numerous, at 9,264 residents: 3,240 from Mexico, and 6,024 from "Caribbean or other Central American countries." Next are 5,948 "other Asian" immigrants – that is, Asians from countries other than China, Japan, Korea, Taiwan, or Hong Kong. Those five nations account for 2,927 Kensington residents. Pakistan, by itself, is the birthplace of 2,351 people in the Kensington neighborhood.

Why does all that matter?

I start with a description of the Kensington neighborhood, its attributes, and connections for several reasons.

In the first place, Kensington offers a case in point illustrating how "24-hour cities" are much more than their heralded "bright lights, big city" downtowns. The idea that so-called "superstar cities" are no more than playgrounds for the rich elite is a broad-brush and superficial criticism. Second, Kensington shows how the idea of residential neighborhoods connecting to places of work are not simply walk-to-work options for the 24-hour downtown, provided that the residential communities themselves partake of the city's 24-hour character. Third, this neighborhood is an excellent reminder that transportation is not simply a mechanism to funnel workers into a central business district, but truly a two-way connection that communicates the energy of the vibrant downtown into a wider region. In the fourth place, Kensington and similar communities show how successful cities refuse to be fixed in time, but are adaptable to change across many dimensions. And, fifth, the diversity of population in Kensington exemplifies the bubbling up of "creative class" innovation, as people of varying ethnicities, religions, sexual orientations, and other distinctive characteristics interact – as all are drawn by the opportunities arising from the economic, social, and cultural vibrancy of the city's 24-hour profile.

There are just a few basic thoughts I seek to pass on to my students each year, regardless of the title of the course they have enrolled in. One was the pithy introduction given to me by my first boss in the real estate industry, on the day I was brought on staff. "Hugh," he said, "you don't have the background in business we usually look for, but you'll do fine. Just remember two kinds of people you'll be meeting in real estate: dreamers and liars. Watch out for them both." The second is, "It is very hard to kill a big city. Cities can stop growing. They can become sick. They can suffer terribly. But death comes only rarely – though it does happen."

The most critical idea, though, is not about real estate or even about cities. It is about research itself, namely, "There is no point in doing research unless you are willing to be surprised."

Conceptions and misconceptions about cities that never sleep

We have all met people about whom we can say, "He might be right and he might be wrong, but he is never uncertain." Futurists such as *Megatrends* author John Naisbitt are nothing but confident in their pronouncements about what lies ahead for us. Big-picture ideas are, by necessity, couched in broad-gauged assertions. This is helpful to the pundits, as they can subsequently select an array of examples that "prove" the correctness of the assertions and dismiss contrary evidence as "exceptions that prove the rule."[6]

Economies now run on ideas, not goods

The first edition of *Megatrends* (1988), for instance, talked of the evolution of an information society replacing the industrial society of the previous two centuries. We have unquestionably seen an explosion of information, though how "tweets" are superior to Montaigne's 16th-century essays is difficult to see. But the idea that we live in a "post-industrial world" certainly may be questioned on many fronts. No knowledgeable person would be surprised to find that industrial production in China and India has been growing by leaps and bounds. China's goods production grew by an average 13.3 percent annually between 1990 and 2012. India's industrial production between 1994 and 2012 has risen 7.2 percent per year on average. South Korea's annual growth over those 18 years has averaged 14.1 percent. That's a lot of "stuff" – material goods produced for world markets. Emerging economies have to become industrial before they can be post-industrial.[7]

But developed economies have accelerated their industrial output since 1994 as well. Even with the ups and downs of cycles, including the Great Recession of the last decade, the US, Canada, Australia, Germany, the Netherlands, and other advanced economies had higher industrial production index levels in 2012 than in 1994. To take another measure, the merchandise trade volume of the US – imports plus exports – was $660 billion in 1988, when *Megatrends* hit the best-seller list. In 2012, merchandise trade for the US was nearly $3.9 trillion, almost a sixfold increase in the amount of "stuff" flowing through the American trade, transportation, and distribution system.

We are still living in a material world, and those who might doubt that are invited to go to our top "services-oriented" cities and count the number of trucks clogging the streets and highways.

The future belongs to the "new"

When the term "24-hour city" was introduced by *Emerging Trends in Real Estate* in 1994, here was the conventional wisdom about urban America:

> Edge City is the crucible of America's urban future. Having become the place where the majority of Americans now live, learn, work, shop, play, pray, and die, Edge City will be the forge of the fabled American way of life well into the twenty-first century.[8]

There is no doubt that Americans have a long-standing love affair with novelty. A random word association test would likely show the pervasive influence of advertising's most popular slogan "new . . . and improved." Newer is better, modern, hip, edgy. What's not to like?

Couple that concept with "frontier" and you have a sure-fire winner. Where Franklin D. Roosevelt promised a New Deal, John F. Kennedy set the American imagination soaring with his New Frontier in the 1960s. Kennedy, who was steeped in history, knew that Frederick Jackson Turner had enunciated his "frontier thesis" as long ago as 1893. Turner argued that democracy, individual liberty, self-reliance, and economic innovation were American virtues stimulated by the frontier experience. Joel Garreau's allusion to a "new frontier" in his subtitle was surely no accident, but an appeal to a deep-seated belief in his readers.

The formal US frontier, of course, was already "closed" by the time the 1890 Census of Population was taken. California had been admitted to the Union in 1850, Oregon in 1859, and Washington State in 1889. What remained was "infill," the gradual conversion of areas with territorial status to full statehood. Thus Idaho and Wyoming achieved statehood in 1890, Utah in 1896. Oklahoma followed in 1907, and, finally, New Mexico and Arizona completed the assemblage of 48 contiguous states by 1912. It would take until 1959 before more stars were added to the American flag, with the addition of Alaska and Hawaii.

The fascination with the new and with the frontier, however, never truly ebbed in American discourse. But "new" wasn't always "improved." New Coke, for instance, was remarkable only as a marketing failure. The New Math, introduced to accelerate US science education after Sputnik, was widely derided and was pedagogically dead by the time *Why Johnny Can't Add* was published in 1973.[9] And, over the years, we've had many sightings of the so-called "new economy," which has yet to repeal such fundamental laws as supply and demand, or alter the preference of investors for profitable operations above all other economic attributes.

As for "frontier," that too has retained a popular ring. *Star Trek* reminded us at the beginning of each episode that space constituted "the final frontier." Less elegantly, Kenneth Jackson described the suburbanization of America in *The Crabgrass Frontier* – whose outposts evolved into Edge Cities.[10] To remind us that lawlessness is often a state of being on the edge of civilization, today we have the Electronic Frontier Foundation, "defending your rights in the digital world," suggesting that *cyber*space may be challenging for the role of "final" frontier.[11]

Imagine the puzzle, then, that the so-called 24-hour city might have claims to superiority. Why a puzzle? Contrary to the claims for new urban forms, the 24-hour cities were a set of *old* places: New York, Boston, San Francisco, Chicago, Washington, DC. Moreover, most of these cities had been shaken by urban unrest and had been subject to significant depopulation trends, enervating their economies from the late 1960s into the 1980s. They were dark, dangerous, and putatively ungovernable. They were high-cost, high-tax places, losing competitive advantage to the Sunbelt. Yet *Emerging Trends* not only heralded these older cities, but the cities began justifying that confidence in the late 1990s and in the first decade of the new century.

Why did that happen? And, how?

Claim: "Back to the city" is for the elite, the young, and the single

Joel Kotkin, in a *Wall Street Journal* op-ed piece in 2007, took on what he called the "myth of the superstar cities." Kotkin's targets were the usual suspects: New York, San Francisco, and Boston – the elite cities of (in his characterization) Ferraris, planes, multimillion-dollar condos, and $200 lunches. His contrasting and favored set of cities include Atlanta, Dallas, Phoenix, and other places considered more affordable to middle managers, technicians, and skilled laborers.[12]

Even an urban booster (and top-of-the-line scholar) such as Harvard's Edward Glaeser, author of *The Triumph of the City*,[13] wonders about the divide between rich and poor in the "superstar cities" – a term coined, by the way, by a trio of academics (Joseph Gyourko and Todd Sinai of Wharton, and Christopher Mayer of Columbia).[14] Glaeser, like Kotkin, argues for greater laissez-faire in urban planning, especially in encouraging less-restrictive development regulation. Their approach favors supply-side corrections to perceived market imbalances. Investors, however, appear to prefer markets that are supply-constrained, with dampened development cycles. "Let the market rule" is not as simple a prescription as efficient market theorists believe it to be.

Glaeser, interestingly, proposes The Woodlands, a development near Houston, as an exemplary alternative to the old cities that form the "superstar" cadre and the core of the 24-hour-city roster. The Woodlands won the Urban Land Institute's 1994 special award for outstanding development. Glaeser correctly notes that housing prices at The Woodlands are far less expensive than in the major cities on the East and West Coasts. He spends less attention (actually no attention at all) on the fact that 92.4 percent of the residents of The Woodlands were white, as recorded in the 2000 Census, an amazingly high level of de facto segregation for 21st-century America.

Kotkin, too, extols the preferred lifestyles of his preferred population – smaller cities with low density – claiming that higher growth rates demonstrate that such places win in the "voting with your feet" category. Queried about claims that seniors and the young Gen Y cohort are anchoring a "back to the city" movement, Kotkin asserted, in a 2013 interview, "There is no data to support that."[15]

Even though he vociferously proclaims his allegiance to "the data," I have my doubts. He is a prime example of what Nate Silver describes as a "hedgehog" in his marvelous *The Signal and the Noise: Why So Many Predictions Fail – But Some Don't*.[16] Kotkin selects data to refine his established model of the world, rather than evaluate whether that model needs alteration, updating, or a thorough reconfiguration. Kotkin is very order-seeking and so tends to abide by overly simple governing relationships. And he rarely changes his basic story: any exceptions are merely outliers that prove the orthodox general rule.

Unfortunately for Kotkin's denial about the "back to the city" movement, there is plenty of statistical evidence affirming renewed strength in urban neighborhoods. The issue is, "which cities are we talking about?" Eugenie Birch of the University of Pennsylvania has developed an interesting matrix (which we will see in Chapter 7) clustering cities across the US by their downtown residential patterns. Since the 1980s, 17 major US cities have seen steadily increasing downtown populations. These urban residents fit the pattern of being young, college-educated individuals in creative-class industries – far more so than the "average" city across America and exceptionally more so than the nation's suburbs.[17] Such a picture is far more consonant with Richard Florida's research[18] than with Kotkin's. Moreover, even prior to the housing bubble, downtown residents were tilting toward home ownership, belying the assumption of footloose singles/mingles. My own research (presented in Chapters 7–9) indicates that the 24-hour cities have been assembling the mix of schools, shopping, restaurants, and amenities (social, cultural, recreational) that retain households in the child-rearing years. The old assumption that families will flee to the suburbs once the kids start school sounds more like a look in the rearview mirror than an accurate prospect – for a subset of cities.

I wonder what Joel Kotkin would have to say about the Kensington neighborhood of Brooklyn – as a former Brooklynite himself.

Claim: There is one model for US cities – and it looks like LA

One of the glories of the United States is also one of its most frustrating aspects for analysts: the nation refuses to be pinned down by simple categorization. It is diverse, heterogeneous to a degree that becomes more apparent the longer one looks at it. That's true of people, and it is true of cities.

This book proposes that we look at a taxonomy of American cities, through the lens of history, by means of socioeconomic variables and by the "score-keeping quality" of real estate prices. If there is a genus *Urbs Americanis*, I propose that the 24-hour city and the 9-to-5 cities have emerged as the key species competing for dominance.[19] But this is not a winner-take-all competition. Americans have a wide range of preferences and very particular distinctions in what they are seeking in their life experiences. We vote with our feet and we vote with our wallets. The configuration and the operations of cities are part of this sorting-out process.

This is not to say, though, that, from an economic perspective, we should be indifferent to the choices. If one of the species of American city turns out to be more productive, more energetic, more efficient, more environmentally sustainable, it would be quite valuable to know which species that is.

This is not just a question of counting. It is not enough to see whether there are more cities of Type A than Type B. There is certainly more kudzu growing wild across the US than asters or fleabane, but kudzu is merely prolific, whereas asters and fleabane benefit a wide range of species, from insects to birds to mammals, including rabbits and deer. There is a difference.

Different can just mean "different," but the claim for 24-hour cities is that they are in some way "better." This book examines the evidence for that claim across a variety of disciplines, by statistical evaluation surely, but also by tapping the experience of decision-makers in both the public and the private sectors. The interaction of those two sectors is very much at the heart of the subject and well worth discussing.

Claim: Private enterprise favors small government, low taxes, and low business costs

From the point of view of the private sector, there is a very important principle guiding financial investment. Termed "the principle of anticipation," it teaches that, "value is the present worth of expected future benefits."[20] When you purchase any item, you are paying for the "goodies" it will provide over some future time. For many items, that benefit is very short-run: a nice dinner, an evening at the movies, a weekend getaway. For other items, the benefit comes over time: buying a home, going to college, a diamond engagement ring.

Those who invest in commercial real estate live by the principle of anticipation, the expectation of return on investment. Hundreds of billions of dollars are invested in commercial properties each year. Investors have a full and free choice of where to place their money. Prices paid and amounts committed over time should tell us a lot about the advantages gained up to now in particular cities and also about the sustainability of those advantages.

If, as I will argue, 24-hour cities disproportionately capture this mobile financial capital by the attractiveness of a form of physical capital that is essentially immobile, office buildings, that is significant. If it further turns out that what attracts the financial capital is the presence of a particularly productive agglomeration of human capital, that takes the significance to a vastly higher plane.

Joel Kotkin and Edward Glaeser are right that the so-called 24-hour cities have higher taxes, higher wages, and higher overall business costs. Nevertheless, businesses continue to elect to pay the higher rents in 24-hour downtowns. I believe this is a rational choice, based upon management's evaluation of the comparative productivity and profitability of these urban locations.

I remember a long-ago TV commercial quoting coffee magnate John Arbuckle[21] as saying, "You get what you pay for," as a justification for premium pricing. Arbuckle may not have originated that phrase, but it resonates as an entrepreneurial maxim. Places with a strategy of being "low-cost providers" are engaged in a race to the bottom. Entrepreneurs and established companies alike do not appear inclined to bet on such a strategy as a favorable signal of a city's future. Neither do commercial property investors.

Public policy makers should care deeply about this, because high and sustained valuations for commercial real estate are a potent and extremely reliable source of public finance. If the value of the real estate is driven by productive human capital, cities should think long and hard about what attracts such high-output persons. The productivity of property (its capitalized rents) can enhance the quality of life and the quality of public services for the population as a whole, given appropriate public policy. And this, as it turns out, is not about corporate giveaways, beggar-thy-neighbor economic development efforts, or trickle-down economics. It is about the dynamics that link the bright lights/big city core of Manhattan with my Kensington neighborhood in Brooklyn.

Testing the 24-hour city hypothesis

It is not easy to be a 24-hour city. The set of such cities is quite small. It is hard to put together all the ingredients, and even then an ingredients list does not constitute a gourmet recipe. Many cities proclaim a desire to become 24-hour places. Some will progress toward their goals, but many will be frustrated in their ambitions. That's OK: there is room for a diversity of urban forms in America. Cities, by definition, are in the game for the long haul. Those seemingly disadvantaged now can, perhaps, take some hope from the realization that today's 24-hour cities were largely written off as dinosaurs in the 1970s. Change, transformative change, is certainly possible for cities.

This book attempts to look at that possibility, to examine the claims made for (and against!) 24-hour cities in a serious way, and to sort out the hows and whys that emerge as we consider, not only the differences between cities, but their comparative advantages. As cases in point, 14 large US cities are examined in detail.[22] Furthermore, to make sure that the selection of those cities is not biased or arbitrary, the data for the "test" cities (seven hypothesized as 24-hour markets, and seven as 9-to-5 markets) are compared with the balance of the 25 largest urban areas in the country. And the observations of senior business executives and public officials are taken into consideration, observations elicited in personal interviews conducted from 2009 to 2011, and in a formal survey conducted in 2010.

The first three chapters confront the questions about why cities apparently started to disintegrate in the decades following World War II. Carl Sandberg lauded Chicago as "the city of the big shoulders," but America's largest cities seemed to become pitiful helpless giants in the late 1960s. Downtowns hollowed out, and economic energy sprawled. The tools of city planning appeared counterproductive, and certainly ineffective in managing urban change. Frank Lloyd Wright's 1974 quip that, "The outcome of the cities will depend on the race between the automobile and the elevator, and anyone who bets on the elevator is crazy," nicely captures the informed judgment about cities versus suburbs for most of the late 20th century.

Chapters 4–6 examine a crisis period, roughly the four decades from 1960 to 2000. This was a turbulent time, not just for cities, but for the nation as a whole. Social revolutions and counterrevolutions roiled America. Politics took on an increasingly confrontational attitude, shaping public policies that had nasty urban consequences. Finance – including real estate finance – was radically altered, with enormous consequences for institutional risk. Technology exploded, and serious scholars – as well as popular commentators – spoke about "the death of distance," the assumption that the world was becoming flat and the historical advantages of urban centers were being permanently eroded by telecommunications and the Internet.

Chapters 7–9 outline the surprising rebound of cities, a process I see as very Darwinian in nature. I review the complex set of attributes that shape how city dwellers (and suburbanites) experience daily life – their residential environment, how they get around for work and leisure, what they are concerned about in terms of crime and safety, the opportunities for education, the guilty pleasures of consumption, the whole panoply of options that fall under the rubric of "quality of life." The data, and the observations of interviewees and survey respondents, are indications that, from the mid-1990s onward, both the general public and real estate professionals have been "voting with their feet and voting with their wallets" in some surprising ways. And, like the "red state/blue state" electoral map, the voting patterns point to a very heterogeneous set of preferences. But the empirical measures are pointing to the emergence of the 24-hour cities as the dominant species in the taxonomy of "fittest" American cities.

In Chapter 10, I pull together some observations about how the 24-hour cities have found a recipe that lets urban ingredients present in most places interact in a particularly satisfying way, creating a live–work–play environment that has made them advantageously competitive in attracting physical, financial, and human capital. And, if the history narrated in the earlier chapters is a messy one, I argue that such messiness should be appreciated. No one should think the best cities are problem-free. What the best cities excel at is problem-solving.

There is still a lot to be learned. History has not stopped, and many questions warrant deeper investigation. Is there a difference between "good density" and "bad density"? How do various cities convert resources into economic energy, and do 24-hour cities have higher metabolisms to accomplish this? What role does leadership play, and how do we take the critical step from mere calculation to wise judgment?

I believe we should care deeply about this. It is my hope that this book prods our thinking and our discussion.

Notes

1 Data available at http://web.mta.info/nyct/facts/ffbus.htm#routes for bus, and http://web.mta.info/nyct/facts/ridership/ridership_sub_annual.htm for subway.
2 Partly in recognition of de facto shunning by yellow cabs, New York City created a new class of green "Boro Taxis" in 2013.
3 Data cited for ZIP code 11218 accessed at http://city-data.com/zips/11218.html
4 Data cited for the United States accessed at http://census.gov/topics/population.html
5 For a marvelous introduction to the history and impact of the public library system in the US, see, Robert Dawson, *The Public Library: A Photographic Essay*, Princeton Architectural Press (New York, 2014).
6 John Naisbitt, *Megatrends. Ten New Directions Transforming Our Lives*, Warner Books (New York, 1982).
7 See UNIDO (United Nations Industrial Development Organization), *Industrial Development Report 2013. Sustaining Employment Growth: The Role of Manufacturing and Structural Change* (United Nations, 2013).
8 Joel Garreau, *Edge City: Life on the New Frontier*, Doubleday (New York, 1991).
9 Morris Kline, *Why Johnny Can't Add: The Failure of the New Math*, Random House (New York, 1973).

10 Kenneth Jackson, *The Crabgrass Frontier: The Suburbanization of the United States*, Oxford University Press (New York, 1985).
11 Accessible at www.eff.org/about
12 Joel Kotkin, "The Myth of 'Superstar Cities,'" *Wall Street Journal*, February 13, 2007, p. A 25 (op-ed).
13 Edward L. Glaeser, *The Triumph of the City: How our Greatest Invention Makes Us Richer, Smarter, Greener, Healthier, and Happier*, Penguin Press (New York, 2011).
14 Joseph Gyourko, Christopher Mayer, and Todd Sinai, "Superstar Cities: Why Do Housing Prices Rise Faster in Some Cities?", NBER Working Paper 12355, National Bureau of Economic Research (Cambridge, MA, 2006).
15 See http://livability.com/blog/community/joel-kotkin-livability-qa
16 Nate Silver, *The Signal and the Noise: Why So Many Predictions Fail – But Some Don't*, Penguin Press (New York, 2012). Silver credits Isaiah Berlin and Philip E. Tetlock with the "hedgehog versus foxes" distinction, and note that "foxes . . . tend to be more tolerant of nuance, uncertainty, complexity, and dissenting opinion." These are excellent traits for serious researchers, but not for one-theme ideologues.
17 Eugene Birch, "Who Lives Downtown?" Living Cities Census Series, The Brookings Institution (Washington, DC, 2005).
18 See Richard Florida, *The Rise of the Creative Class*, Basic Books (New York, 2002), *Cities and the Creative Class*, Routledge (New York and London, 2005), and *Who's Your City?* Basic Books (New York, 2009).
19 For the purposes of the analysis in this book, I posit a "bright line" distinction between the two groups of cities and test that in the quantitative study in Chapter 7–9. However, it is apparent to me that cities are arrayed along a spectrum, and are always evolving. With Dr. Emil Malizia of the University of North Carolina, Chapel Hill, I have been exploring urban vibrancy as a phenomenon across many US cities of varying size and believe that an intervening category – the 18-hour city – also deserves attention.
20 See www.allbusiness.com/barrons_dictionary/dictionary-anticipation-principle-of-4962107-1.html. But we might prefer the more capacious understanding expressed by Edward Alsworth Ross:

> Any established or known policy, whether of government, or an association, or of an individual which affects people favorably or unfavorably according to their conduct, will come to be anticipated and will result in modifying behavior. A favorable reaction will call forth more of the conduct, condition, or type of character favored, while an adverse reaction will tend to repress it.
>
> (*American Journal of Sociology, 21:5* (March 1916), University of Chicago Press)

21 See www.brooklynwaterfronthistory.org/story/where-coffee-was-king/ for more about this "King of Coffee," a century before Starbucks.
22 The "sets" of cities are as follows: 24-hour cities – Boston, Chicago, Las Vegas, Miami, New York, San Francisco, Washington, DC; 9-to-5 cities – Atlanta, Dallas, Los Angeles, Minneapolis, Philadelphia, Phoenix, Seattle; balance of large metros – Baltimore, Cincinnati, Cleveland, Denver, Detroit, Houston, Portland (OR), Riverside (CA), St. Louis, Sacramento, San Diego, Tampa.

1 The dynamism of the American city

Triumphs and troubles

Love them or hate them, people care about cities.

Boosters talk about "superstar cities" that are magnets for the creative class.[1] They note that civilization itself is rooted both etymologically and historically in the city. Urbanity is a much desired attribute in an individual, denoting culture, education, and style. Theodore Parker, a Massachusetts divine of the mid-19th century, said that cities "have always been the fireplace of civilization, whence light and heat radiated out into the dark."[2] Kevin Lynch, in *The Image of the City*, observed, "Looking at cities can give a special pleasure, however commonplace the sight may be."[3] Such laudatory sentiments about cities date at least as far back as Sophocles, who wrote, "The highest achievements of man are language and wind-swift thought, and the city-dwelling habits."[4]

Bosh and bunk, the detractors retort. Cities attract and concentrate all that is worst in human experience. They are the home of degeneracy, crime, and corruption, the locus of disease and discontent. "When [Americans] get piled up upon one another in large cities as in Europe," predicted Thomas Jefferson, "they will become corrupt as in Europe."[5] Even so urbane an individual as Brooks Atkinson, the notable mid-20th century *New York Times* theatre critic, wrote, "All cities are superb at night because their hideous corners are devoured by darkness."[6] Those looking to find examples of dysfunction in urban America do not have to look hard: Detroit; Camden, NJ; Gary, IN; East St. Louis, IL. But, even in apparently prosperous cities, we find dangerous slums: East Brooklyn and South Jamaica in New York City; Houston's south central Third Ward; the Ashburn, Woodlawn, and South Lawndale neighborhoods in Chicago; areas between downtown Atlanta and Hartsfield-Jackson airport.

When William Julius Wilson noted that, "in 1959 less than one-third of the poverty population in the United States lived in metropolitan central cities; by 1991 the central cities included close to half of the nation's poor," the statement resonated with Americans.[7] By 2012, the numbers looked like this: 46.5 million Americans lived below the official poverty line: 38 million of them lived in the nation's metropolitan areas. Whereas rural poverty gripped 8.5 million people, 19.1 million poor resided in the nation's principal cities, and 18.1 million poor were in the suburbs surrounding those cities.[8]

Even so ardent a supporter of cities as Benjamin R. Barber acknowledges that we must "examine the pitfalls of the city . . . taking into account urban injustice, inequality, and corruption." Barber maintains that cities exacerbate "many of modernity's most troubling features," and that they do so "in every domain, from education, transportation, and housing to sustainability and access to jobs."[9] As he celebrates "the triumph of the city," Edward Glaeser worries that the city's great strength, its density:

> makes it easier to exchange ideas or goods but also easier to exchange bacteria or purloin a purse. All of the world's older cities have suffered the great scourges of urban life: disease, crime, congestion. The fight against these ills has never been won by passively accepting the way things are or by mindlessly relying on the free market.[10]

It should not be considered a "spoiler alert" that this book will be arguing in favor of a particular urban configuration termed "the 24-hour city." Nevertheless, there will be no attempt to ignore the problems of any American city. This is a story of differentiation, change, and the interrelation of urban economies and the built environment. The so-called 24-hour cities are not the "average" American city. In fact, they will be seen to be distinctively different by a number of socioeconomic and real estate-market measures. It is not a simple story, and it is one where history plays an important role.

In this chapter, we will start the story in the middle, at the mid-20th century, with a look at two of the largest American cities, New York and Los Angeles, as they were depicted in popular culture on broadcast television. We will look at the countervailing forces, powerful forces that were shaping cities in the decades prior to World War II, and the singular effects that the war itself exerted. Toward the end of the chapter, we will examine the concept of change itself, unpacking what seems to be a very simple notion and seeing that change comes in at least five varieties.

A look from our living rooms

Long before the pall of "reality television" descended over American popular culture, network TV reflected, and to some degree shaped, the way people lived their lives across the United States. Families gathered around 21-inch black and white TV sets in their living rooms, sets made in the USA by Motorola, RCA, Philco, GE, Zenith, and other "household names." The news of the day was delivered by trusted names such as Douglas Edwards, John Cameron Swayze, Walter Cronkite, Chet Huntley, and David Brinkley. TV was one of those unanticipated but massive shifts that occurred following World War II, as technological advances emerging in military applications were translated into the private sector, and US manufacturing refocused on civilian applications.

The enormous demographic wave that became known as the Baby Boom provided a vastly expanding market. Television's penetration into that market was breathtaking. Whereas only 1 out of every 200 US households had a television set in 1946, more than 55 percent had one in 1954. By 1962, nine out of ten households owned a TV set, and since then TV ownership has become practically universal (Figure 1.1). The way Americans spent their time also changed. In 1950, the average household watched television 4 hours and 35 minutes per day. By 1975, that time had expanded to 6 hours and 7 minutes. By 2005, the time allocation was 8 hours and 11 minutes, as 24-hour programming options and technologies such as the video cassette recorder (VCR) and digital video recorder (DVR) became available and increasingly affordable.[11]

Television not only used more and more of the time available for all US households' activities – reducing time available for social and civic interactions outside the home[12] – but it also conveyed information about life beyond the walls of the homestead. In this, there was both good and bad.

The early decades of television are celebrated as a "golden age" that brought original dramas by Alfred Hitchcock, Rod Serling, and Paddy Chayefsky. Leonard Bernstein and Arturo Toscanini presented classical music and had a vision of bringing its cultural heritage to a mass audience.[13] *The Ed Sullivan Show*, which at its peak in 1957 had a weekly audience of nearly 15 million viewers, is remembered for its presentation of pop stars, but also brought scenes from the Broadway stage – including *West Side Story*, *Oklahoma!*, *Gentlemen Prefer Blondes*, and *Man of La Mancha* – to a coast-to-coast viewership.

Public opinion was increasingly shaped by the immediacy of televised events. The 1954 Army–McCarthy hearings began the unravelling of the power and reputation of Wisconsin Senator Joseph McCarthy, with the drama capped by Army attorney John Welsh's parting response, "Senator: you've done enough. Have you no sense of decency, sir? At long last, have you left no sense of decency?"[14] The Kennedy–Nixon debates of 1960 brought live political discourse directly into American living rooms.

Support for the ambitious US space program was galvanized by the countdown-to-splashdown coverage of the Mercury, Gemini, and Apollo

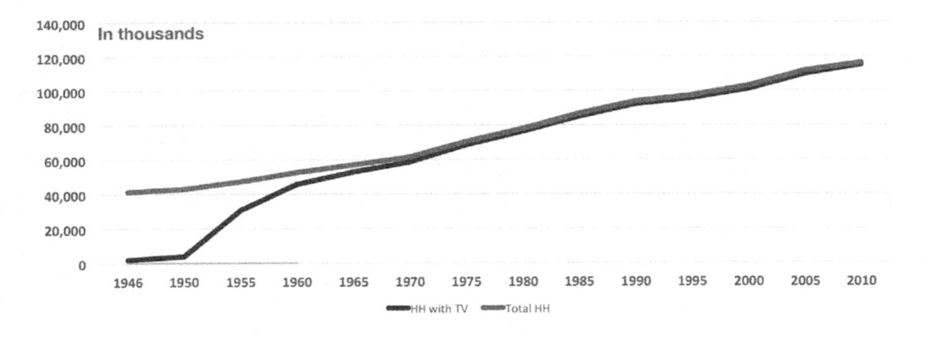

Figure 1.1 Penetration of television into US homes

Source: TV Basics, based on A.C. Neilson data (households with TV); US Census Bureau (total households)

missions – with their moments of triumph and of tragedy. News of the civil rights and Vietnam War protests and counterreactions sharpened awareness of social divisions shaped by economics, politics, race, and age.

The shocking round of assassinations of President John F. Kennedy, Senator Robert F. Kennedy, and Dr. Martin Luther King Jr. provided a national experience of shared grieving. The impeachment proceedings against President Richard Nixon, the ensuing Watergate hearings, and his eventual resignation from office, covered "live from Washington," offered a lesson in civics unparalleled in US history.

For all this, however, the most remembered judgment about television in this era is that of Newton N. Minow, Chair of the Federal Communications Commission, in a speech to the National Association of Broadcasters. He challenged the network executives in 1961 to watch an entire day of their own programming, without distraction, and declared, "Keep your eyes glued to [the] set until the station signs off. I can assure you that what you will observe is a vast wasteland."[15]

Minow's characterization undoubtedly referred to the array of sitcoms, westerns, game shows, and cop shows that were the staples of network programming, especially in the evening "prime time" hours. From today's perspective, though, there is a fair amount to be learned from that programming. Not the least interesting feature is how cities and American life were portrayed on the airwaves, both reflecting what audiences identified with and shaping how viewers perceived their contemporary environment.

Depictions of New York and Los Angeles

Let's look at a couple of classic blue-collar sitcoms, *The Honeymooners* and *The Life of Riley*.

The Honeymooners was introduced as a comedic sketch as early as 1951 and was expanded into a 30-minute program for the 1955–1956 season. It was set in Brooklyn, one of New York City's five boroughs, and featured two young couples living in walk-up flats. Both wives had vaguely defined careers prior to marriage, but are mostly "stay-at-home" spouses, whose characters are largely foils to their husbands. Ralph Kramden (Jackie Gleason) is a New York City bus driver, and Ed Norton is a New York City sewer worker – interestingly, both municipal employees. They both make the same wage: $62 a week. The set was spare: most action occurs in the Kramden's main room, which served as living room, kitchen, and dining room. A chest of drawers, stove, and icebox were the furnishings. A window, without curtains, gave a view onto the fire escape and neighboring tenements. This was a no-frills, working-class slice of life – and realistic enough that, as a child of a Brooklyn bus driver myself, I imagined I could actually identify where the Kramdens and Nortons lived.[16]

This lifestyle was much in contrast with other popular family sitcoms, which were decidedly more upscale. The emblematic sitcoms of the era were *The Adventures of Ozzie and Harriet* (1952–1966), *Father Knows Best* (1954–1963), and *Leave It to Beaver* (1957–1963). Whereas *Father Knows Best* and *Leave It to Beaver* were set in unidentified Midwestern suburbs, *Ozzie and Harriet* featured the Hollywood home of the actual Nelson family, a 5,214 square-foot home set on a half-acre lot just a block from the 160-acre Runyon Canyon Park, at the eastern end of the Santa Monica Mountains. (The tract was in private hands, owned by the supermarket magnate Huntington Hartford, during the time the TV show ran.) The TV families were all two-parent households with several children, whose fathers were white-collar workers. Ozzie Nelson, who played himself, was an entertainer and bandleader; Jim Anderson, the dad in the *Father Knows Best* household, was an insurance agent. Ward Cleaver, father of "the Beaver," was simply known as a white-collar office worker in the fictional suburb of Mayfield, whose particular job was never specified.[17]

But a more telling contrast to *The Honeymooners* is another California-situated household comedy, *The Life of Riley*, which was broadcast on NBC from 1953 to 1958. Chester A. Riley was the father in a suburban Los Angeles family, but, like Ralph Kramden, was a blue-collar worker. His employer was the fictional Cunningham Aircraft Company, where Riley worked as a riveter. His wife, Peg, was the long-suffering parallel character to Alice Kramden. Unlike the Brooklyn couple, Chester and Peg had the stereotypical two-child household of son Chester Jr. and daughter Babs. And, mirroring the Ed Norton foil character of *The Honeymooners* (and Ed's wife Trixie), the *Riley* program featured Chester's scheming neighbor and co-worker, Jim Gillis, and his wife Honeybee.[18]

But the differences in the settings were much more striking than the similarities in the casting. Where the Kramdens and Nortons were in cramped, spartan flats, the Rileys and Gillises enjoyed detached, single-family houses. Where the Brooklyn couples had aging, depression-era appliances, the L.A. households learned, often comically, to cope with the array of gadgets that steadily advanced American consumer spending – devices that were a core source of advertising revenue for the developing TV industry. And, despite the common scheming of the principal characters at home and at work, there was little hint of upward economic mobility in Brooklyn, but an underlying theme that hard work would pay off with economic and material gratification for the L.A. families, who already had achieved a measure of the American Dream of house, property, and middle-class comfort.

The images and the *sub rosa* message of older and denser cities caught in a mire of urban stagnation and younger, more sprawling cities as a setting for upward mobility set a tone for expectation and a script for urban discussion that would shape popular attitudes and, to some degree, politics and urban studies for a generation or more.

Cities: Their growth and their problems – not a laughing matter

Since the days of Aristophanes (446–386 BCE), comedy has held up a mirror to society that has been disarming and occasionally critical. Plays such as *The*

Frogs, *The Clouds*, and *Lysistrata* are also some of the best evidence we have about the perspectives of common people (the *demos*), as well as notables such as Socrates, Alcibiades, Athenian negotiators, and Spartan ambassadors. Furthermore, the Greek city (or *polis*) actually gave rise to the performances of comedies and tragedies. Plays were events staged in the context of public festivals and served civic functions, cultic as well as cultural.[19] Although 1950s TV programs are unlikely to last for 25 centuries, they do serve as cultural markers of some importance.

The 1950s stand as a key transition decade for US demography and urban life. As Figure 1.2 shows, the rise of urban areas was well underway by the mid-20th century across the United States. About 1920, urban residents surpassed the nation's rural population. World War II accelerated migration to the cities, as these centers of America's Industrial Revolution were transformed into production machines for war matériel and logistic support.

What triggered and sustained the growth of large cities in the first 150 years of the nation's history? Technology, trade, and transportation all played critical roles. Events around the world, in Europe primarily, but also in Asia and Latin America, provide early examples of global connectedness. Money and banking cannot be ignored. Neither can the arts and culture, in both "high" and "popular" expressions.[20]

Real estate in all its forms – housing, workplaces, stores, warehouses, hotels, and land itself – emerged as the tangible, visible instantiation of all those factors. Real estate provided the locus of manifold human activity, activity that took on distinctively urban characteristics over the course of the 19th and early 20th centuries. Real estate became emblematic of cities, a source of identification and frequently of pride. Real estate, truthfully, also became a discriminator in

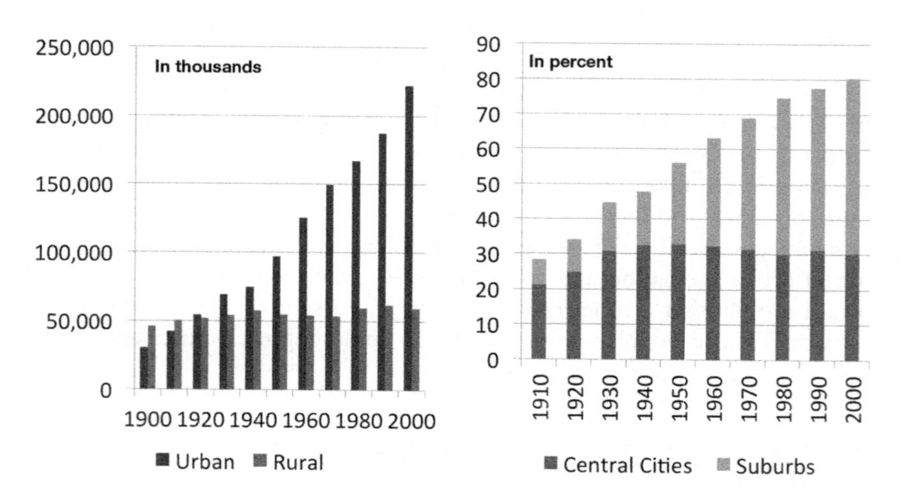

Figure 1.2 US population trends in the 20th century
Source: US Census Bureau

the urban setting, a marker of class, of education, of ethnicity, and of race. And real estate became a repository of physical capital, as well as a source of fungible wealth. In all these ways, too, real estate became a way to "keep score," a point of comparison between past and future, between one place and another. The metrics of cities became bound up with the metrics of real property.

But, even as the urban–rural balance tipped toward the cities, the suburban ring around the central city began to assert greater dominance. This, too, was not accidental and had its origins in both new visions of urban life and deliberate public policy, closely related to strategic military aims. And, as we shall see, even the physical form of the "tract suburb" owes much to planning and development by the Department of War (the official name of The Pentagon until 1947). Again, as seen in Figure 1.2, the share of population living in the central cities began to decline as the suburbs accounted for virtually all of the growth in metropolitan areas.

Some basic background

The historical evidence is that cities and their regions have had evolving relations for centuries. Physical form, social structure, and economic functions have shifted over time, and America's cities have certainly seen such shifts in dramatic fashion. The results were mixed.

The Industrial Revolution introduced scale issues hitherto unexperienced in American urban life. Mass production dissolved the links of home to workshop, while simultaneously creating unprecedented demand for labor. Despite Luddite fears of machines replacing workers, the factory generated an enormous requirement for workers to tend the tireless machines that multiplied labor inputs many times over, creating huge gains in factor productivity and, consequently, high returns to capital. Workers flocked to major cities, in both a rural-to-urban migration within the US and a great wave of international immigration to the US. With the limitations of 19th-century transportation, those workers typically resided close to the factories where they worked.

Quantitatively, as measured by population, total employment, aggregate wealth, and vital statistics such as mortality rates, the nation improved as urbanization intensified. To briefly sketch the picture, real per capita GDP is estimated to have been $2,445 in 1870, prior to the era of explosive US urban growth. That figure more than doubled to $5,079 by 1906, and doubled again to $11,518 by 1943. (It would take until 1992 to double again to $23,059.[21])

In 1900, annual deaths ran at 17.2 per 1,000 of population. By 1940, the mortality rate had dropped to 10.8 deaths per 1,000 of population. That represented real progress.[22]

In the largest cities in the most industrialized states, however, mortality ran above the national average on the eve of World War II. As Figure 1.3 suggests, the large cities (above 100,000 in population – the threshold used by the Bureau of Vital Statistics) in New England, the Mid-Atlantic, and Midwestern US had marginally higher mortality rates than the nation as a whole in 1940.

US average	10.8
Connecticut	10.8
Illinois	11.1
Maryland	12.6
New Jersey	11.4
New York	10.4
Ohio	11.5
Pennsylvania	12.2
Rhode Island	11.1

Figure 1.3 Deaths per 1,000 population in 1940 (cities of 100,000 or more residents) in industrialized states of Northeast US

Source: Vital Statistics Rates in the United States 1900–1940; Federal Security Agency – US Public Health Service

The numbers merely hint at the lived experience of many city residents, though. Students of urban society have long recognized the deleterious impact explosive industrialization had for quality of life in cities. Mumford describes the dreary streets, rubbish-strewn alleys, darkened houses, and fire-trap layouts of worker housing in 19th-century Philadelphia, Chicago, and Brooklyn.[23] Even as early as 1890, Riis had exposed the squalid conditions of Manhattan's Lower East Side and Five Points slums in *How the Other Half Lives*, an example of the power of photojournalism.[24] Veiller describes a city block of 200 × 400 feet crowded with 605 tenement units, housing 2,781 people, with 264 water closets in total and not a single bath. Writing as secretary of the Tenement House Commission, he concluded that these were breeding grounds of "disease, poverty, vice, and crime."[25]

The stress on the urban fabric spread across virtually all cities. Boston experienced waves of immigration in the 19th century, even as its economy enjoyed an industrial boom. A Harvard study of lodging houses depicted the sordid conditions of the South End, with its "disorderly houses" and streets "heavy with the evil odors of degradation." The Back Bay, now Boston's great urban jewel, was then a fetid lowland of tidal mud flats and breeding ground for disease-spreading insects.[26] Warner's path-breaking 1978 study, *Streetcar Suburbs: The Process of Growth in Boston 1870–1900*, shows how improvements in transportation hastened the outflow of population from the city center to Roxbury, Brighton, Dorchester, and other formerly rural outposts.[27]

Sandberg conferred upon Chicago the epithets "hog butcher for the world" and "city of the big shoulders."[28] As the 19th century drew to a close, Chicago had accomplished remarkable feats. It battled its way back from the Great Chicago Fire of 1871, which destroyed more than 18,000 structures over 3.5

square miles in the heart of the city. The hub of the US long-distance railroad system after the Civil War, Chicago became the center of the nation's meat-packing industry, with the assembly-line-based Union Stock Yards employing 25,000 workers. The railroads and Great Lakes shipping allowed Chicago to become a center of the iron and steel industry, with some 10,000 men employed by the Illinois Steel Company in 1900, just before its acquisition by J. Pierpont Morgan and merger into Morgan's US Steel Corporation.[29] The 1893 Columbian Exposition was known as "The White City," an ironic contrast to the general first impression offered by the city itself, called "a miasma of din, anthracite, and putrefaction" by Larson.[30] Seeking to provide escape from such conditions, Frederick Law Olmstead designed the upscale suburb of Riverside, whose early architecture featured the work of Frank Lloyd Wright, Louis Sullivan, and William Lebaron Jennings. The dense network of railroads connected agricultural centers such as Roselle, Argo, and Elgin to the city, as well as linking satellite industrial cities such as Waukegan, Gary, Cicero, and Aurora to the center. Chicago, in a matter of a few decades, evolved into a vast and fully articulated metropolitan area.

Lincoln Steffens, one of the journalists who achieved prominence in the early 20th century as "muckrakers," linked the appalling living conditions of big-city residents to the political corruption rife at the municipal level. In *The Shame of the Cities*, Steffens presented his investigations of "boodling" (institutionalized payoffs) in St. Louis, police graft in Minneapolis, the political/industrial machinations that governed franchising, vice, public contracts, and outright bribery in Pittsburgh, and general civic debasement in Philadelphia ("the worst-governed city in the country").[31]

It is against this background that alternative visions of urban life and/or the geographical alternative of suburban life were developed. Choosing the 1950s as a particular starting point for discussion is somewhat arbitrary – as any historical starting point would be. But this decade was a watershed in many ways, as unheralded as it has been when compared with the war decade that preceded it or the socially turbulent decade that followed. One measure of its significance is that, by 1960, the urban (metropolitan area) population was already 70 percent of the US total, and almost half of that number already lived in the suburbs. The momentum of population growth had shifted. The important story was no longer farm-to-city. It was city-to-suburb. Although downtowns were not yet "hollowing out" by 1960, they were no longer dominating the choices of America's households.

Tug of war: Centralization versus dispersion

Practically speaking, the second half of the 20th century was a period of stress for most large cities and a procession of triumphs for suburban America. Most analysis of US suburbanization has focused on the post-World War II spread of population and the consequent reconfiguration of metropolitan areas in physical form, social structure, and economic function. Yet the urban–suburban relationship is anything but a "new phenomenon" of the late 20th century.

Frequently, the starting point for discussion of metropolitan change is the largely monocentric configuration of downtown central business district (CBD) and residential suburbs that epitomized urban form in the United States at the dawn of World War II. As Robert Beauregard notes, this form energized the economy, epitomized its power, and held its social fabric in coherence for a century. Suburbs and CBDs functioned as complements to one another. The city had not yet shed its manufacturing base, and so the urban center had a multiform economy, with white-collar office functions, blue-collar production, and retail districts contributing to its commercial base. Walk-to-work housing and mass transit tied residential neighborhoods to jobs. Civic, entertainment, and cultural facilities were integrated into the core, giving citizens familiarity and identification with the downtown.[32]

Suburbs were typically linked with the core by streetcars or railroads. Warner described the "naturally occurring" streetcar suburbs surrounding Boston. The advance in transportation technology permitted the middle class to extend the distance to work, much as the upper income groups had previously enjoyed. The wealthiest households had multiple transportation options and could afford country homes as well as townhouses, utilizing the former as rural retreats and as summer residences in the era before air conditioning. Electrified streetcars allowed the middle class to have convenient access to downtown jobs from residential areas up to six miles from Boston's City Hall. As these workers' jobs were exceptionally stable, fixed rail connections by streetcar nicely accommodated their journey-to-work needs as well as contributing to stability in their residential neighborhoods. Lower-middle-class workers, on the other hand, had somewhat lesser job security and depended on the flexibility of Boston's "cross-town" streetcars, which supported residential options within three miles of the center. Lower-income workers, for whom marginal transportation costs were critical, remained tied to walk-to-work neighborhoods. Warner termed the resulting configuration "a selective melting pot," as it effectively segregated neighborhoods by income level, in roughly concentric circles. The concentration of the poor (in Boston's case, immigrant Irish and Italians) in the city center set the stage for more serious urban economic and social problems in later years.[33]

Philadelphia's suburban growth featured similar patterns. Michael McCarthy notes the influence of Andrew Cassatt, a railroad executive who was also supervisor of Lower Merion Township, in the development of "Main Line" suburbs such as Bryn Mawr and Villanova as middle- and upper-income suburbs in the late 19th and early 20th centuries.[34] Robert Fishman identifies the key institutions of the Main Line suburbs: the railroad hotel, the Episcopal Church, and the country club. These imprinted the residential desirability of the suburb, while the railroad itself provided a durable link with the CBD as the commercial core of the metropolitan area.[35] McCarthy indicates a similar pattern in railroad suburbs located throughout New York's Westchester County.

In the pre-World War II era, several attempts were made to put into practice the concepts of the "utopian" theorists such as Ebenezer Howard and Frank Lloyd Wright. Several of these sprang up in the New York metro area, including Radburn (in Fairlawn, New Jersey), Forest Hills and Sunnyside (in the New York City borough of Queens), and Garden City (in New York's Nassau County, just beyond the city's eastern border). Efforts to accommodate a range of income groups notwithstanding, development costs made them too expensive for working-class families, thwarting the intentions of project sponsors such as the Regional Plan Association and Russell Sage Foundation.[36]

These attempts at "model" decentralization did inspire later efforts, but illustrated the difficulty of putting the planners' ideals in practical form. The New Deal-era US Resettlement Administration (USRA), tried again in the 1930s with satellite new towns Greenbelt, Maryland, Greenhills, Ohio, and Greendale, Wisconsin. The efforts ran afoul of political and private economic obstacles – seeing the value of "prime suburban land" at risk if lower-income households were to be resettled outside the cities. Private developers and the savings-and-loan industry attempted legal obstruction. After the three small experiments were completed in 1938, the USRA was abolished (with the lesson that government policy should favor "safer" programs such as mortgage insurance for suburban housing and subsidized public housing for the poor in urban centers where they were already concentrated).[37]

Even before World War II, then, governmental and market responses to the troubled condition of central cities, where stresses were magnified by the economic crisis of the Great Depression, set the stage for a particular pattern of metropolitan restructuring. A centrifugal pattern of population resettlement expanded the perimeter of urban areas, assisted by transportation access. The initial stage, however, kept the role of the employment center downtown – especially for upper-middle-class and upper-income households. The dispersal of homes from jobs meant that, for white-collar office workers, especially those with executive and managerial functions, departure from the central city at the end of the working day became more and more the norm, giving rise to more 9-to-5 business districts. Less-wealthy households then became relatively more residentially concentrated in the heart of the city. Income segregation placed higher social costs upon the core city, making the suburbs more of a refuge. Such dynamics were already in place by 1940. The following decades would see their power intensify.

For many Americans, World War II is viewed largely as a disruptive departure from the trends shaping the domestic social and economic landscape. But the 1941–1945 period has been identified by scholars as a critical era in US urban development patterns. The war effort was a powerful de-concentrating influence. Part of this was a "spread the risk" strategy, and part was a realization that the older industrial cities (such as Hartford and Bridgeport) had limited capacity to handle an influx of defense workers.[38]

The risk-spreading consideration was acutely felt in the aftermath of the destruction of the Pacific Fleet in just a few hours at Pearl Harbor and

Figure 1.4 Defense Housing prefigures post-war suburbs

This site plan for Federal Works Agency housing during World War II illustrates the prototype of post-war suburban housing development: a residential monoculture, with cul-de-sac local circulation feeding an arterial collector road connecting to work centers and shopping outside the housing community. The low-density housing, with its dependence upon the automobile, was "ahead of its time," in modeling what the suburbs of the 1950s and 1960s would epitomise.

Figure 1.4 Continued

widespread concern about German U-boat activity in the shipping lanes proximate to New York City. The War Production Board appointed by President Roosevelt in January 1942 elected to disperse matériel production by developing new plants in suburban areas, rather than retool existing and underutilized plants in the central cities. This stimulated new investment opportunities and expanded suburban economies around New York, Detroit, Baltimore, and Pittsburgh.

With the collapse of homebuilding during the Great Depression, and the diversion of resources to the war effort in the 1940s, urban housing shortages were endemic in the nation's large cities. One symptom of the shortage was the imposition of rent control as an anti-war-profiteering measure in many places. Neil Leibowitz reviews the determination by the wartime Office of Price Administration that stabilizing rents would lower the cost of production inputs, positively affect labor turnover, check popular unrest, and contribute to improved national morale. Pressure on the housing stock was immense. The counties surrounding the major California cities experienced two-year population increases of 130,000 in Los Angeles, 97,000 in San Diego, and 95,000 in San Francisco. Chicago's Cook County grew by 149,000. Fears were expressed that, unless the migrants were directed to the "geographic fringe" of the metropolitan area, the requirement for municipal services would break the cities financially.[39]

Real estate boards across the nation mobilized to block rent control at the same time that they opposed the construction of "non-competitive public housing" in the cities. The Lanham Act in 1940 authorized some 700,000 units of housing to be built, and, in response to the siting of the defense industry plants and the resistance within established urban centers, much of this housing took the form of low-density suburban development, as illustrated in the Federal Works Agency pamphlet "Defense Housing 1941" (Figure 1.4). Focusing on the short-term effects of wartime price controls and subsidized workforce housing, the property industry encouraged conditions by which residential and job dispersions would, over time, compromise demand for its own fixed assets.

The reality of a two-ocean war shifted the gravity of industrial activity westward. Los Angeles and San Francisco, in particular, became key defense industry locations, along with San Diego and many southern and western areas where military bases were situated. In this case, housing followed defense jobs and military facility construction as well. Hall details how the US military had identified the strategic importance of the West Coast's ports as early as the 1920s.[40]

Military contracting favored the "Sunbelt regions." During the course of the war, approximately 65 percent of all federal investments flowed to defense industries such as shipbuilding, ordnance, aircraft and parts, and related industries such as petroleum refining. Gregory Hooks and Leonard Bloomquist argue that federal investment expanded and modernized industrial plant and equipment in industries ranging from energy and munitions to aeronautics and electronics. At the end of the war, the federal government owned 40 percent of the nation's capital assets, much of which it then divested into the private sector.[41]

The process shifted the location of economic growth in the post-war period. These authors identify Los Angeles, Houston, Dallas–Fort Worth, Memphis, and Miami as the major beneficiaries over the quarter-century commencing 1947. The biggest "losers" in the regional rebalancing of this period were New York, Chicago, Pittsburgh, Boston, and Philadelphia. Markusen et al. present a mild corrective to the Hooks and Bloomquist analysis, noting that research and development spending boosted economic growth in states such as Maryland and Massachusetts. But, overall, the secular shift of advantage toward California, Texas, Florida, and other Sunbelt states – a process imprinted on the nation's economic geography by World War II-era decisions – is validated in the research by Markusen and her colleagues.[42]

Thus, even before the much-analyzed post-World War II suburbanization, there was considerable momentum toward expansion at the perimeter of cities, and a direction of growth favoring the newer regions of the South and West over the large but aging cities in the nation's northeastern quadrant. The commuter framework of 9-to-5 downtowns was being erected, and the stage was being set for larger, automobile-oriented, multinucleate metropolitan areas in regions that were remote and sparsely populated in the first half of the 20th century.

Modalities of change

Cities are evolving organisms. Critical changes in recent decades have affected the relative competitiveness of particular cities, but the entire phenomenon of change is rarely considered in its full complexity. Every metropolitan area experiences, and must expect to continue experiencing, turbulent cross-currents of change. I propose that five fundamental forms of change are worthy of special attention, with the understanding that these forms of change are typically at work simultaneously and that they interact with each other. The five basic "flavors" of change are *cycles*, *trends*, *maturation*, *changes of state*, and *disruption*.

Briefly, the various forms of change can be described as follows:

- *Cycles* are periodic fluctuations around some equilibrium point, where the critical measures are amplitude and duration of the fluctuation. The measurement and mathematics are modeled on signal theory. Its driving dynamic in market and general economic theory is the presumption of reversion to the mean when systems move out of equilibrium.
- *Trends* are long-term shifts in system variables, including durable changes in the quantity and quality of system components. The measurement and mathematics of trends are fundamentally modeled on classical Newtonian physics, using concepts such as inertia and momentum. Demography, technology, and industry/occupation structure are examples of variables subject to trend-like changes pertaining to cities and real estate markets.
- *Maturation* refers to the organic unfolding of system potential and adaptation to habitat. The normal path of change follows laws of biological and ecological science, frequently described by the sigmoid or logistics curve. Key concepts are resource availability and utilization, systems development and interaction, and sustainability. Like trends, maturation conceives of change as telic, i.e., directional; like cycles, maturation anticipates that change is not a "constant." Distinct from both cycles and trends, maturation is conceived of as inherently finite.
- *Change of state* denotes fundamental alterations of system conditions, which may be short term and reversible (but not necessarily so). The scientific model is drawn from molecular structure and chemistry, as for instance when a molecule retains its elemental structure while passing from a solid to a liquid to a gaseous state. Behavior is state dependent. Key concepts

for changes of state are critical thresholds, boundary conditions, triggers, and external pressures. Volatility characteristics may vary significantly depending upon state condition.

- *Disruption* involves a sudden, profound shock to a system that constitutes a radical challenge to system integrity. Disruptions may resolve into some degree of return to *status quo ante*, but may issue into some permanent and irreversible change requiring systemic reconceptualization. Risk managers sometimes refer to disruptions as "event risk." Following Taleb, disruptions can be considered low-probability, high-impact events, dubbed "black swans."[43] A coherent scientific model may follow the principles of particle physics, along the lines of nuclear fission or fusion. Accommodating disruptions into neoclassical economics may require reconceptualization at the level of Kuhn's "paradigm shift."[44]

The five basic "flavors" of change are illustrated in the following figures. These are idealized forms, of course, but they depict the underlying patterns clearly. Associated with each graph are a few examples of how the changes appear in urban economies and real estate markets. As we proceed in our analysis, we shall see empirical evidence of these forces of change at work. Because of interactions, though, each basic form of change rarely follows the idealized shapes in these graphs, even if the driving force is often one dominant type of change.

Each of these forms of change will be seen as relevant to cities. We will trace change as it influenced the emergence of 24-hour cities in particular, and deal with the related question about why most cities have not evolved to 24-hour

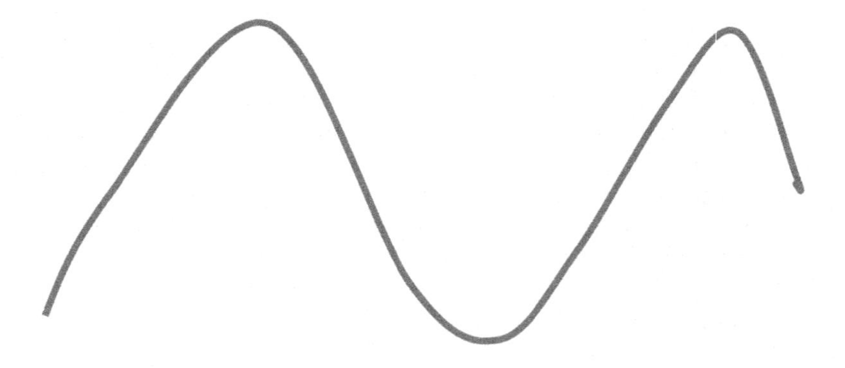

Cyclical change is a feature of all market-based economies. GDP, employment, interest rates, the stock market, and many other macroeconomic measures conform to this pattern. Real estate markets display a similarly recurrent exposure to cycles, measured by rents, prices, construction, occupancy rates – property market measures that are related to, but not identical with, the underlying economic cycles themselves.

Figure 1.5 Cycles

status. Some may object that "evolved" is not the proper word, as it implies that, over time, cities as economic entities are inexorably moving away from 9-to-5 configurations toward more round-the-clock activity. Not only is there no evidence that such is the case, there are both theoretical and practical reasons to believe that such a universal tendency would not be expected.

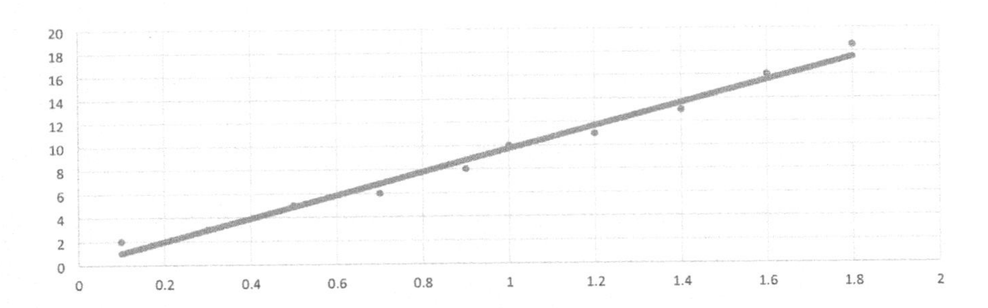

Long-term changes that display a persistent direction are common in economies and in real estate. Population growth, the move from blue-collar to white-collar work, the impact of inflation on pricing levels, the expansion of the inventory of residential and commercial property, and the aggregate value of real estate as an asset class are examples of long-term trends.

Figure 1.6 Trends

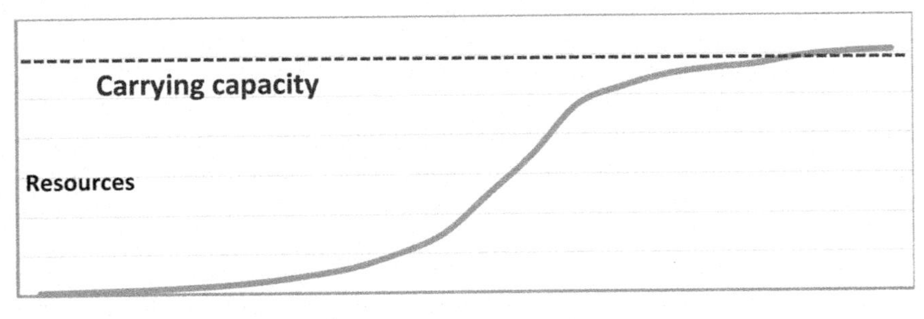

Maturation recognizes the relationship between internal dynamics and external constraints. Just as human beings (and all other organisms) balance a propensity for growth with genetic and environmental limits, so do cities and real estate markets. This can also be seen in the introduction of new products – whether consumer goods, financial products, or forms of real estate. Periods of slow growth, followed by acceleration, and finally a levelling off are a third fundamental pattern of change.

Figure 1.7 Maturation

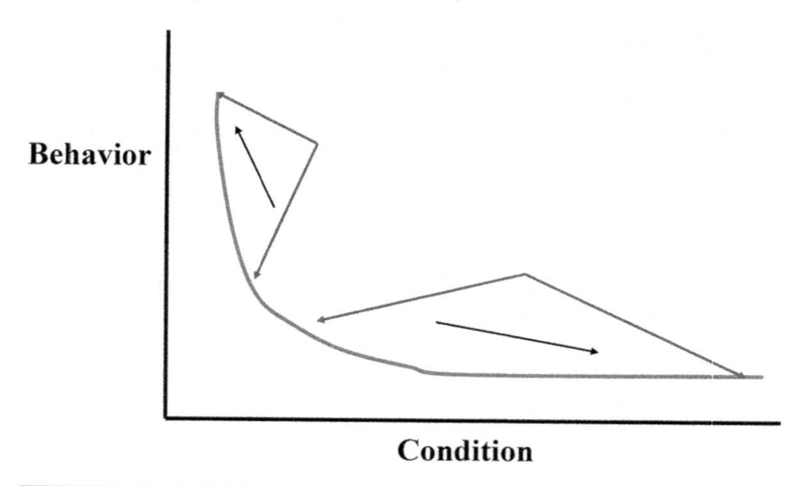

Economies and real estate markets show variable, not constant, elasticity over time. There has been much frustration in recent years that employment and earnings have not responded proportionately to monetary stimulus, for instance. In the real estate markets, we often note far different rent and price movements, dependent upon where the market stands relative to an "equilibrium range."

Figure 1.8 Changes of state

Some reflections on theory

From a theoretic standpoint, evolution is not "telic", or ends-determined. Evolutionary biologists, ever since Darwin, have pointed out that we live in a world where contingency and randomness are inescapable features. Evolution as classically understood proceeds by mutation (another word for "change"), with subtly differing organisms seeking to survive. Hence, the notion of "the survival of the fittest" remains the watchword for evolution in the popular mind, with overtones of inevitability, determinism, and even a kind of superiority for the "winners."

All three of those telic presumptions must be called into question. Inevitability has been a feature of theories for centuries and has enjoyed something of a resurgence in our own time. Most of these theories might be characterized as forecasting out of the rearview mirror, or extrapolating pre-existing trends indefinitely into the future. One of the more famous examples of this, in urban planning circles, was an international conference gathered in New York in 1898 on the subject of horse manure. Transportation in the pre-automobile era was primarily horse-drawn. The mathematics of population growth, demand for goods and services requiring delivery, and the need for ever-enlarging cities to connect people to jobs led to an astonishing, disturbing, and yet apparently inescapable conclusion: by 1950, the pile of horse mature in the streets of New York would rise to a height of nine feet.[45]

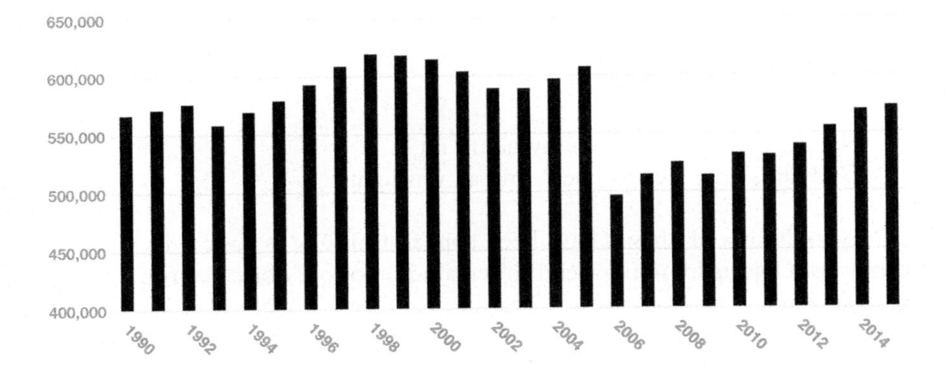

Singular events may suddenly and profoundly alter pre-existing patterns of change. Natural disasters such as storms and earthquakes, human interventions such as a terrorist attack and systemic financial failures such as the crisis of 2007–2010 are examples of such disruptive change. However, there are positive examples as well: the impact of the internet, the "green revolution" in food production, and the introduction of the standardized shipping container might be cited as "positive disruptions."

Figure 1.9 Disruption

Malthus, in 1798, had also looked at two apparently inexorable facts: population growth proceeded on a geometric curve, while food production expanded in a linear fashion. His conclusion? Without the moral resolve to lower birthrates (by a variety of means), the growth of population would inevitably be checked by negative forces, namely, war, famine, and pestilence.

In our own times, critics of globalization have pointed to technology, communications, consumerism, transportation efficiency, and brand-driven capitalism as a confluence of forces inevitably driving toward worldwide homogeneity. Those more accepting of globalization as a phenomenon speak of the world as converging as a single market where the law of supply and demand drives advantage inexorably to low-cost producers. *New York Times* columnist Thomas Friedman wrote a best-selling book, *The World is Flat*, and followed up with *Hot, Flat, and Crowded*. Although Friedman earnestly appeals for changes in attitudes, policies, and behaviors, his basic thesis is that the path of history is one of convergence, a basic condition we are largely unable to alter.[46]

Despite some resurgence in popularity, theories of inevitability are now mostly viewed with deserved skepticism. This is not just owing to the failure of confidently announced and apparently scientifically based predictions. It is owing to a change in the very way we think of science.

The Age of Enlightenment undergirded the optimism of "inexorable progress" as the basic arc of history. The sense of certainty, and the determinism that was its intellectual foundation, owes much to one of the true geniuses of Western civilization, Sir Isaac Newton. The poet Alexander Pope proposed a couplet as Newton's epitaph:

> Nature and Nature's Laws lay hid in night.
> God said, "Let Newton be." And all was light.[47]

Newton's laws of physics have been applied fruitfully as basic principles for centuries. Objects in motion tend to remain in motion; objects at rest tend to remain at rest – absent some external force. Force is equal to the mass of an object times its acceleration. For every action, there is an equal and opposite reaction. Newton famously articulated the principle of gravitation and used it to compute planetary orbits. His work in optics revealed, through the prism, that white light is composed of a spectrum of colors. His understanding of the principles of refraction led him to construct a telescope that uses mirrors, now known as a "Newtonian telescope," still being built today by civilian and military researchers. And Newton was, along with Leibnitz, the developer of calculus as a mathematical tool.

Applying Newton's work to the problems of thermodynamics, later scientists derived laws showing the conservation of energy and the irreversible pattern of entropy toward a constant state, entropy referring to the tendency of a system toward a condition of uniformity.[48]

The power of Newton's work, and that of those who built upon it, was that it allowed predictability and hypothesis-testing to rule in the scientific realm. Laws of nature were truly laws: mathematically rigorous and objective tools derived from the empirical study of the world (the universe) that were powerfully explanatory. With the proper information and the proper application of the laws, outcomes could be anticipated and the truth of the laws could be validated by the ability to find events predicted by theory in the test of experiment.

At base, this meant that systems were deterministic.

In important ways, and with implications across all fields of knowledge, the Newtonian consensus in physics was compromised by scientific advances throughout the 20th century. In 1905, Albert Einstein published the theory of special relativity, with its famous equation $E = mc^2$, which announced the convertibility of energy and matter, a concept beyond Newton's theoretical boundaries. Time and space are seen as relative to each other, a particular departure from the classical Newtonian assumption of absolute space and time.

By 1916, Einstein had advanced a more general theory of relativity, which rethought the notion of gravity, recalculated such assumed "knowns" as the orbits of planets, and conceived of an expanding universe with characteristics incompatible with the mechanical balances assumed in Newton's order of things.

A generation of physicists pushed science in the direction of acknowledging that the same element of the material world sometimes exhibited the character of a particle and sometimes the character of a wave. They also wrestled with the idea that so-called "objective knowledge" disturbed objects in the act of observation. So measurement itself had to be considered inexact. Werner Heisenberg and Neils Bohr, between 1924 and 1927, articulated what came to be known as "the uncertainty principle" so closely identified with Heisenberg. In its weak form, that principle says we cannot simultaneously know the position and the momentum of a particle at any moment. As we cannot know all characteristics at any time, properties must be described as probabilities, or (somewhat negatively) as uncertainties. The strong form of the principle goes further: uncertainty not only resides in our capacity to measure, it is a feature of particles themselves. The statistical property of the standard deviation was introduced at the heart of the description of the physical world, and the familiar determinism of the earlier physics was eclipsed.[49]

No one who has looked at, much less tried to work through, post-Newtonian physics would ever say it was neither mathematical nor rigorous.[50] On the contrary, even if indeterminate, the physics of Planck, Einstein, and Heisenberg, as continued by later scientists such as Murray Gell-Mann, Richard Feynman, Stephen Hawking, and Roger Penrose, have remained true to the standard that theory should be expressed in rules-bound math and be tested by its ability to anticipate as yet unobserved phenomena (i.e., predicting discoveries ranging from the bending of starlight around the sun (1919) to the recent (2012) confirmation of the existence of the previously hypothesized Higgs boson by scientists at CERN).

Applying theory to our practical issue: Cities, their forms and their performance

Why bother with this long discussion of changes in the physical sciences?

It is because physics has been the "hard science" model that is the gold standard that the "softer" social sciences – including economics and urban studies – have striven to emulate. The neo-classical economic models of Alfred Marshall, John Maynard Keynes, and even Milton Friedman attempt to create prediction of future observations by means of point estimates derived by analyzing variance from some point of equilibrium, in a mathematical extension of Newtonian physical science into the economic realm. Similarly, the standard model of urban spatial patterns for the second half of the 20th century was the Alonso–Mills–Muth model,[51] whose conceptual substrate was the pattern of entropy drawn from the Second Law of Thermodynamics.

Under the post-Newtonian way of thinking, the outcomes of models are not deterministic but probabilistic. We see the results of economic thinking as encompassing a variety of behaviors, not a single correct answer. In particular, we can note that any model that incorporates the standard deviation as an element in its mathematics should result in not one, but two correct answers. Why? Elementary statistics teaches us that the standard deviation is the square root of the "variance" in a set of data. And variance is the sum of the *squared* differences of each observation from the average. So equations employing these statistical concepts always implicitly consider terms at the power of two (x^2), the form of quadratic equations. And, in elementary algebra, such equations are understood to have as many solutions as the "power" of their variables.

If this all seems complex, that's exactly the point. Physical science is increasingly appreciated as the study of complex systems, and economies – including urban economies and their physical manifestations in the built environment – are likewise conceived of as complex systems. Components are no longer analyzed as stand-alone elements best appreciated by being isolated for study. Rather, elements are considered as a network, where the individual components not only mutually define each other, but are defined, not as nouns (things), but as verbs (activities) that only reveal themselves in their interactions.[52]

Such a perspective allows us to take a particularly open-minded approach to urban studies, to the characteristics of cities, to their trajectories of change, and to the properties of real estate markets that help us to measure physical elements in monetary terms. When we examine the changes that have resulted in 24-hour cities, we neither claim that such a result was inevitable, nor that all cities will (or even should) tend toward 24-hour-ness as the correct outcome of urban evolution. Moreover, though we will make claims that 24-hour cities are in some important ways superior to 9-to-5 cities, we must be careful to acknowledge that, in a complex urban network, we may find a "two-solution" result of our analysis – useful places for both 24-hour and 9-to-5 cities in a sustainable urban system.

We will, in other words, be taking seriously the idea of a taxonomy of cities. That means recognizing differences, but also recognizing enormous similarities. The similarities do not negate the differences, and biological taxonomy explicitly illustrates this difference–similarity combination. Take the degree of DNA shared between a variety of mammals. Domestic cats, for instance, share 90 percent of their DNA with humans. They share the same percentage of DNA with their feline cousins the leopard and the cheetah. But, taxonomically, it is quite easy to distinguish between *Homo sapiens* and any member of the family of *felidae*, the scientific name for cats. So, similarity and difference should be expected in our urban taxonomy, just as we find them coexisting in nature.[53]

Within each type of city, we should also expect a reasonable amount of variation, just as we would expect that members of the same family would differ in height, weight, and the color of eyes and hair. Members of the class of 24-hour cities are not clones of each other; neither are the members of the class of 9-to-5 cities.

Finally, as we think about the evolution of cities, as the 24-hour character has been an emergent phenomenon, we should not discount the possibility that this has morphed from a form that was once a 9-to-5 urban configuration. But, if that is a possibility, it should represent an evolutionary step of some significance, a large rather than a small step, and one that represents a "change of state" rather than merely a point along a general trend line.

Keeping this foundational discussion in mind, let's proceed to apply the basic concepts to sets of actual US cities and their commercial real estate markets.

Notes

1 Joseph Gyourko, Christopher Mayer, Todd Sinai, "Superstar Cities: Why Do House Prices Rise Faster in Some Cities?", NBER Working Paper 12355, National Bureau of Economic Research (Cambridge, MA, 2006). Reference to the "creative class" alludes to the work of Richard Florida, *The Rise of the Creative Class*, Basic Books (New York, 2002).
2 Cited in James A. Clapp, *The City: A Dictionary of Quotable Thoughts on Cities and Urban Life*, Center for Urban Policy Research, Rutgers University (New Brunswick, NJ, 1984) p. 190. Hereafter cited as "Clapp."
3 Kevin Lynch, *The Image of the City*, Joint Center for Urban Studies, Harvard University and Massachusetts Institute of Technology (Cambridge, MA, 1960).
4 Sophocles, *Antigone*.
5 *Letter to James Madison*, December 20, 1787. In Clapp, p. 128.
6 *The New York Times*, March 17, 1964. In Clapp, p. 16.
7 William Julius Wilson, *When Work Disappears: The World of the New Urban Poor*, Knopf (New York, 1996).
8 Data are from the 2012 American Community Survey of the US Bureau of the Census, published in *American Community Survey Briefs: Poverty 2000 to 2012* and in the *Current Population Reports, Income, Poverty, and Health Insurance Coverage in the United States: 2012.*
9 Benjamin R. Barber, *If Mayors Ruled the World: Dysfunctional Nations, Rising Cities*, Yale University Press (New Haven, CT, and London, 2013), pp. 41, 178.
10 Edward Glaeser, *The Triumph of the City: How Our Greatest Invention Makes Us Richer, Smarter, Greener, Healthier, and Happier*, Penguin Books (New York, 2011), p. 10.
11 Data from the Television Bureau of Advertising, *Televisions Basics*, accessed at www.tvb.org/trends/95487
12 See Robert D. Putnam, *Bowling Alone: The Collapse and Revival of American Community*, Simon & Schuster (New York, 2000), especially Chapter 13, "Technology and Mass Media," pp. 216–246.
13 James Roman, *From Daytime to Primetime: The History of American Television Programs*, Greenwood Press (Westport, CT, 2005), Chapter 8, "Television Drama: From the Golden Age to the Soap Factory," pp. 149–154.
14 Ibid., pp. 242–243.
15 Cited in Ella Taylor, *Primetime Families: Television Culture in Postwar America*, University of California Press (Berkeley and Los Angeles, CA, 1989), pp. 28–29.
16 A introduction to and summary of the series is at www.avclub.com/article/how-the-honeymooners-invented-the-domestic-sitco-83668
17 Taylor, op. cit., pp. 24–26, 40.
18 An introduction to and summary of the series is at www.imdb.com/title/tt0045406/
19 "The great trinity of Greek tragedians [Aeschylus, Euripedis, and Sophocles] . . . reveal how closely the fortunes of the arts were tied to the fortunes of the community.

They show us the poet as the public man." Daniel J. Boorstin, *The Creators: A History of the Heroes of the Imagination*, Vintage Books (New York, 1993), p. 211.

20 The historical development of the urban United States is cogently discussed in Jon C. Teaford, *The Twentieth Century American City* (2nd ed.), Johns Hopkins University Press (Baltimore, MD, 1993). Peter Hall's masterful *Cities in Civilization*, Pantheon Books (New York, 1998), explores the interconnections of commerce, technology, culture, and structural organization in cities around the world.

21 Angus Maddison, *The World Economy: Historical Statistics*, OECD Publishing (Paris, 2003). Data for the 19th century come from Nathan S. Balke and Robert J. Gordon, "The Estimation of Prewar Gross National Product: Methodology and New Evidence," *Journal of Political Economy 97:1* (1989), pp. 38–92. Accessed at http://socialdemocracy21stcentury.blogspot.com/2012/09/us-real-per-capita-gdp-from-18702001.html

22 www.cdc.gov/nchs/nvss/mortality_historical_data.htm

23 Lewis Mumford, *The City in History: Its Origins, Its Transformations, and Its Prospects*, Harvest Books/Harcourt Brace Jovanovich (New York, 1961), pp. 458–474.

24 Jacob A. Riis, *How the Other Half Lives*, Charles Scribner & Sons (New York, 1890).

25 See Howard P. Chudacoff, *The Evolution of American Urban Society* (2nd ed.), Prentiss Hall (Englewood Cliffs, NJ, 1981), pp. 113–118, for more on the tenements. The Tenement Museum in New York City is a rich resource for information on high-density housing in the 19th century; its website is www.tenement.org/. The complete report of the Tenement House Commission (1903) is available online at https://archive.org/details/tenementhousepro01deforich

26 Walter Muir Whitehall, *Boston: A Topographical History* (2nd ed., abridged), Belknap Press/Harvard University Press (Cambridge, MA, 1968). Chapter VII: "The Filling of the Back Bay," pp. 141–173.

27 Sam Bass Warner, *Streetcar Suburbs: The Process of Growth in Boston 1870–1900* (2nd ed.), Harvard University Press (Cambridge, MA, 1978).

28 Carl Sandberg, "Chicago," *Poetry* (March 1914). Accessible at www.poetryfoundation.org/poetrymagazine/poem/2043

29 For a wealth of information on Chicago's urban development, see www.encyclopedia.chicagohistory.org/

30 A historical novel by Erik Larson, *The Devil in the White City: Murder, Magic, and Madness at the Fair That Changed America*, Vintage Books (New York, 2004).

31 Lincoln Steffens, *The Shame of the Cities*, McClure, Phillips & Co. (New York, 1904), republished in unabridged form by Dover (Mineola, NY, 2004).

32 Robert A. Beauregard, *When America Became Suburban*, University of Minnesota Press (Minneapolis, MN, 2006).

33 Warner, op. cit.

34 Michael P. McCarthy, "Corrupt and Contented? Philadelphia's Stereotypes and Suburban Growth on the Main Line," in B. Kelly, *Suburbia Re-examined*, Greenwood Press (New York, 1989), pp. 111–118.

35 Robert Fishman, *Bourgeois Utopias: The Rise and Fall of Suburbia*, Basic Books (New York, 1987), pp. 134–154.

36 See Eugenie L. Birch, "Radburn and the American Planning Movement: The Persistence of an Idea," *Journal of the American Planning Association 46:4* (1980), pp. 122–155; Donald A. Kreuckeberg, *Introduction to Planning History in the United States*, The Center for Urban Policy Research, Rutgers University (New Brunswick, NJ, 1983).

37 Chudacoff, op. cit., pp. 246–250.

38 Peter Drier, John Mollenkopf, and Todd Swanstrom, *Place Matters: Metropolitics for the Twenty-First Century*, University of Kansas Press (Wichita, KS, 2001), pp. 105–107.

39 Neil H. Lebowitz, "'Above Party, Class, or Creed': Rent Control in the United States, 1940–1947," *Journal of Urban History, 7:4* (August 1981), pp. 439–469.
40 Peter Hall, op. cit., pp. 423–433.
41 Gregory Hooks and Leonard E. Bloomquist, "The Legacy of World War II for Regional Growth and Decline: The Cumulative Effects of Wartime Investments on US Manufacturing, 1947–1972," *Social Forces, 71:2* (December 1992), pp. 303–337.
42 Ann Markusen, Peter Hall, Scott Campbell, and Sabina Dietrick, *The Rise of the Gunbelt: The Military Remapping of Industrial America*, Oxford University Press (New York, 1991).
43 Nassim Nicholas Taleb, *The Black Swan: The Impact of the Highly Improbable*, Random House (New York, 2007).
44 Thomas Kuhn, *The Structure of Scientific Revolutions*, University of Chicago Press (Chicago, IL, 1962).
45 Eric Morris, "From Horse Power to Horsepower," *Access 30* (Spring 2007), pp. 2–9.
46 Thomas L. Friedman, *The World Is Flat: A Brief History of the Twenty-First Century*, Farrar, Straus & Giroux (New York, 2005); *Hot, Flat, and Crowded: Why We Need a Green Revolution*, Farrar, Straus & Giroux (New York, 2008).
47 Cited in James Gleick, *Issac Newton*, Pantheon Books (New York, 2003), p. 178.
48 See Peter Coveny and Roger Highfield's discussion in *The Arrow of Time: A Voyage through Science to Solve Time's Greatest Mystery*, Fawcett Columbine (New York, 1990), pp. 147–158.
49 John L. Casti, *Searching for Certainty: What Scientists Can Know About the Future*, William Morrow & Company (New York, 1990), pp. 48–76. The change of perspective from determinism to indeterminism is an excellent of example of "paradigm shift," discussed in Kuhn, op. cit., pp. 43–52.
50 As an illustration, try reading Richard P. Feynman's *Six Easy Pieces: Essentials of Physics Explained*, Helix Books/Addison Wesley (New York, 1995). These pieces are *not* easy!
51 See, for example, James Heilbrun, *Urban Economics and Public Policy* (3rd ed.), St. Martin's Press (New York, 1987), pp.124–134.
52 An excellent introduction to a difficult subject can be found in Melanie Mitchell, *Complexity: A Guided Tour*, Oxford University Press (New York, 2009).
53 A discussion of genetic (DNA) similarity across species can be found at www.quora.com/What-percentage-of-human-DNA-is-shared-with-other-things

2 The 1950s

A revolutionary decade?

Often viewed as the calm before the storm, the 1950s set in motion trends as revolutionary for cities as any decade before or since. Physically, mentally, culturally, aspirationally, competitively, American urban life in 1960 was vastly different from what it was in 1950. Although the gradual pace of change masked its profound implications, and although each individual component could be viewed as benign, the sum of the differences set the stage for a redefining of cities that would last until nearly the end of the 20th century.

The 1950s were the heart of the Baby Boom years, by many accounts the most significant demographic event in American history. During the decade, US births averaged 4 million per year, an unprecedented population increase. Total population grew 18.7 percent between 1950 and 1960, from 151 million to 180 million. The share of US population living in central cities edged downward, from 32.8 percent to 32.3 percent, and the share for the suburbs grew from 23.3 percent to 30.9 percent. With the large overall population gain, of course, the absolute population of central cities was up by 8.6 million residents, and few saw any cause for alarm in what seemed to be a "win–win" proposition. Such complacency would soon be confounded.[1]

Between 1900 and 1940, the home ownership rate for the United States fluctuated narrowly between 43.6 percent and 47.8 percent. Largely owing to the 1944 GI Bill's support for veterans' mortgages, home ownership became the majority form of housing tenure by 1950, at 55.0 percent of US households. (The Veterans Administration had backed 2.4 million home loans by 1952.) US home ownership surged to 61.9 percent by 1960 and has remained between 60 percent and 70 percent ever since.

The GI Bill also provided education benefits for servicemen returning from World War II. In 1947, 5.6 percent of Americans aged 25 or older had 4-year college degrees. But, with the new government support, veterans accounted for 49 percent of all college admissions, and, by 1950, 7.7 percent of the 25-and-over population held bachelor's degrees or higher. By 1959, that figure was 11.1 percent, roughly double the 1947 ratio. As the Boomers hit college age themselves, the ratio would redouble again, hitting 24.0 percent by 1977.[2]

US living standards were on the rise throughout the 1950s, and the gains are, if anything, understated by real per capita income figures showing inflation-adjusted growth of 18.7 percent over the course of the decade. The growth figure is misleading, to a degree, because of the Baby Boom impact: there were far more children under the age of 15 in 1960 than in 1950, and, therefore, many more "dependents" in the denominator of the per capita income calculation.

Some sense of the extent of the Boomers' impact can be gleaned from the dependency ratio (those aged 14 and under plus those 65 and older, divided by the number of those between the ages of 15 and 64). That ratio leapt from 55 percent in 1950 to 67 percent in 1962, when it hit its all-time peak for the United States. As the Boomer generation began to enter the workforce later in the 1960s, the dependency ratio began a long-term decline, falling back to 55 percent by 1975 and dropping as low as 48.8 percent by 2007. But now, as the Boomers reach retirement age, that ratio is beginning to edge upward once more.[3]

These variables will continue to be part of the discussion of how the nation and its cities have changed over time. However, we should first devote attention to a few important topics worth discussing in depth at this point.

Federal investment in highways a key policy driver for the decade

At the beginning of the 1950s, the governing transportation plan for the nation dated to 1927, mapping a system of US highway routes that would eventually grow to 162,200 linear miles of road. Even into the 1950s, though, there were unpaved roads on that system. Although organized by a nationally conceived route numbering system, highways were largely built, maintained, and paid for by state transportation agencies. West of the Mississippi River, US highways were sometimes little better than during the days of the 19th-century wagon trains.[4]

In the 1950s, US automakers rolled out approximately 8 million vehicles annually, two-thirds of the global total. The number of vehicles on the road grew from 43.2 million to 67.9 million over the decade. Cars per capita jumped by 32 percent, with nearly 38 percent of Americans owning cars by 1960, or 1.3 cars per household. The number of cars per employed worker jumped from 0.74 in 1950 to 1.03 in 1960, a rise of nearly 40 percent that testifies not only to increasing affluence but also to increasing automobile dependence for commutation.[5]

As with the social measures of home ownership and educational attainment, the changes in automobile use in the 1950s were just the launch pad for even greater shifts in future decades. The 1960 Census was the first to track systematically Americans' journey-to-work patterns by mode of transportation. In 1960, 41.4 million workers used personal vehicles to get to their jobs. This represented 64.0 percent of all employed persons. By 2000, automobile commutation would soar to 112.7 million workers, 87.9 percent of the total. Meanwhile, mass transit commutation dropped from 7.8 million (12.1 percent of the total) in 1960 to 6.1 million (4.7 percent of the total) in 2000. And those walking to work each day fell even more drastically, from 6.4 million (9.9 percent share) in 1960 to just 3.8 million (2.9 percent share) at the turn of the millennium.[6]

Unsurprisingly, as the suburbs grew from 1950 onward, so did the nation's automobile dependency. The shifts were facilitated by massive public investment in highways, the infrastructure that was required to accommodate – and to spur – Americans' change in residential location and modality of journey to work.

American President (1952–1960) Dwight D. Eisenhower had experienced the pre-World War II highway system as an officer managing the US Army's first transcontinental convoy rolling from Washington, DC, to San Francisco. This 1919 event was marked by vehicles stuck in sand or mud, trucks and other equipment crashing through wooden bridges, and extremes of weather conditions where roads iced over or sagged in desert heat. The journey took 2 full months and is recounted by Eisenhower in his 1967 memoir, in a chapter entitled, "Through Darkest America with Truck and Tank." The convoy averaged 6 miles per hour, covering 58 miles per day.[7]

That experience help prompt Eisenhower to declare, just after his election in 1952, that:

> The obsolescence of the nation's highways presents an appalling problem of waste, danger and death. . . . A network of modern roads is as necessary to defense as it is to our national economy and personal safety.
>
> We have fallen far behind in this task – until today there is hardly a city of any size without almost hopeless congestion within its boundaries and stalled traffic blocking roads leading beyond these boundaries. A solution can and will be found through the joint planning of the Federal, state and local governments.[8]

This long-standing commitment led Eisenhower to convince Congress to pass the 1956 National Interstate and Defense Highways Act, which would eventually add 46,876 miles of improved routes across the country, at a cost of $129 billion, funded 90 percent by the federal government. The key changes were technological. Interstate highway lanes had to be 12 feet wide and have shoulders 10 feet wide. Bridges had to have 14 feet of clearance. Grades had to be less than 30 percent, and the highway had to be designed for travel at 70 miles an hour. The most notable attribute of the system was the limited access concept, promoting safety as well as speed. Along this system, the trip that took the Army's convoy 60 days in 1919 was reduced to 5 days.

Although it has never been necessary to use the interstate system for massive troop movements, modern highways have certainly facilitated population migration. There is a 68 percent correlation between the number of interstate highway miles in a given state and its population increase in the half-century between 1960 and 2010. Leading the parade (or, perhaps, the auto rally) are states such as California (2,457 interstate miles, 21.5 million population increase, or growth of 137 percent over 50 years), Florida (1,472 miles, 13.8 million population gain, or 280 percent), and Texas (3,333 miles, the most of any state, 15.6 million new residents, 163 percent population increase). Although they did not see as much absolute population gain as the three leading states, there were other notable beneficiaries as well. Arizona, for instance, received 1,169 miles of new interstate routes and added 5.0 million to its population, a stunning 386 percent expansion. Georgia, similarly, saw 1,253 miles of new highway and grew its population by 5.7 million, or 146 percent. And Utah leveraged its 977 new interstate miles into a gain of 1.9 million residents, or 210 percent.[9]

Percentage-wise, there has been no faster-growing state than Nevada since 1960. Nevada has increased its resident population nearly tenfold, from 265,278 in 1960 to 2.4 million as of 2010. Though Nevada received an allocation of just 560 miles of interstates, this was 211 miles for every 100,000 state residents in 1960. Only Alaska and Wyoming enjoyed higher interstate allocations per capita.

A thorough look at highway mileage allocation, normalized by population at the beginning of the interstate era, generates some enlightening numbers to contemplate. Across all 50 states, the interstate system was developed at 26.6 miles per 100,000 of 1960 population. The sparsely populated states garnered federal highway allocations far above average: Arizona (90 miles), Idaho (92 miles), Montana (178 miles), New Mexico (105 miles), and Utah (110 miles). By contrast, the more densely populated states had substantial under-allocations from the highway trust fund (the mechanism by which tax dollars were collected and distributed for the interstate system). For instance, California's highway mileage per 100,000 was just 15.6, Massachusetts' was 11.0, New Jersey's was 7.1, New York's was 10.1, Ohio's was 17.8, and Pennsylvania's was 16.3.

The net effect, then, was a significant redistribution of capital within the United States. Financial capital, in the form of taxes collected in states with large, dense populations, along the coasts and in the industrial Midwest, was converted into the physical capital of modern roadways supporting growth in the Sunbelt and the rural West. When combined with other technological, economic, and social forces, the national investment in the interstate highway system exerted a powerful influence on the reshaping of America's economic geography.

Long-term changes in macro-level social and economic variables, together with the massive investment in automobile-centered infrastructure, began to affect cities and their suburbs during the 1950s. The most dramatic impacts might not be seen until later, but the 1950s were the turning point. We can turn to one specific industry to watch how the socioeconomic statistics played out over time. That industry is professional sports.

Sports franchises: Lessons in economic geography

Figures 2.1 and 2.2 illustrate the location of teams in four major professional sports leagues in 1950 and 2014, respectively. Taken together, these two maps tell intriguing and important stories about American (and, to some extent, Canadian) economic evolution since the midpoint of the 20th century.

Perhaps the most striking feature of the 1950 map is that, with the exception of two football teams on the West Coast, the Mississippi River set the western

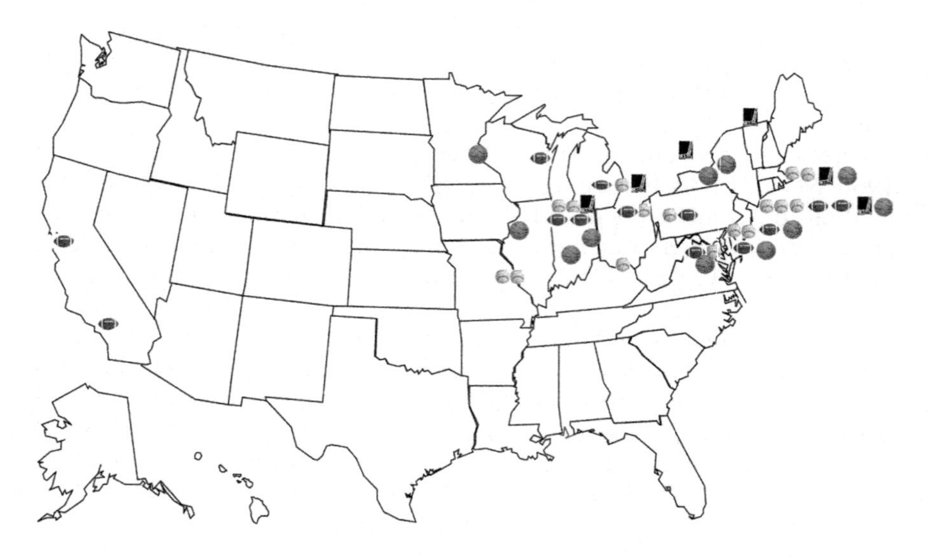

Figure 2.1 Major professional sports teams: 1950

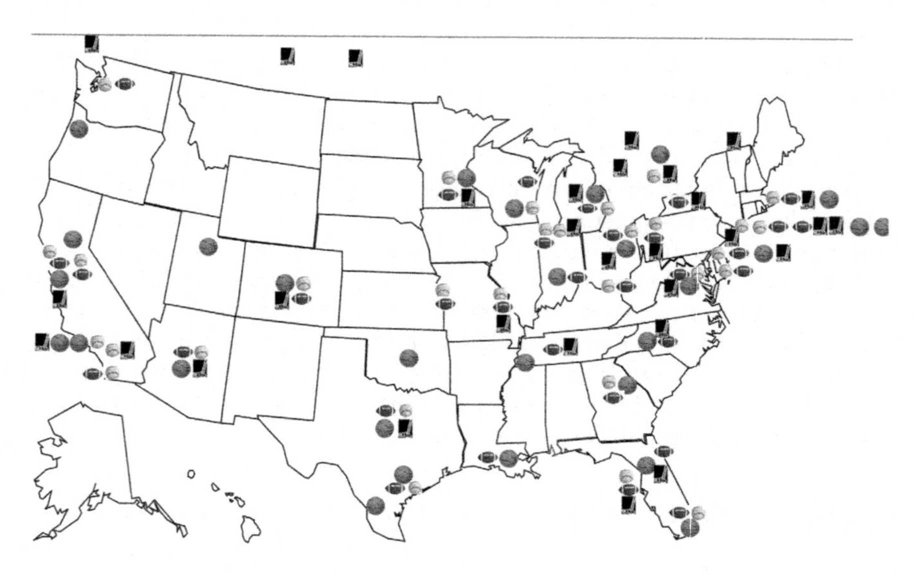

Figure 2.2 Major professional sports teams: 2014

boundary for professional sports. And a line running from Washington, DC, through Cincinnati, to St. Louis describes the southern boundary for the cluster of professional sports leagues. Canada hosted two franchises in ice hockey, its national pastime: the Montreal Canadians and the Toronto Maple Leafs. But, otherwise, the map shows a muscular concentration of big leagues sports in the northeastern United States.[10]

This is hardly an accident. Intercity travel in the early 1950s was predominantly by rail, and the Northeast had a dense network of rail connections for both passengers and freight. The New York Central Railroad's "20th Century Limited" made the run from New York to Chicago in a scheduled 16 hours. By contrast, the Santa Fe Railroad's "California Limited" express train took 36 hours for the run from Chicago to Los Angeles. Chicago was the primary hub for the nation's rail service, which in 1950 had 224,000 miles of track. Prior to the development of the interstate highway system, railroads accounted for 65 percent of all freight tonnage and virtually all long-distance passenger movement.

By 1950, transcontinental air travel certainly was available. In the era of piston-driven planes, however, the trip was relatively long and hardly quiet. TWA's Lockheed Constellation aircraft made the journey from New York to Los Angeles in about 11 hours, including a refuelling stop in Chicago. Weather conditions made travel relatively unreliable, and vibrations contributed not only to noise but to mechanical problems as well. Flying was expensive: adjusting for inflation, it was five times as expensive as in 2014. And safety was questionable: in 1950, there were 2.4 fatal accidents per million departures (that figure has been reduced presently to 0.01 per million departures during the first decade of the 21st century, as indicated in Figure 2.3). All things considered, sports teams – and most businesses – preferred the rails in the 1950s.[11]

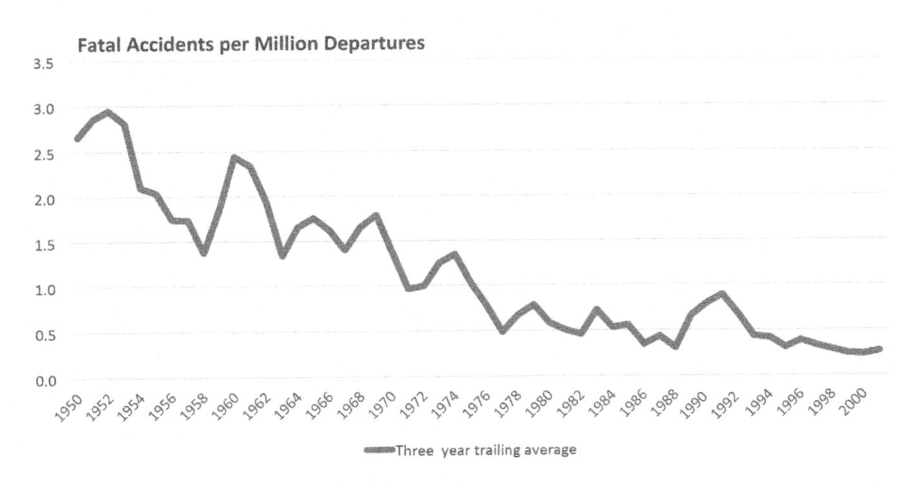

Figure 2.3 Economic expansion propelled by air travel safety as well as speed

Source: Air Transport Association of America

The concentration of sports teams reflected the nation's concentration of population and commerce, of course, and this was the economic foundation of 1950's dense network of rail connections. The US population centroid in 1950 was located in Clay County, Illinois, about 240 miles south of Chicago.[12] Of the ten largest American cities, nine were located in the northeast quadrant depicted in Figure 2.1. Los Angeles ranked 4th, and San Francisco 11th, in the 1950 Census. Pittsburgh was more populous than Houston. Buffalo outranked Seattle. Newark was larger than Dallas. Fort Wayne (Indiana) was home to more people than Tampa or Phoenix.

To put this in context, in 1950, the United States represented 27.3 percent of world GDP, the total output of goods and services in the global economy.[13] Much of that production was controlled in the historical manufacturing industries of the Northeast. In 1955, *Fortune* magazine published its first list of "the top 500 industrial companies in the United States." Figure 2.4 shows their revenues that year and the location of their headquarters – all in the Northeast.

Sports teams often adopted names reflecting their hometown's economic power. Some are obvious. We find football teams such as the Pittsburgh Steelers and the Green Bay Packers, baseball teams such as the Boston Red Sox (hearkening back to New England's textile industry) and Washington Senators, and basketball teams such as the Fort Wayne Pistons. Brooklyn's baseball team, the Dodgers, was a backhanded tribute to its ubiquitous streetcar system – the nickname being a shortened version of "the trolley dodgers." So sports teams were expressions of local pride, as well as being economic entities in their own right.

Firm name	Industry	Revenue ($ million)	Location
General Motors	Automobiles	9,823.5	Detroit
Esso	Energy	5,661.4	New York
US Steel	Steel	3,240.4	Pittsburgh
General Electric	Equipment manufacturing	2,959.8	New York
Swift	Foods	2,510.8	Chicago
Chrysler	Automobiles	2,071.1	Detroit
Armour	Foods	2,056.1	Chicago
Gulf Oil	Energy	1,705.3	Pittsburgh
DuPont	Chemicals	1,687.7	Wilmington

Figure 2.4 Top ten US industrial companies: 1955

Source: Fortune Magazine

As much as the 1950s are viewed, in retrospect, as an era of relative stability, an examination of development in sport shows considerable flux. As in the macro-level socioeconomic variables introducing this chapter, changes that surfaced in the 1950s were the leading edge of some trends that accelerated in the 1960s and later decades. One of those changes was in the movement of franchises, relocations that affected cities' economies and their morale.

The illusion of home team permanence was shattered early in the decade. Places as large as New York City and as small the Tri-City cluster of Moline and Rock Island (IL) and Davenport (IA) saw their teams pick up stakes. The actual number of cities with big league teams actually declined from 21 in 1950 to 20 in 1960, as four towns (Fort Wayne, Tri-Cities, Rochester, and Minneapolis) saw basketball franchises relocate to larger urban areas. Milwaukee, Dallas, and Kansas City were added to the roster of franchises. During the first half of the decade, historic baseball franchises moved from Boston, Philadelphia, and St. Louis. All three of those cities found it difficult to support two baseball teams, and so the Boston Braves, Philadelphia Athletics, and St. Louis Browns departed, leaving the cities to the stronger Boston Red Sox, Philadelphia Phillies, and St. Louis Cardinals.

The most stunning move, however, was the paired relocation of the Brooklyn Dodgers and New York Giants to Los Angeles and San Francisco, respectively, leaving the enormous New York market to the American League team, the Yankees. Negotiations between the Dodgers and Giants franchise owners and the National League required the teams to move simultaneously, thus spreading the costs of travel to the West Coast. With the exception of Kansas City's addition, the geography of the pre-1950 clustering of teams was unaltered. This shows the vital importance of the transportation network, especially intercity rail, during that era.

The leakage of teams from their historic locations became a flood in the 1960s. Chicago, Dallas, Los Angeles, Minneapolis, Philadelphia, St. Louis, Syracuse, and Washington, DC, all saw sports teams decamping in search of greener pastures – greener, as in the color of money.

But relocation was not the only story, by any means. The US population of about 150 million in 1950 grew to 180 million in 1960 and to 205 million in 1970, as the Baby Boom climaxed. This meant that markets of all kinds were expanding, and professional sports were no exception. Opportunities for league expansion were abundant.

Some cities that lost teams were awarded replacement franchises (New York and Washington, DC, for example). But other cities were simply targets of opportunity for baseball, football, basketball, and hockey. Baseball placed new teams in Houston, Montreal, San Diego, and Seattle. Football opened for business in Dallas, Minneapolis, Atlanta, and New Orleans. Professional basketball came to Chicago, San Diego, Seattle, and Phoenix. And the National Hockey League doubled its size in 1968 by adding teams in Los Angeles, Minneapolis, Oakland, Philadelphia, and St. Louis.

The age of jet airline travel surely facilitated the coast-to-coast and border-to-border (and beyond, into Canada) spread of big league sports. Travel time was cut approximately in half when jet airliners replaced propeller planes. Nevertheless, the high capital costs of conversion from a piston-driven fleet to a jet fleet made the dawn of the jet age a gradual one, with passenger enplanements rising only 27.8 percent (from 53.1 million to 67.8 million) between 1958 and 1962. By 1965, air travel volume had soared another 51.8 percent, to 102.9 million, and it jumped another 65.1 percent by 1970, to 169.9 million.[14]

Fueled by demographic expansion and greater travel ease, the number of cities hosting big league teams reached 31 by 1970 and has roughly increased in tandem with overall population growth ever since. As the US was growing, so was Canada, albeit from a much smaller base. In 1950, the total Canadian population was just under 14 million, and Statistics Canada estimated the national population at 35.3 million at the beginning of 2014.[15] Currently, we find major sports franchises in Ottawa, Winnipeg, Calgary, Edmonton, and Vancouver, in addition to the historic franchises in Montreal and Toronto.

With each succeeding decade, the number of "big league cities" increased. This relatively exclusive club admitted nine new members in the 1970s: Indianapolis, San Antonio, Denver, Salt Lake City, Hartford, and Buffalo in the US, and Quebec City, Winnipeg, and Edmonton in Canada. Orlando and Charlotte, as well as Calgary, joined the list in the 1980s. By 2000, the Canadian capital city of Ottawa had its NHL team (the Senators), and Sacramento, Tampa, Nashville, Raleigh, San Jose, and Columbus were boasting about new teams. However, the fickle nature of sports competition was also dropping cities. Quebec lost its hockey team to Denver, Winnipeg to Phoenix, and Hartford to Raleigh. In football, the Houston Oilers moved to Nashville as the Tennessee Titans.

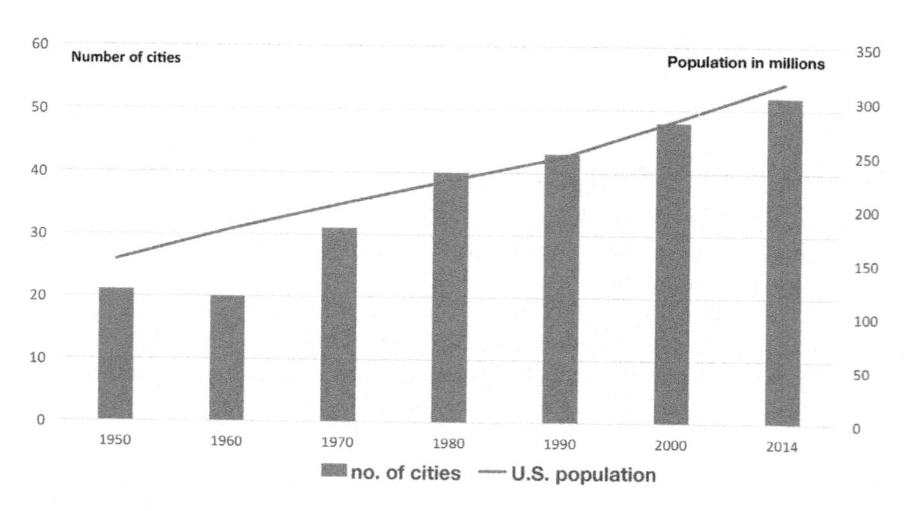

Figure 2.5 US population and number of big league cities

Since the turn of the millennium, Winnipeg has been awarded a replacement hockey franchise, and Newark has become home to the NHL New Jersey Devils. The NBA saw the Seattle Supersonics morph into the Oklahoma City Thunder and the Vancouver Grizzlies relocate to Memphis.

Although there has been unquestionably an absolute increase in the number of teams and cities, the economic geography is not just about "more." Even with greater breadth across the continent, the total number of teams supported in the northeast quadrant dominant in 1950 has grown from 43 to 56. Yet that gain of 13 teams pales before the 61 teams added in the Sunbelt, the Pacific Northwest, in Kansas City, Denver, and Salt Lake City.

It is nearly impossible to imagine that this particular configuration of growth could have been achieved had not one additional technology been developed to supplement the transportation improvements already alluded to. That essential technology was air conditioning.

The Sunbelt's indispensible technology

US Energy Information Agency data show that about 100 million US homes have air conditioning (in a 2009 survey). That's 87 percent of all American households. Currently, 90 percent of all new homes are built with central air conditioning installed. In the South, 85 percent of homes now have central air conditioning, and in 67 percent of those it is in constant use throughout the summer (many have programmable thermostats that cycle down the system to reduce consumption when intensive cooling is not needed). About three-quarters of households in the West and in the Midwest have central air conditioning. The more temperate Northeast sees only 44 percent of homes with central air.[16]

That technology, then, contributes to our understanding of how the residential population of Sunbelt cities could explode over the past half-century. The Sunbelt's share of the national population expanded from just 28 percent in 1950 (about 42 million) to 40 percent in 2000 (or 112 million), nearly a tripling in absolute terms. That's quite an impressive impact for a technology that saw its first shipment of a million units in 1953. And it offers another example of an invention from the Northeast (with William Carrier generally credited as its pioneer, and a printing factory in Brooklyn as the first commercial user) contributing to growth in far-flung areas of the country.

It is not only homes, of course, that rely on air conditioning (and other climate control included in the rubric "HVAC," standing for heating, ventilating, and air conditioning). Office buildings and shopping centers, hotels and computer centers, indeed virtually all forms of income-producing property are only marketable with high-quality HVAC systems. Sports stadiums and arenas should not be omitted.

Basketball and hockey arenas, naturally, have long hosted "indoor sports," where climate control systems are integral to facilities design. But, with the opening of Houston's Astrodome in 1965, baseball and football stadiums – once considered "outdoor" venues only – became either pure "indoor" fields (in the case of true domed buildings) or "hybrid" facilities, where retractable roofs could permit play either under natural weather conditions or sheltered from inclement weather if necessary. Domes were the norm in the 1970s (Seattle, New Orleans, and Pontiac, Michigan – a Detroit suburb) and 1980s (Indianapolis and Minneapolis). With the opening of the Rogers Center (Toronto, 1989), retractable roofs began to make their appearance. Domes remained the most common design during the 1990s, though, with new fixed-roof facilities in Atlanta, St. Louis, San Antonio, and Tampa-St. Petersburg. SAFECO field in Seattle became the first US retractable roof stadium in 1999 and has since been emulated in Miami, Arlington TX (Dallas), Houston, Phoenix, and Milwaukee. Since 2000, Lucas Oil Stadium in Indianapolis and Ford Field in Detroit have been the only "true dome" construction projects in the US.[17]

The effect of bringing traditionally "outdoor" sports under a roof has been to provide fans and players with greater environmental comfort (although with some additional physical wear and tear on athletes due to artificial surfaces). Better-controlled environments make for superior production values for televised sporting events as well. Some feel a greater "home field advantage" in domed arenas, owing to crowd noise from fans – long a feature in the indoor arena sports – but this may be somewhat negated when "indoor" teams need to play in true outdoor stadiums for road games.

Whatever the competitive pluses and minuses, though, adopting a dome or retractable roof design has assisted cities subject to extreme hot and cold temperatures, or frequent or excessive rainfall. In this, sports growth has been consonant with evolving building technologies that have permitted residential, commercial, and industrial property to expand into previously inhospitable locations.

Intrametropolitan choices

Both in the core Northeast markets and in the relocation/expansion markets that became more numerous in the past half-century, sports teams reacted to, and to a degree also influenced, urban patterns that ultimately affected cities in their evolution toward 24-hour or 9-to-5 status. Transportation policies, land-use regulation, demographic segmentation, and economic incentives all played key roles. The business of sport provides a salient example of decision-making that influenced industries and cities more generally for decades.

When the Dodgers played in Brooklyn and the Giants in northern Manhattan, transportation to their games was largely by subway, bus, and streetcar. That was generally true for other vintage sports fields, such as Fenway Park in Boston, Forbes Field in Pittsburgh, Wrigley Field in Chicago, Sportsman Park in St. Louis, and Griffith Park in Washington, DC. However, that changed as new stadiums needed to be built to accommodate team relocations and expansion franchises. The Giants, for example, found their home in San Francisco out by the airport, at Candlestick Point, 8 miles from downtown. The New York Mets, the replacement franchise for the relocated Giants and Dodgers, also had their new stadium built near an airport, LaGuardia, in northern Queens.

As new venues were built during the 1960s, the locations tended to be suburban. Oakland's Alameda Stadium was out near the Oakland airport along the Nimitz Freeway. The Minnesota Twins located in Metropolitan Stadium in suburban Bloomington, 15 miles from downtown Minneapolis. The locations near airports, and the highways that served them, were not coincidental. Cheap land at the perimeter of cities was needed in abundance, not only for the footprint of the stadium itself, but for the thousands of cars that became the preferred transportation option for fans. Anaheim Stadium (Los Angeles Angels) was situated to be convenient to no fewer than five interstate highways. Milwaukee opened County Stadium for the Braves, "surrounded by oceans of asphalt" in the suburbs, and the Dodgers' purpose-built stadium, though only a few miles from downtown Los Angeles, was accessible only by automobile, rendering the team's "trolley dodger" nickname totally anachronistic.

Kansas City situated its baseball and football complex at the intersection of I-70 and I-435. Dallas's football and baseball venues are along I-30. Boston's New England Patriots play in suburban Foxborough, between I-95 and US Route 1. The Pontiac Silverdome became home to the Detroit Lions, and the basketball Pistons play in the "Palace at Auburn Hills" near I-75, an arena 32 miles from downtown Detroit. The New York Islanders hockey team called Uniondale, Long Island, home. And first the New York football Giants (1976) and then the New York Jets (1984) shifted their home games to East Rutherford, NJ, hard by the New Jersey Turnpike in the Meadowlands west of the Hudson River.[18]

Cities attempted to counter the trend toward dispersal by subsidizing facilities in the urban core. A debate has raged about the effectiveness of such subsidies in achieving public aims, and we will return to this topic when we consider the desiderata of 24-hour downtowns. But, for now, it is enough to note that one secondary effect of federal interstate development was to draw economic activity – not just sports, but the attendant multiplier effect, as well as more conventional business – away from the urban center and out toward the metropolitan periphery where inexpensive land by the scores of acres could be devoted to surface parking lots. The centrifugal force pushing activity to the edges, and the correlative devaluation of the urban center, became a dominant trend in economic geography for decades. In its own way, this was as significant as the "Frostbelt to Sunbelt" population trend over the same period.

Impact of design

The "cookie cutter" or multipurpose stadiums dotted the landscape as the most common arena configuration. Circular in design, these ballfields were at first lauded for their functionality, with unobstructed sight lines, symmetrical playing areas, and the ability to serve both baseball and football homogeneously, and offering the then-latest electronic features, including animated scoreboards and pre-set musical exhortations to fans. Atlanta, St. Louis, Pittsburgh, Philadelphia, and Cincinnati are best known for the standardized stadium design, but New York's Shea Stadium also stands as an early example, as does Washington's RFK stadium.

Tastes in stadium design have changed over the past two decades, and, especially as teams have sought to return to downtown venues, new arenas have become more idiosyncratic and linked both to urban history and neighborhood context.

In fairness, sports were not alone in adopting a bland modernism in building design. Skyscrapers of the post-World War II period were heavily influenced by the so-called "International Style," which featured a rectilinear footprint, simple cubic form, grid-like fenestration, and above all the dominance of the 90° angle. It was a clean and functional architecture, and some classic buildings such as Manhattan's Lever House and Seagram Building epitomize its strengths. The principle of "form follows function," attributed to Louis Sullivan and associated with Frank Lloyd Wright, is powerful in its clarity and, in the hands of masters, it produced some excellent urban configurations. But sheer repetition and imitation eventually made the style banal. City skylines began to resemble one another, just as much as the circular sports stadiums did. What was at first "modern" soon became "tired and old."[19]

The same turned out to be true for retail properties. The concept of an "exemplary design" for shopping centers, where the very architecture promoted the efficiency of the center in encouraging consumption, became a norm. Two 1950s-era malls – Northgate in Seattle and Southdale in Edina, Minnesota – were widely emulated. Southdale, which opened in 1956 and is still in operation, was America's first fully enclosed and climate-controlled shopping center. As the concept of the "brand" was reaching its apex in the advertising world of the 1950s and 1960s, it was perhaps inevitable that shopping center design would stress familiarity as much as it did efficiency. More than the external "envelope," whose variation was largely a function of how many department store "anchors" a mall might have, the insistence of mall developers and financing sources on "national credit tenants" for the so-called "in-line" shops meant that the same retailers occupied shops in virtually every regional mall in the country. Additionally, the mall formula was very similar to the ballpark formula of the era: highway access, cheap land, proximity to consumer dollars, and acres upon acres of parking lots.[20]

Corporate real estate users also began to go with the flow toward the suburbs, as the 1950s passed into the 1960s. Although the emergence of the major corporate relocation and development trend did not turn into a downtown exodus until later, the 1950s did see low-rise suburban offices pop up along the highway system. Unsurprisingly, Los Angeles was an early leader in this trend. Edward Durell Stone's Stuart Pharmaceutical building in Pasadena opened in 1958, and CBS had already established Television City in Hollywood earlier in the decade. Southfield, Michigan, was maturing as an office location in the early 1960s, and vintage photographs show a typical configuration of rectilinear mid-rise towers surrounded by ample surface parking.[21]

The recurrent theme of standardized and highly functional (for its period) design was not out of character with the era. Although it might seem obvious in retrospect, the tendency toward an easy acceptance of conformity was well noted even during the 1950s itself. *The Organization Man*, by William H. Whyte, was published in 1956.[22] Sloan Wilson's novel *The Man in the Gray Flannel Suit* came out that same year.[23] In an era of generally rising affluence, the phrase "go along to get along" had resonance and was a byword of the long-time Speaker of the House Sam Rayburn, as he mentored a generation of politicians, including the future President, Lyndon B. Johnson.

So, although the "cookie cutter" stadium may now be scorned by sports enthusiasts, like so many other features of sports as a business, such stadiums reveal much about the trends and preferences of society over time.

Conventional truths and unexpected consequences

Human beings pride themselves on their rationality. Since the days of the Enlightenment, this has been a point of special pride. But the identification of man as the "rational animal" dates back to the Greeks, Aristotle and Porphyry in particular. In our own century, the theory of rational markets has claimed that the combined information available to market participants makes collective decision-making, expressed in prices, so far superior to that of any individual that it is fruitless to out-think or otherwise rein in market forces: the market, in some profound sense, is "right" and should ideally be left to its own devices. Rationality, in our time, has largely become synonymous with metrics, objective analysis that is reducible to mathematics. In this, the progress of science since the age of the Enlightenment has had a particularly forceful influence on our self-understanding.[24] But, even in this, the Greeks can claim a priority. It was the pre-Socratic philosopher Pythagoras (yes, the same Pythagoras that gave us the geometry theorem $a^2 + b^2 = c^2$ that we all learned in high school) who taught, "The whole world is made of numbers."[25] From physics to sociology to econometrics, the spirit of Pythagoras has helped shape our perspective on the world, and on ourselves.[26]

It is likely that no branch of math has made the rational link of mathematics to the world of objects more directly than geometry, beginning with Euclid. Euclid's book *The Elements* has been called the most influential non-religious text of all time. Quite a claim! But it is hard to dispute that the logical presentation of Euclid has shaped the thinking, not only of mathematicians, but also of philosophers, scientists, and political thinkers, including Jefferson and Lincoln, over the centuries.

There is, therefore, a prima facie predisposition to consider as logical and rational a system of urban organization that has come to be known as Euclidean zoning. Unfortunately, it is not for the rational geometer Euclid that this zoning approach is named. Instead, it is a suburb of Cleveland, Ohio, bearing Euclid's name. In a historic test case of a municipality's ability to organize its domain by particular land uses, the Supreme Court ruled in *Village of Euclid* v. *Ambler*

Realty, Inc. (1926) that Euclid had the right to specify by statute the permissible uses, heights, and densities of land parcels in its jurisdiction, and that Ambler did not have a valid claim that the zoning constituted a "taking" requiring compensation for a loss of value based upon its intended alternative use of the land. Importantly, the court agreed (actually in a close four–three decision) that the zoning was not an arbitrary fiat, was not discriminatory against Ambler, and had a rational basis for its permitted and excluded uses.[27]

The basic idea of zoning was separation of functions, based upon an understanding that certain forms of land use were incompatible with others, and their contiguity diminished the desirability of each. For example, the smokestack industries that dominated manufacturing in the early 20th century created such noise, dirt, traffic, and physical hazard that it was unwise to situate those industries in residential areas. In America's older cities, such contiguity was absolutely the norm. But that was considered a problem to be resolved, rather than a model to be continued.[28]

Noxious influences were prevalent and severe. Sanitary reform, in the interest of public health, was widely viewed as an urgent priority. Philadelphia's public records note that, well into the 20th century, disease was synonymous with urban living, with numerous complaints from residents about noxious odors and gases from nearby manufacturing operations.[29] As noted in Chapter 1, Chicago's Columbian exhibition in the 1890s was popularly called "The White City," but Chicago itself was described as "a miasma of din, anthracite, and putrefaction." That description is paralleled by Thomas Wolfe's depiction of the area around Brooklyn's Gowanus Canal in the 1930s, in *You Can't Go Home Again*, most particularly its "stunningly symphonic stink cunningly compacted of unnumbered separate putrefactions."[30] In those days, the residential neighborhood around the canal was known simply as South Brooklyn, but now has more exalted local names such as Carroll Gardens and Cobble Hill – though the industrial canal remains. Manhattan itself had its peculiarities of contiguous uses. The site of the United Nations complex on the East River occupies an area that was once known for its slaughterhouses, coal-fired electric generating plants, and breweries, just blocks from the residential areas of Murray Hill, Turtle Bay, and Sutton Place.[31]

If, today, the "hollowing out" of the urban core is lamented, it must in fairness be remembered that center cities had these and related dystopic features. Once the dominating disruptions of the Great Depression and World War II were concluded, urban planners seized upon the powers conferred upon cities to reshape land-use patterns via zoning codes to address the critical issues of health and safety. Public perception of the center as a place to live and work suffered as these debates raged in city councils and in the popular press. For planners, the easiest approach was to treat the urban map as a tool to shape the future. Where existing uses were undesirable, they might be permitted to remain or, if possible, they might be acquired using the power of eminent domain and razed. But, whichever of the two paths was taken, the future would be determined by Euclidean zoning, with its segregation of uses.[32]

Growing up, I had personal experience of that mixing of uses. Up until the age of 7, I lived on Fenimore Street in Brooklyn. It was a street of attached and detached homes, but right down the block was a yard housing construction cranes. We'd often watch the derricks being loaded and moved on flatbed trucks, heading off to building sites. Just a couple of blocks away were the gas tanks of the local utility, enormous structures we could watch rise or fall depending upon the volume of natural gas present at any given time. Right next to the tanks was a sanitation department garage, a depot for scores of trucks. The garage's distinctive diesel smell is a vivid olfactory memory even today. Amazingly, a Parks Department playground was on the very next corner from the tanks and the garage, and one of New York City's largest medical facilities, Kings County Hospital, was right across New York Avenue from the garage. Despite the predominantly residential uses in the neighborhood, there was a two-square-block concentration of industrial uses – including some light manufacturing and auto repair shops – around the gas tanks.

It was the "old normal," and we were quite used to it. The neighborhood had just about all you would need within walking distance (depending upon your tolerance for walking – which I remember was somewhat limited when I was 4 or 5 years old). We walked back and forth to grammar school and church, for instance, about four blocks away. There were grocery stores and a small supermarket, an Army–Navy surplus store where we bought dungarees, tee-shirts, and shoes, candy stores and barber shops. And, of course, bars could be found on every block, where the men would spend their hours and which were, for many, the places where television was first encountered. The IRT subway ran under Nostrand Avenue, linking the area to the downtown department stores on Fulton Street, to the salt-water swimming pool in the basement of the Hotel St. George in Brooklyn Heights, and thence to Manhattan and the Bronx. My father drove the Nostrand Avenue bus, which ran the length of Brooklyn from the East River to Sheepshead Bay and the beaches on the Atlantic Ocean. Sometimes, my mother would bring us up to the avenue to wait for his bus, and we would pile on for a round-trip ride. In my imagination, I could pick out the very tenements where Ralph Kramden and Ed Norton lived with Alice and Trixie.

In retrospect, my experience was emblematic of the striving and transitioning working class of that era. Our family were renters on Fenimore Street. We were on the second floor of a row house. And we were an extended family in those seven rooms – my grandparents, my unmarried uncle Frank, my mom, dad, and three of us kids. My dad was a US Marine veteran from World War II and remained a blue-collar worker all his life. My grandfather (my mother's dad) was a manager for the American Locomotive Company. My uncle was a secretary for a Manhattan law firm. Another uncle, living a few miles away, worked in the front office of Todd Shipyards. Down the block lived my godmother, Agnes McHugh, and her husband John, a New York City cop. White-collar and blue-collar workers, Brooklyn-based or Manhattan commuters, all mixed in the neighborhood.

We lived just 1 mile, a 15-minute walk, from Ebbets Field, where the Brooklyn Dodgers played. No one in our extended household owned a car, and, in fact, automobiles were so uncommon that it was safe and accepted for the kids on our block to play games of touch football, stickball, and various forms of tag right in the street. Sometimes a lookout would be posted and, if an automobile came down that block, the lookout would call "Car Time!" to halt play. Parents, sitting on the stoops or at chairs near windows, kept an eye on us. Occasionally, the distinctive bell of the "grinder truck" would sound, bringing mothers out to have knives and scissors sharpened. And even better, from our perspective as kids, was the clop-clop sound of the horse pulling the Italian Ices cart in the summertime. All that outdoor activity, day and night, was especially intense in the summer months, as absolutely no one on our block even dreamed of home air conditioning. If you wanted air conditioning, you went to the movies (or some of the bars, to tell the truth). Although I'm sure Jane Jacobs never made her way to our neighborhood, her analysis of the Greenwich Village of her day closely mirrors my growing-up experience. (Jane Jacobs is discussed in the section on urban renewal below, and her contributions are frequently mentioned throughout this book.)

If this all seems idealized and nostalgic, the forces of change were definitely at work, even in my limited experience. My uncles on my father's side lived in a slightly more affluent neighborhood nearby, but were already looking outside New York. My Uncle Mitch owned a business in Manhattan that made dentures, and Mitch relocated his family to Teaneck, New Jersey, and became a daily commuter into New York. I remember spending a leafy summer in northern New Jersey after the death of my paternal grandmother, when my parents were moving us from Fenimore Street to my grandmother's house on the other side of Kings County Hospital. And my Uncle Matt took the move from Brooklyn even further. Matt was a lineman for the telephone company and moved his family 100 miles north, "to the country" in Dutchess County, New York, where I found out about horses, hay, and creeks one other summer. So, dispersal to suburbs and exurbs is part of my own family history.

All this might suggest that organic change was going to alter urban form in any event, and I believe that there is substantial truth to that presumption. But, as the forces of demographic, economic, and public policy impacted cities, organic and incremental change was both accelerated and redirected.

Zoning and planning

As early as 1939, the Federal Housing Administration (FHA) published a comprehensive study, authored by land economist Homer Hoyt, that looked into how neighborhoods configured themselves in a variety of large and small cities, 64 in all, across the country.[33] Hoyt was clear about the intended usefulness of his work. "The selection of areas for slum clearance, the determination of mortgage lending policies by area, and decisions in regard to zoning or rezoning" made Hoyt's examination of city structure (in his words), "not merely an academic matter." He described his goal, at the highest level, as a "search for an orderly arrangement of land uses . . . a series of techniques . . . designed to bring order out of chaos."[34] Presciently, before the outbreak of World War II, Hoyt suggested:

> In years to come, continued eradication of slum areas or a cessation of population growth may foster conditions favoring a doubling back of high grade residential areas rather than a continuation of growth in line with past experience. Or, it is possible that common use of aircraft or other means of rapid transit may result in a greatly accelerated decentralization.[35]

Hoyt unquestionably advanced the theoretical understanding of how cities structure themselves, and did so on the basis of exhaustive empirical study. Unfortunately, he, and especially lesser minds who followed him, blurred the line between what is descriptive – the way things are – and what is normative – the way things should be. For instance, Hoyt's study of *The Structure and Growth of Residential Neighborhoods in American Cities* rigorously described the racially segregated neighborhoods that characterized the urban areas of the 1930s. He also noted the social taboo of interracial marriage prevalent at the time. But then, in language that is jarring to the contemporary ear, he accepted that such conditions were structurally determined and embedded them in analyses of urban growth patterns and property values, setting the stage for (and actually statistically undergirding) future patterns of mortgage discrimination.[36]

The sectoral theory of urban growth, which indicates that cities grow asymmetrically rather than in concentric rings, did in fact fit the facts of actual cities rather well. This was one of Homer Hoyt's many contributions to a science of land economics. And it was put to practical use by at least two generations of city planners in the development of zoning plans, infrastructure plans, and community service allocation. The expectation of growth toward the perimeter of the urban area, and the anticipation of high value improvements in some sectors and the erosion of value in others, became something of a self-fulfilling prophecy in city after city.[37]

Euclidean zoning was one of the principal means of applying sectoral theory in the service of city planning goals. Those goals, in turn, were expressions of both economic and political interests. Those interests, however, were veiled by a scrim of "rationality," reflecting a judgment that ordered differing land uses into separated, single-use zones that not only segregated commercial/industrial property from residential uses, but even sought to set aside distinct areas for single-family houses apart from multifamily dwellings. It was an engineering approach to cities, an extension of Newtonian deterministic science that saw cities as functional machines, with standardized parts that presumably could be tuned for maximum efficiency. The alternative view, that cities could be construed as social organisms whose growth might follow biological rather than mechanical principles, was hardly considered.[38]

Although it was a real estate company, Ambler Realty Co., that was the counterparty in the Euclid cases, the real estate industry seized upon the opportunities presented by a readily identifiable "target market" for housing, where a relatively standardized product could be developed on relatively inexpensive land, shielded from the competitive pressure of alternative land uses. Following World War II, Levittown and its emulators across the country made the tract-housing suburb the staple of housing development. As Chapter 1 has already suggested, a variety of federal incentives helped shape this market, to the benefit of the urban perimeter and to the detriment of the urban core. The GIs and their families were a potent political constituency, and a grateful nation was eager to reward their heroic service in Europe, Northern Africa, and the Pacific.

It is easy to idealize "The Greatest Generation," as the World War II veterans have come to be called. Tom Brokaw's book on the subject is a paean to them:

> It is a generation that, by and large, made no demands of homage from those who followed and prospered economically, politically, and culturally because of its sacrifices. It is a generation of towering achievement and modest demeanor, a legacy of their formative years when they were participants in and witness to sacrifices of the highest order.[39]

Yet divisions of race and class were still very much a part of the ethos in those years when the veterans were making their imprint upon American society and upon its cities. It may be difficult to recall that, although President Harry Truman's Executive Order 9981 abolished racial discrimination in the Armed Forces in 1948, it did so with the proviso of "taking due regard to the time required to effectuate any necessary changes without impairing efficiency or morale." In fact, the last all-black unit in the Army was phased out in 1954, and implementing Executive Order 9981 required the forced resignation of the Secretary of the Army, who refused to comply with the desegregation policy.[40] The year 1954 also saw the landmark Supreme Court decision *Brown* v. *Board of Education*, which overturned a system of separate schools for whites and blacks. As Figure 2.6 displays, at the time of *Brown*, segregated schools were *required by law* in 16 states, while forbidden by law in 16 others. The regional division of the country along segregationist lines is apparent.[41]

What is not so apparent is the urban profile of race and the underlying rationalizations that supported racial divisions. At the turn of the 20th century (1900), the African–American population was represented in white urban neighborhoods in approximately the same proportion as in the national statistics – even in the cities of the South. The Supreme Court's 1896 ruling in *Plessey* v. *Ferguson* upheld the right of individual states to have "separate but equal" facilities to keep races apart. This, in practice, meant segregated transportation, segregated washrooms, segregated drinking fountains, segregated lunch counters, as well as segregated schools. When combined with Euclidean zoning, it also meant a vast increase in segregated urban neighborhoods. Harvard Law

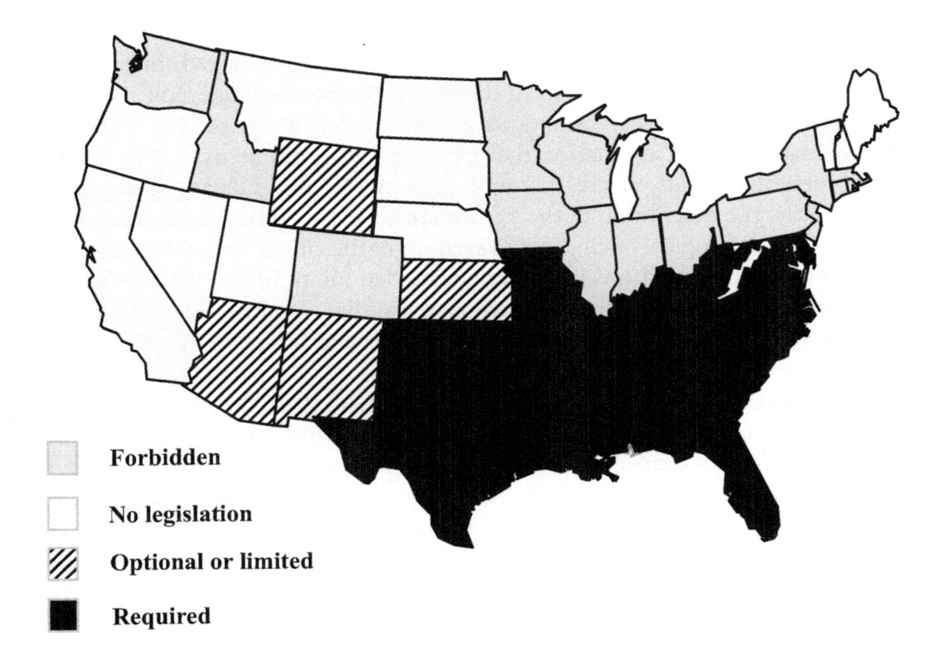

Figure 2.6 Educational segregation in US before *Brown* v. *Board of Education*

School's Gerald Frug notes that, "neighborhoods . . . have been organized, through zoning and other policies, in a way that has divided them into racially identifiable spaces."[42] By 1939, Homer Hoyt was able to observe:

> The presence of a certain proportion of members of an inharmonious race in a block affects the characteristics of the entire block. . . . in most of the areas there is an abrupt transition from blocks occupied by Negroes [*sic*] to those in which the entire population is white. These concentrated Negro areas . . . tend to fall in those same . . . sections of the city where, as we have seen, rents are lowest, buildings are oldest and in the poorest condition, and the largest percentage of dwelling units lack private baths [and] the sections occupied by the colored race generally lack central heat and are overcrowded.[43]

Redlining

After *Plessey* and *Euclid*, the degree of ghettoization of US cities had intensified greatly. The practice of "redlining" pushed this trend along with a vengeance. The FHA developed "residential security maps" (or HOLC maps, referring to the Home Owners Loan Corporation, another federal agency) to guide home valuation, dividing 239 US cities into areas that were suitable for mortgage lending (green or blue areas on the map indicating rising or stable areas, by FHA's criteria) or unsuitable (yellow, for declining, or red, for already in decline). These maps overtly institutionalized discrimination in housing credit, and the high correlation of African–American populations with substandard housing characteristics meant, in practice, that fresh capital would be denied to whole urban neighborhoods where it could have made a significant difference. Importantly, the description of the "green areas" specified that they were "*new* [emphasis in the original] well planned sections of the city." "Yellow" and "red" areas were characterized specifically as having "an infiltration of a lower grade population." The result, as foreseen by FHA, was that, "some mortgage lenders may refuse to make loans in these neighborhoods and others will lend only on a conservative basis."[44]

The dominant scholarly consensus, after more than three decades of research on these city maps, is that they contributed significantly to disinvestment in the residential stock in all US central cities and particularly disadvantaged African–American neighborhoods. Researchers have also documented that private interests from the financial and real estate industries helped shape and then benefited from the racially biased public policies.

Recently, however, some researchers have sought to argue that the interpretation of the HOLC maps as simply race-based fails to account for the multifactor criteria used to develop the neighborhood ratings. This nuance, although helpful, only confirms the linkage between redlining and zoning. In Chicago, for example, the redlined neighborhoods included those areas where housing was mixed with commercial/industrial uses and where the stock was relatively older and consequently in poorer condition than newly developed and homogeneously residential districts. Homer Hoyt had already seen that African–American concentrations were highly correlated with such characteristics.

Thus, even in the more recent scholarship reviewing the HOLC maps and 1940 Census data, it has been found that *all* predominantly black census tracts in the City of Chicago were classified as "red" areas, whereas all other ethnicities ranged from "red" up to "blue" (Irish, Polish, Eastern European, and Italian) or "green" (Scandinavian and Jewish).[45]

Redlining persisted through the 1950s and 1960s, and its abuses prompted the passage of the 1968 Fair Housing Act (Title VIII of the Civil Rights Act of 1968). By then, however, much of the damage was already done, and remediation would be a matter of decades. The destructive combination of redlining and Euclidean zoning was known well in advance of 1968, however. And one of the most prominent urban voices had already sounded an alarm in 1961.

Urban renewal

That year, 1961, Jane Jacobs published her startling and iconoclastic book, *The Death and Life of Great American Cities.*[46] In this work, Jacobs, who was not herself a professional planner, turned planning and zoning orthodoxy on its head. She held that, "diversity is natural to big cities." Four key principles serve as the preconditions for vibrant urban areas, according to Jacobs. They are high population densities and the activities they spawn; a mixing, rather than a segregation, of primary land uses; small-scale, pedestrian-friendly blocks, rather than streets designed primarily with automobile throughput in mind; and the preservation of old buildings, even as new construction is provided within existing neighborhoods.

Although Jacobs considered Euclidean (or "rationalist") zoning the most prevalent of the city-destroying forces of the 1950s and 1960s, she believed that urban renewal was the most violent of the tools in the planning arsenal.[47] Her position, now widely endorsed, ran smack against powerful public policy (especially in the person of Robert Moses, who once held 12 public offices simultaneously) and such shapers of public opinion as *The New York Times*. Unsurprisingly, the momentum behind urban renewal was enormous, and its very relentlessness to some degree lay behind Jacob's *cri de coeur*. Here is an excerpt from her observations in *Death and Life*:

> Look what we have built [in urban renewal] . . . low income housing projects that become worse . . . than the slums they were supposed to replace. Cultural centers that are unable to support a good bookstore. Civic centers that are avoided by everyone but bums. . . . Expressways that eviscerate great cities. This is not the rebuilding of cities. This is the sacking of cities.[48]

The US Housing Act of 1949 was the legislation enabling urban renewal programs, by extending federal funds to cities to acquire and redevelop areas that the cities deemed to be slums, using the powers of eminent domain. The 1949 Housing Act, and its successor Act of 1954, enabled cities to acquire title to land and subsequently sell that real estate to public authorities or to private developers. In theory, slum clearance was to be the initial step in the provision of public housing, and a one-to-one correspondence between demolition and construction of housing units was supposed to be the norm. That never happened. Although development *did* occur on the newly acquired and cleared land, the amount of local housing created was often just a fraction of what was lost. And the real estate community, which had vigorously lobbied on behalf of the Housing Acts, invoked the principle of highest and best use to construct a wide variety of uses to replace the razed housing.[49]

A few examples may suffice to illustrate how this worked. One of the very first urban renewal projects was undertaken in downtown Pittsburgh. Hundreds of structures were demolished (mostly obsolete railroad buildings, an unused exposition center, as well as small shops and houses). Work began under the leadership of the civic elite, led by banker Richard Mellon and food magnate Howard Heinz, and a 36-acre state park was developed at the point where the Allegheny and Monongahela Rivers flow together to form the Ohio River. The city's new Gateway office district was also developed, with seven office towers, a Hilton hotel, an underground parking garage, and a single apartment building. Using the powers of the Housing Acts, though, Pittsburgh's leaders moved on to the adjacent Lower Hill district, a predominantly immigrant and African–American community. Beginning in the mid-1950s, some 1,500 families and 400 local businesses were evicted from a 95-acre site. A civic arena was constructed, doing double duty as a sports venue and a home for a light opera company. More parking was provided, to accommodate the volume stimulated by a new expressway, which isolated the arena from other downtown districts. Another hotel and two apartment buildings sufficed as "eyewash" for achieving the objectives of the housing legislation.[50]

In Chicago, the street grid in African–American neighborhoods was ripped apart to create "superblocks" for high-rise public housing such as the Cabrini–Green development of 3,607 "no frills" units. On the city's south side, the nation's largest public housing project, the Robert Taylor Houses, was built, clustering 27,000 people in 4,300 units between 39th and 54th Streets. Both Cabrini–Green and the Robert Taylor Houses became synonymous with chaos and violence, and both have since been largely demolished.[51]

St. Louis took advantage of changes in the 1954 version of the Housing Act, which permitted commercial and industrial redevelopment of cleared slums, to tackle the 100-square-block Mill Creek area, a 454-acre, largely residential but severely blighted, district that generated a mere $365,000 in total property taxes. Mill Creek was home to 1,772 families and 610 single-person households. Their living conditions were unquestionably grim. Eighty percent of the existing housing units lacked private bathrooms, and 67 percent had no running water. The infant mortality rate was twice that of St. Louis as a whole, and crime was four times as high. With $10 million from a bond issue, St. Louis condemned and razed 93 percent of the entire 454 acres. Over the course of two decades, some 2,346 units of middle-income, elderly, and low-income housing were built on 83.5 acres. However, these units were largely occupied by households from elsewhere, and the displaced households merely moved the location of blight elsewhere. Meanwhile, the mean income level of the neighborhood actually dropped, the proportion of home ownership declined, and maintenance levels, incredibly, worsened over time.

Of the remaining 370 acres, 26 acres were developed with 643,000 square feet of commercial buildings, 132 acres became the site of 886,000 square feet of non-manufacturing industrial space, and 212 acres were dedicated to parks, schools, and churches. The automobile and truck appear to have been the big winners: 174 acres were reserved for city and state rights-of-way, related to all of the foregoing land uses. In all, $28.2 million in public costs (city, state, and federal) stimulated $97.8 million in private investment over a 15-year period, beginning in 1955. Even with the combined $126 million in investment, the assessed valuation of Mill Creek rose a mere $16.8 million, from $13.3 million to $30.1 million.

St. Louis public housing statistics show 4,733 units constructed during the 1950s, more than replacing the 2,393 units demolished to clear the sites for these housing projects. Quantity did not mean quality, though – 2,738 of those units were in the infamous Igoe–Pruitt complex, whose 33 buildings were demolished in the 1970s, with the implosion broadcast on national TV.[52]

Examples could be multiplied indefinitely, but let's return to Los Angeles and New York for compare-and-contrast purposes.

Los Angeles designated the Bunker Hill area, just to the north of its historic downtown, as a redevelopment area as early as 1951, and adopted a "tentative" urban renewal plan under the Housing Acts by 1956. An area of extremely steep topography, Bunker Hill's 133 acres were described as an aging residential pocket of dilapidated housing, residential hotels, rooming houses, and run-down corner stores constituting a squalid and crime-ridden urban quarter. Implementation of the renewal plan began in the later 1950s and is only now nearly complete. Unlike many other such areas across the nation, Los Angeles has made something of a success in transforming Bunker Hill, though it has taken a half-century to do so.[53]

The Community Redevelopment Agency of Los Angeles reports that, in addition to infrastructure and topographic realignment, 11.4 million square feet of commercial space have been built in the urban renewal zone, as well as 3,255 residential units. There are also public and institutional facilties, including the Museum of Contemporary Art and the Walt Disney Concert Hall. Moreover, project funds have provided for more than 21,000 subsidized housing units, either newly constructed or rehabilitated, throughout the City of Los Angeles.

A 2014 walk through the area showed – with due adjustment for differences of time and place – how prescient Jane Jacobs' prescriptions turned out to be. Although Los Angeles' automobile-dependence remains, with its broad streets serving primarily as a collection network for the freeways still slicing through downtown, Bunker Hill and the flatlands below it are starting to sprout restaurants, pubs, convenience stores, and street-level retailing, all of which are in a mutually reinforcing urban ecology with the offices, apartment buildings, and cultural institutions. Though the rush hour traffic into and out of the Los Angeles downtown remains fearsome, the de facto creation of a mixed-use district – setting aside the strictures of Euclidean zoning – must be acknowledged and applauded.

Jacobs' opprobrium in *Death and Life* was aroused most strongly by the policies and programs of master builder Robert Moses. Moses was the subject of an exhaustive and highly critical biography, *The Power Broker*, by Robert Caro, detailing his extraordinary career and its impact, not only on New York City and State, but also, by his example, on public works nationwide. Moses' reputation is being somewhat rehabilitated in our day, with books such as *Robert Moses and the Modern City*, exhibitions, and even an attempt to emulate Moses' big-vision transformations by economic developers such as Dan Doctoroff in fashioning successive (but unsuccessful) Olympic bids for New York City. Without discounting Moses' positive impacts, such as massive bridge-building connecting New York's five boroughs, and hundreds of parks and playgrounds, it is Moses' housing and highway programs that remain controversial to this day.[54]

For good and ill, Moses' approach was all about scale – grand scale. As New York's Commissioner of Slum Clearance, Moses initiated exactly the kinds of project that Jacobs railed against. The Lincoln Square Urban Renewal Area, for instance, was populated by a black and Puerto Rican community of incredible density – 67 acres averaging 5,000 tenants per square block. Its center, known as San Juan Hill, was the birthplace of the musical form "bebop" and the Roaring Twenties dance "The Charleston," the home of jazz great Theolonius Monk, and the setting for Leonard Bernstein's *West Side Story*. The neighborhood was razed, requiring the forced relocation of all its residents. Most moved to housing in Harlem and the South Bronx, exacerbating the existing overcrowding in those areas and straining their public resources. What replaced this neighborhood was the Lincoln Center for the Performing Arts, a campus for Fordham University, and high-rise luxury apartments.[55]

Moses' biography notes that he was responsible for the development of 150,000 housing units. Many of these were on urban renewal land, under Moses' supervision and control, though built by the New York City Housing Authority (NYCHA). In 1954, the ethnic breakdown of NYCHA tenants was 59 percent white, 34 percent black, and 7 percent Puerto Rican. By 1959, with slum clearance in full swing, black and Puerto Rican tenants accounted for 57 percent of all tenants in NYCHA housing, and the Agency had embarked on a 35 percent expansion (38,000 new units), concentrated in the South Bronx, northern Manhattan, in Brooklyn communities from Fort Greene to Red Hook, Brownsville, East New York, and Coney Island, and in south Queens areas such as Jamaica and the Rockaways. The densities were extraordinary, about 275 persons per acre, or an astounding 175,000 per square mile – about seven times the citywide average population density – overcrowding schools and parks, straining transportation systems, and placing enormous pressure on public safety and sanitation agencies.[56] When Jane Jacobs spoke of density, this is not what she had in mind.

Slum clearance scraped neighborhoods that still lie fallow, incredibly enough. The Lower East Side is just now finding a plan for the Seward Park Urban Renewal Area. Brooklyn's Atlantic Terminal area has just seen, in the last decade, the beginning of Forest City-Ratner's Atlantic Yards. Arverne, on Queens' Rockaway Peninsula, still has acres of vacant beachfront land unutilized 45 years after the bulldozers completed their job.

As for expressways eviscerating cities, we might consider those that were built and those that were thwarted. The Cross-Bronx Expressway linked the George Washington and Throggs Neck Bridges and connected to the New England Thruway by slicing through the Tremont neighborhood. Although the Brooklyn–Queens Expressway was double-decked under the Brooklyn Heights Promenade, which affords spectacular views of Lower Manhattan, it emerges into a trench, cutting the Cobble Hill and Carroll Gardens neighborhood from the Brooklyn waterfront, and (on its Gowanus Expressway leg) isolates Red Hook from other neighborhoods, and its Prospect Expressway extension displaced residents and dissolved community cohesion along its path. Partly as a result of negative civic reaction, in which Jane Jacobs' voice was prominent, two similar proposed expressways across Manhattan were never built, one in Lower Manhattan and the other in Midtown, linking the Lincoln Tunnel to the Queens–Midtown Tunnel.

Frank Lloyd Wright once said, "The outcome of the cities will depend upon the race between the automobile and the elevator. Anyone who bets on the elevator is crazy."[57] For a long time, the odds seemed to be in favor of Frank Lloyd Wright's position, and Robert Moses clearly not only placed his bets accordingly, but tilted the playing fields as much as he could toward the automobile's side. These giants of American design and development were not alone.

Public policy in the 1950s, in such disparate fields as transportation, housing, education, public finance, and law, was a powerful force in shaping cities in the second half of the 20th century. Local influences provided nuance, of course, but by and large cities were more reactive, rather than initiating action. Money is potent, and the federal government was not afraid to spend it. How much the consequences were intended may never be known, but the impact of the federal programs on regional and urban economies was profound. Entropy directed growth toward and beyond the urban perimeter, and the amalgam of influences began the process of hollowing out center cities. Although the vocabulary of "edge cities," "9-to-5 CBDs," and eventually "24-hour cities" had yet to be articulated, the conditions that would add these concepts to urban studies can be found in trends that were already visible in the decade of the 1950s.

Trends: A key form of change

At the conclusion of Chapter 1, the notion of "trends" was noted as one of the principal modalities of change. I distinguish trends from other forms of change mainly because of the longevity of trends. Trends represent a fundamental and persistent shift in some socioeconomic variable, a shift that continues even through the ups and downs of cycles and that follows the Newtonian law of inertia: an object in motion tends to remain in motion, unless acted upon by some countervailing and unbalanced force.

Demography presents us with many of the trends that pertain to our study of cities. For one thing, the US population has not only increased overall since the founding of the nation, but it is projected to increase even further, from about 317 million in 2014 to about 400 million by 2050. With the mechanization of agriculture, a vast rural-to-urban migration has characterized US place of residence and place of work statistics for the past century and a half. As Chapter 1 indicated, this saw more than half the population in urban areas by 1920. Since then, the urban share of the population has pushed above 80 percent (as of the 2010 Census). The enormous expansion in the number of major sports franchises discussed earlier in this chapter is, in part, a consequence

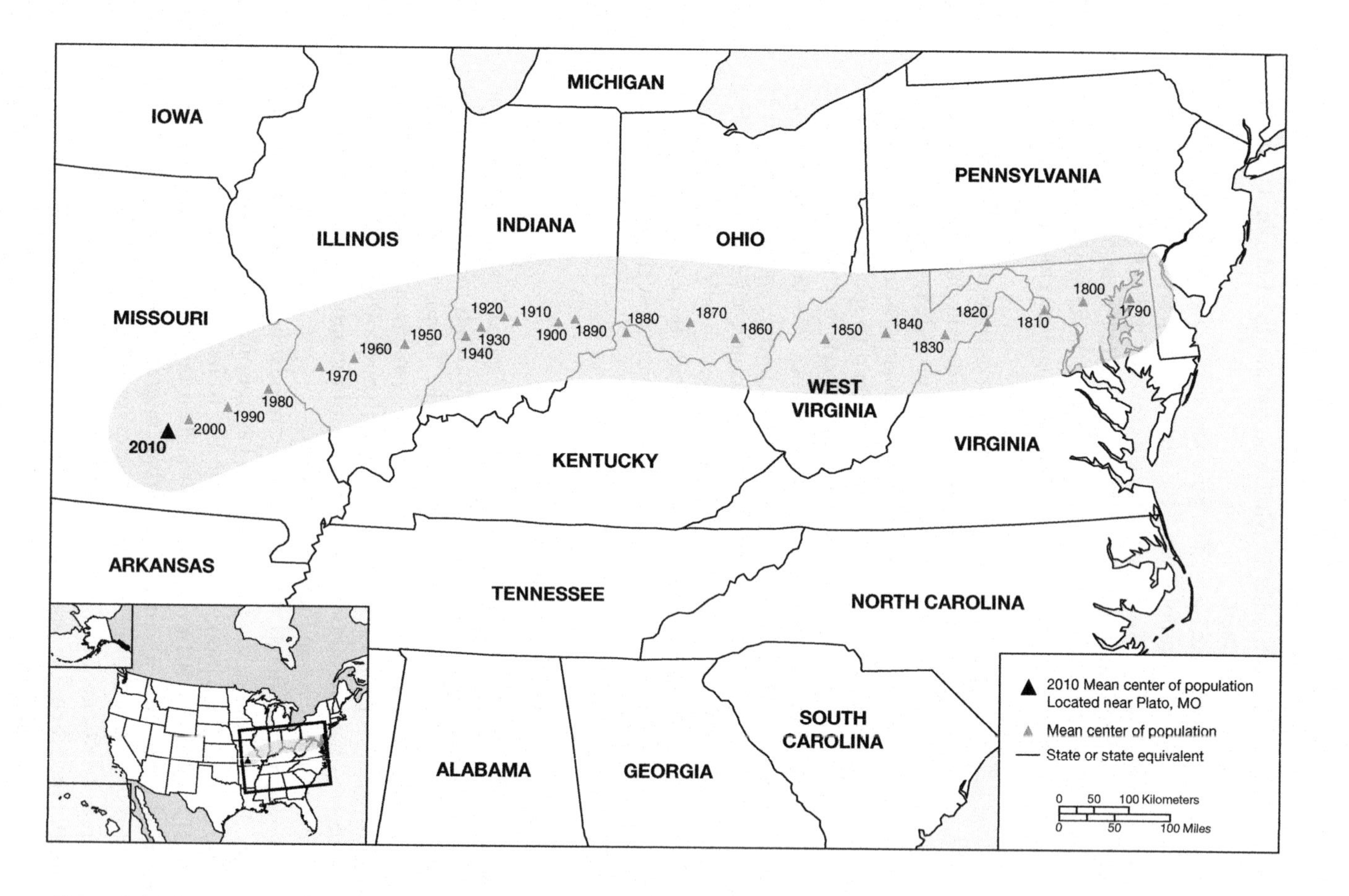

Figure 2.7 Mean center of population for the United States: 1790–2010

Source: US Department of Commerce, Economics and Statistics Administration, US Census Bureau

of this fundamental demographic trend. And what is true of sports has been true of business generally – economic growth follows in the wake of population growth, all things being equal.

Population growth has not been evenly distributed, however, and this has created a long-range expansion of population westward throughout US history and, since 1920, southward. For much of our history, this was a reflection of the expansion of our national frontier, but more recently (as discussed earlier in this chapter) has been influenced by technological advances in transportation and climate control.

Alterations in the level of educational attainment in the United States have also factored into urban change in profound ways over time. In 1995, Peter Drucker wrote a thought-provoking essay in *The Atlantic* entitled, "The Age of Social Transformation."[58] Drucker argued that, in the age of the knowledge worker, it was workers, rather than business owners, who controlled the means of production. If this is correct, urban areas with the capacity to attract and retain knowledge workers, or what Richard Florida has termed "the creative class," have an advantage in concentrating and growing total economic output. Drucker argued, nearly 20 years ago now, that, where government and the business organization were once relied upon as the core forces for economic cohesion, that role is devolving in a more pluralistic society.

> Organizations must competently perform the one social function for the sake of which they exist – the school to teach, the hospital to cure the sick, and the business to produce goods, services, or the capital to provide for the risks of the future. They can do so only if they single-mindedly concentrate on their specialized mission. But there is also society's need for these organizations to take social responsibility – to work on the problems and challenges of the community. Together these organizations are the

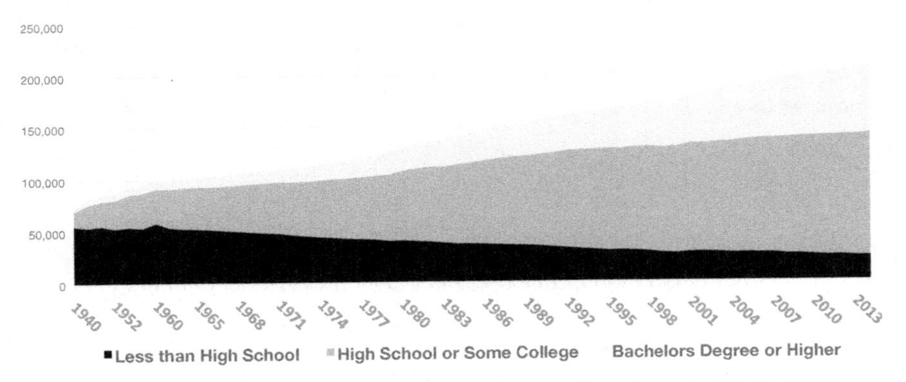

Figure 2.8 Population aged 25 and over by educational attainment: 1940–2013

Source: US Census Bureau, 1947, 1952–2002 March Current Population Survey, 2003–2013 Annual Social and Economic Supplement to the Current Population Survey, 1940–1960 Census of Population

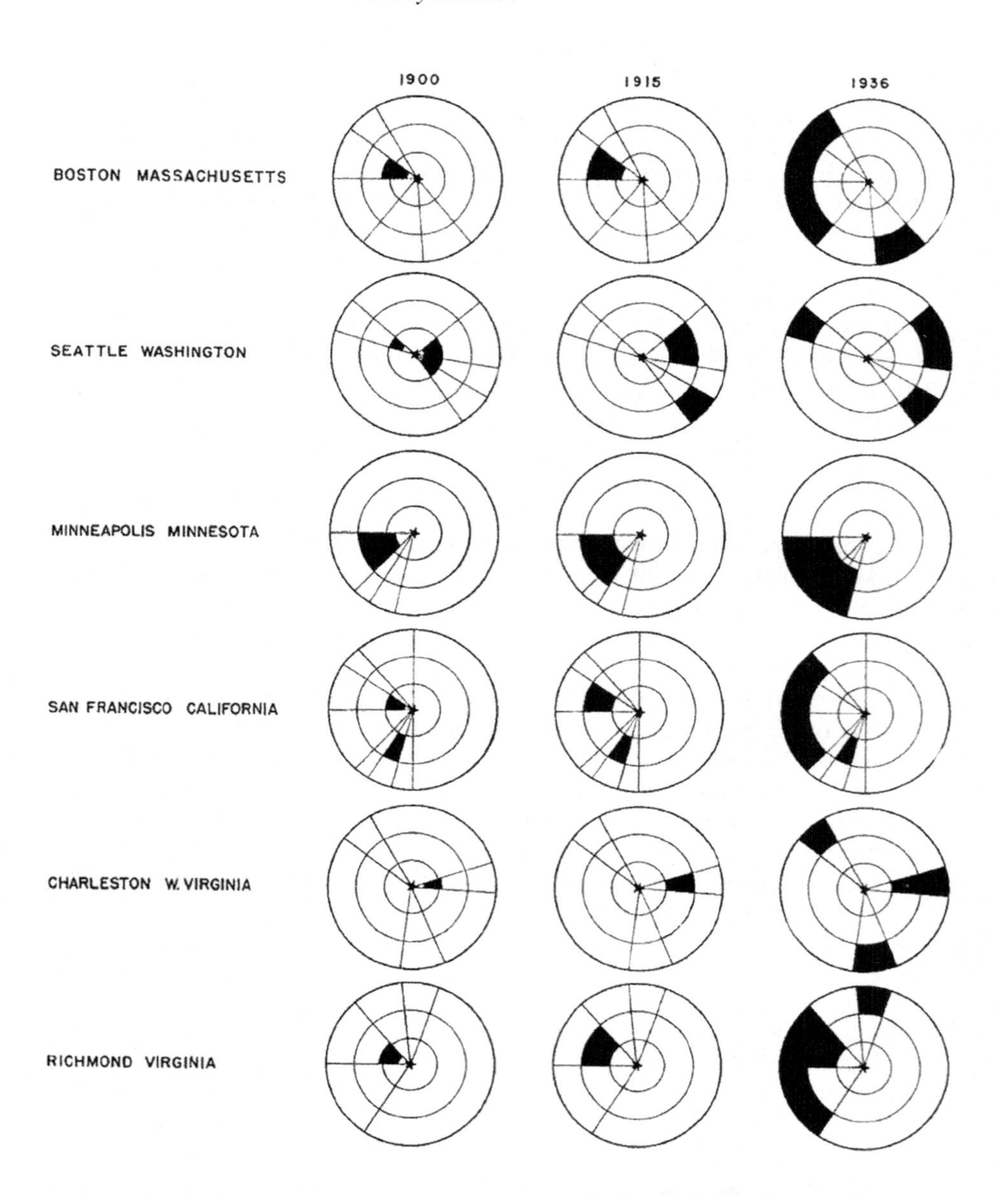

Figure 2.9 Shifts in location of fashionable residential areas in six American cities: 1900–1936 (fashionable residential areas indicated by solid black)

Source: Federal Housing Administration Division of Economics and Statistics

> community. The emergence of a strong, independent, capable social sector – neither public sector nor private sector – is thus a central need of the society of organizations. But by itself it is not enough – the organizations of both the public and the private sector must share in the work.[59]

If the interpenetration of functions – the cross-fertilization of effort and of ideas – is such a vital feature for our economic future, those cities where structures promote dense networking within and across groups of industries and occupation can take greater advantage of the long-term trend in higher-educational attainment than those where separation of functions and districts remains the norm.

Nevertheless, alterations in urban structure are not very easy to achieve, nor are they typically inexpensive. Once economic, social, and political forces have established strong constituencies, there is a huge incentive to keep existing advantages – and disparities – in place. Institutionalizing this in zoning codes helps fix patterns and encourages trend lines to follow their pre-existing paths. Homer Hoyt's 1939 study for the FHA already saw this occurring. Where a vector of growth first establishes itself, as in so-called "fashionable residential areas," the tendency in city after city is for that vector to perpetuate itself, altering only in its momentum toward the periphery of the urban area, as those who can afford to do so can "get more house and more land for less money." Only when the time value of unproductive commutation begins to bite, or when conditions closer to the core become qualitatively better in ways that are economically persuasive to more affluent households, can such trends be reversed. That may be occurring in a minority of US cities, but is not generally the case at present. The centrifugal forces pushing cities outward were powerfully at work in the 1950s and continued to be so for decades afterward. Drucker's observation about the transformative effect of the knowledge worker, published in 1995, might very well be seen as one of the countervailing forces that can disturb the inertial energy of cities' outward push. It may also be more than a coincidence that Drucker arrived at this conclusion at just about the same time that the real estate industry came to focus attention on 24-hour cities.

Notes

1 Frank Hobbs and Nicole Stoops, *Demographic Trends in the 20th Century*, US Bureau of the Census, Census 2000 Special Reports (Washington DC, November 2002).

2 Dayne D. Batten, "The G.I. Bill, Higher Education and American Society," *Grove City College Journal of Law & Public Policy 2:1* (2011), pp. 13–29. Sarah Turner and John Bound, "Going to War and Going to College," *Journal of Labor Economics* (2002), pp. 784, 806–807. *Digest of Education Statistics 2001*, National Center for Education Statistics (2001), online at http://nces.ed.gov/pbs2002/2002130.pdf

3 World Economic Forum: United States' Population Structure, 1950–2000, online at www.weforum.org/pdf/Initiatives/pension_us.pdf; Grayson K. Vincent and Victoria A. Velkoff, *The Next Four Decades: The Older Population in the United States*

(2010 to 2050 Population Estimates and Projections), US Bureau of the Census, *Current Population Reports* (Washington, DC, 2010).

4 R.V. Droz, *US Highways of 1956*, and www.us-highways.com/1927us.htm

5 www.carhistory4u.com/the-last-100-years/car-production; *The Automobile Industry, 1940–1959*, online at http://web.bryant.edu/~ehu/h364/materials/cars/cars%20_60.htm; Oak Ridge National Laboratory, *Transportation Energy Data Book*, Table 8.2 Vehicles and Vehicle-Miles per Capita, 1950–2012, online at http://cta.ornl.gov/data/chapter8.shtm

6 Data can be found at www.census.gov/hhes/commuting/data/commuting.html; navigate to selected census years.

7 Discussed on the website of the Dwight D. Eisenhower Presidential Library, online at www.eisenhower.archives.gov/research/online_documents/1919_convoy.html; Dwight David Eisenhower, "Through Darkest America with Truck and Tank," in *At Ease: Stories I Tell to Friends*, Doubleday (1967).

8 The text of Eisenhower's post-election statement on the need for highway improvement can be found online at www.fhwa.dot.gov/infrastructure/hearst.cfm

9 Federal Highway Administration, *Interstate Routes in Each of the 50 States, District of Columbia, and Puerto Rico*, online at www.fhwa.dot.gov/reports/routefinder/table3.htm; population data from US decennial censuses, online at http://census.gov/topics/population.html

10 Location of the sport franchises in 1950 and their subsequent movements and league expansion drawn from the websites of Major League Baseball (www.mlb.com), the National Football League (www.nfl.com/), the National Basketball Association (www.nba.com/), and the National Hockey League (www.nhl.com/). A book-length discussion of the phenomenon can be found at Michael N. Danielson, *Home Team: Professional Sports and the American Metropolis*, Princeton University Press (Princeton, NJ, 2001).

11 Andrew R. Goetz, "Air Passenger Transportation and Growth in the US Urban System: 1950–1987," *Growth and Change 23:3* (Spring 1992), pp. 217–238. See also www.smartertravel.com/photo-galleries/editorial/six-golden-ages-of-air-travel.html?id=556&photo=55021&max_photos=7

12 See Figure 2.7. The US Census Bureau provides an animated map of the movement of the population centroid online at www.census.gov/geo/reference/centersofpop/animatedmean2010.html

13 Angus Maddison, *Monitoring the World Economy, 1820–1992*, Organization for Economic Cooperation and Development (Paris, 1995).

14 US Department of Transportation, Bureau of Transportation Statistics, online at www.rita.dot.gov/bts/sites/rita.dot.gov.bts/files/subject_areas/airline_information/air_carrier_traffic_statistics/airtraffic/annual/1954_1980.html

15 www.populstat.info/Americas/canadac.htm

16 US Energy Information Agency, online at www.eia.gov/consumption/residential/reports/2009/air-conditioning.cfm

17 See discussion by Mark Byrnes (May 25, 2012) at *The Atlantic's* City Lab website, online at www.citylab.com/design/2012/05/end-domed-stadium/2101/

18 Christopher R. Lamberth, "Trends in Stadium Design: A Whole New Game," *Implications 4:6* (n.d.), online at www.informedesign.umn.edu; see also, www.ballparks.com

19 Jared Shlaes and Mark A. Weiss, "The Evolution of the Office Building," in John R. White (Ed.), *The Office Building from Concept to Investment Reality*, jointly published by the Counselors of Real Estate, the Appraisal Institute, and the Society of Industrial and Office Realtors Educational Fund (Chicago, IL, 1963), pp. 1–20.

20 Natasha Geiling Esri, "The Death and Rebirth of the American Mall," online at www.smithsonianmag.com/arts-culture/death-and-rebirth-american-mall-180953444/?no-ist
21 Louise A. Mozingo, *Pastoral Capitalism: A History of Suburban Office Landscapes*, MIT Press (Cambridge, MA, 2011).
22 William H. Whyte, *The Organization Man*. Doubleday/Anchor (New York, 1956).
23 Sloan Wilson, *The Man in the Gray Flannel Suit*. Simon & Schuster (New York, 1956).
24 The evolution of this mathematical understanding of our world, across a variety of domains, is well covered in Peter L. Bernstein, *Against the Gods: The Remarkable Story of Risk*, John Wiley (New York, 1996), especially in Chapter 6, "Considering the Nature of Man," pp. 99–115.
25 Although Pythagoras himself left no writings, Aristotle (*Metaphysics* A 5. 985 b 23) reports the Pythagorean teaching that, "the principles of mathematics were also the principles of all things that be . . . [so] the elements of numbers were the elements of all that there is, and that the whole world was harmony and number."
26 I have discussed the importance, and the limitations, of this approach in Hugh F. Kelly, "Can Universities Teach Real Estate Decision-Making?" *Real Estate Review 20.2* (1990), pp. 78–84, online at www.hughfkelly.com/papers-and-journal-articles/
27 See William H. Wilson, "Moles and Skylarks," in Donald A. Krueckeberg (Ed.), *Introduction to Planning History in the United States*. Center for Urban Policy Research: Rutgers University (New Brunswick NJ, 1983), pp. 90–98, which cites Alfred Bettman's persuasive argument to the Supreme Court that zoning "represents the application of foresight and intelligence to the development of the community."
28 This perspective was already well articulated in the early days of real estate scholarship, as in Arthur Weimer and Homer Hoyt, *Principles of Real Estate*, The Ronald Press Company (New York, 1939), at pp. 330–332, and 463–465.
29 US Department of Labor, *Proceedings of the Employment Managers Conference*, Philadelphia, PA, April 2–3, 1917, pp. 40, 47. See also Jon A. Peterson, "The Impact of Sanitary Reform upon American Urban Planning: 1840–1890," in Krueckeberg, op. cit., pp. 13–39.
30 Thomas Wolfe, *You Can't Go Home Again*. Harper & Row (New York, 1940).
31 "Fact sheet: History of United Nations Headquarters," a PDF online at www.un.org
32 Urbanist and planner Andres Alexander Price notes:

> Euclidean zoning is the most popular form of zoning in the United States and Australia – with a few exceptions. Euclidean zoning is also known as use-based zoning, single use zoning, or exclusionary zoning, and is known for its emphasis on how a property may be used rather than its form.
>
> (online at www.andrewalexanderprice.com/blog20140515.php#.Vbzm5_kYHjI)

33 Homer Hoyt, "The Structure and Growth of Residential Neighborhoods in American Cities," Federal Housing Adminstration (Washington, DC, 1939).
34 Ibid., p. 3.
35 Loc. cit.
36 Ibid., p. 64.
37 Ibid., pp. 73–78.
38 Eliza Hall, "Divide and Sprawl, Decline and Fall: A Comparative Critique of Euclidean Zoning," *University of Pittsburgh Law Review 68* (2007), 915–952.
39 Tom Brokaw, *The Greatest Generation*, Random House (New York, 2001), p. 11.
40 Douglas J. Gilbert, "Truman's Order Begins Long Process of Desegregation," American Forces Press Service (July 17, 1998), online at www.defense.gov/news/newsarticle.aspx?id=41900

41 Map online at https://en.wikipedia.org/wiki/Brown_v._Board_of_Education#/media/File:Educational_separation_in_the_US_prior_to_Brown_Map.svg
42 Gerald E. Frug, "City Services," *NYU Law Review* (April 1998), p. 91.
43 Hoyt, op. cit., p. 42.
44 Federal Housing Administration (FHA), *Underwriting Manual: Underwriting and Valuation Procedure Under Title II of the National Housing Act.* Government Printing Office (Washington, DC, 1938).
45 James Greer, "Race and Mortgage Redlining in the United States," paper presented at the Western Political Science Association Meetings, Portland, Oregon (March 22–24, 2012).
46 Jane Jacobs, *The Death and Life of Great American Cities*, Vintage Books (New York, 1961).
47 Jay Wickersham, "Jane Jacob's Critique of Zoning: From Euclid to Portland and Beyond," *Boston College Environmental Affairs Law Review 28:4* (2001), pp. 547–563.
48 Jacobs, op. cit., p. 4.
49 Chudacoff, op. cit., pp. 284–289.
50 Edward K. Muller, "Downtown Pittsburgh: Renaissance and Renewal," in Joseph L. Scarpaci and Kevin Joseph Patrick (Eds.), *Pittsburgh and the Appalachians: Cultural and Natural Resources in a Postindustrial Age*, University of Pittsburgh Press (Pittsburgh, PA, 2006), Chapter 1, pp. 7–20.
51 Discussed online at www.encyclopedia.chicagohistory.org/pages/1295.html
52 Wayne Leidwanger, "History of Renewal," A Technical Report of the Development Program – St. Louis, St. Louis City Plan Commission, 1970.
53 John D. Weaver, *Los Angeles: The Enormous Village (1781 – 1981)*, Capra Press (Santa Barbara, CA, 1980), pp. 166–170, 195–199. The Community Redevelopment Agency of the City of Los Angeles, "Bunker Hill Urban Renewal Project: Amended Redevelopment Plan" (October 30, 1976). William T. Fujioka, "Report to the Board of Supervisors, County of Los Angeles, on the Grand Avenue Project" (September 24, 2013).
54 Timothy Mennel examines this controversy in the *Journal of Urban History 37:4*, pp. 627–634, in a review of three books: Hilary Ballon and Kenneth T. Jackson (Eds.), *Robert Moses and the Modern City: The Transformation of New York*, Norton (New York, 2007); Anthony Flint, *Wrestling with Moses: How Jane Jacobs Took on New York's Master Builder and Transformed the American City*, Random House (New York, 2009); and Samuel Zipp, *Manhattan Projects: The Rise and Fall of Urban Renewal in Cold War New York*, Oxford University Press (New York, 2010).
55 For a generation, the dominant biographical treatment of Robert Moses has been the critical book by Robert Caro, *The Power Broker: Robert Moses and the Fall of New York*, Vintage Books (New York, 1975).
56 New York City Housing Authority, "Detailed Project Statistics" (June 30, 1960).
57 Cited in Clapp, op. cit, p. 264. Wright made the comment on *The Chrome-Plated Nightmare*, a television program aired on May 27, 1974.
58 Peter F. Drucker, "The Age of Social Transformation," *The Atlantic Monthly 274:5* (November 1994), pp. 53–80, online at www.theatlantic.com/past/docs/issues/95dec/chilearn/drucker.htm
59 Drucker, art. cit., p. 79.

3 Buying and selling the American Dream

The pull to the perimeter in the 1960s

To speak about the "forces" – social, economic, and political – that propelled the outward movement from center city to suburb in the quarter-century after World War II is to cast an overly deterministic light on the changes in America. Suburban growth was, without question, affected mightily by the trends identified in Chapter 2, but it is just as true that the reshaping of America's metropolitan areas was a matter of choice. The single-family house (and its white picket fence) with a green plot of land on a cul-de-sac was effectively sold as "the American Dream" because it so powerfully appealed to a wide swath of the public.[1] The moves were aspirational, widely understood as a search for a better life. If those aspirations were shaped by a powerful marketing machine, the advertisers were tapping into something deep in the American psyche.

Even at the time, there were voices of dissent and skepticism. Malvina Reynolds composed "Little Boxes" in 1962, inspired by the vista of Daly City, California, south of San Francisco. Memorably, she limned the cookie-cutter sameness of the "ticky-tacky" houses mass produced around virtually all American cities.

Allan Sherman, highly underrated as a social commentator in my opinion, offered his perspective in 1963, in "Here's to the Crabgrass." The migration of families to suburbia brought mortgage, garden clubs, fruit-of-the-month deliveries, and the daily commute back downtown to the lives of millions of households.

Gerry Goffin and Carole King wrote "Pleasant Valley Sunday" for The Monkees in 1967, archly commenting on the "keeping up with the Joneses" pressure on outward evidence of success in what they labeled "status-symbol land."

It was not just songwriters reacting to the trend. Already in 1956, John Cheever had begun to establish his reputation as "the Chekhov of the Suburbs," with scornful short stories in *The New Yorker* about suburban Shady Hill, where everybody has "a nice house with a garden and a place outside for cooking meat." John Updike's 1960 novel *Rabbit, Run* chronicled the vacuity of a suburban couple, the Angstroms, in a fictional Pennsylvania locale. Ira Levin forever imprinted suburban conformity on our literary consciousness with *The Stepford Wives* (1972).[2]

Ridicule did nothing to stem the tide. Even into the 21st century, scholars and commentators have remarked on the length and depth of America's infatuation with the suburb.

Robert Beauregard considers suburbanization a self-expression of US identity. Joel Kotkin believes that the post-war suburban trend is, in fact, a normative phenomenon.[3] George Sternleib and James Hughes considered the population dynamic underlying the growth of suburbs so powerful as to be irreversible.[4]

Just as television shaped the image of cities in *The Honeymooners* and *The Life of Riley*, so too did it market the suburbs as the aspirational destination of the 1950s, 1960s, and 1970s. America's most popular show of "the golden age" was *I Love Lucy*, another program where the interaction involved two couples, the Ricardos (Ricky and Lucy) and their friends and landords, Fred and Ethel Mertz, set in a brownstone apartment house on East 68th Street in Manhattan. There were some groundbreaking elements to the show during its long run in the 1950s. One was the marriage of the redheaded Lucy (née McGillicuddy) to the Cuban–American Ricky. Another was the portrayal of Lucy's pregnancies, as her real life was written into and adapted for the sitcom. And a third was the decision to relocate the family to suburban Westport, Connecticut, as the kids grew and Ricky's career brought them more affluence.

The commuter-based family became a staple of the primetime TV schedule over time. *The Dick Van Dyke Show* (1961–1966) had as its main character Rob Petrie, a New York City comedy show writer who takes the train to work each day from his home at 148 Bonnie Meadow Road in suburban New Rochelle. The home's living room is one of the primary "sets" for the comedic action. Mary Tyler Moore played the stay-at-home mom, and Larry Matthews played their son, Ritchie. In keeping with the standard formula, their next-door neighbors, dentist Jerry Helper and his wife Millie, are the solid, suburban best friends – in contrast with the zany crew of Manhattan-based writers and performers that are Rob's 9-to-5 wisecracking sidekicks.

On *Bewitched* (1964–1972), we find a Madison Avenue "ad man," Darren Stephens, shuttling back and forth to Manhattan from suburban 1164 Morning Glory Circle, variously identified as being in Westport (like the Ricardos) or in Patterson, New York, north of New York City. There are some similarities and some odd reversals vis-à-vis the *Dick Van Dyke* and *I Love Lucy* sitcom formulas. Like the Ricardos, the Stephens have a "mixed marriage." However, on *Bewitched*, it is "mortal man" and "supernatural wife," rather than Scottish–Irish and Cuban–American. And where the zanies are in Manhattan for the Petries, Darren Stephens works with profit-hungry "squares," but comes home to the weird doings of witches and warlocks. Also, we see the Stephens family growing, with both a son and daughter born during the program's eight-season run. The kids take after their mother, supernaturally.[5]

Marketing and motivation

As noted in Chapter 1, television was accessible to nearly 100 percent of American households by the time the 1960s were complete. Although the content of the programming itself had a powerful effect in shaping (and reflecting) public sentiment, we should not forget that the economics of the broadcast TV industry considered the entertainment component simply the vehicle for how money was made: the advertising popularly (and revealingly) known as "commercials." It is no coincidence that the retrospective look at the 1960s that has most recently captured the popular imagination is *Mad Men*. Advertising turns out to have great insight into human motivation and behavior and was fully capable of putting that insight to work 50 or more years ago.

Vance Packard's *The Hidden Persuaders* (1957) identified eight human basic needs that, if addressed, enabled the artful advertiser to sell virtually anything. They are:

- emotional security
- reassurance of worth
- ego-gratification
- creative outlets
- objects of love
- a sense of power
- a sense of roots
- a feeling of immortality.[6]

The commercials that accompanied the suburban sitcoms stroked all of those human needs and, in doing so, told a story of the satisfactions of "the American dream." There was a sense of accomplishment in shifting from urban renter to suburban owner. Along with a sense of spaciousness in low-density tract developments, there came a sense of "I'm worth it." Ironically, ownership housing in the decades after World War II was less expensive than in any era before or since – one reason that the home ownership rate soared as it did. But, in the prism of advertising, this fact was refracted into the dual components of "I have earned this and deserve the credit" and "This is proof of the higher quality of life I've attained, better than the lives of those still trapped in the city."

Hearth and home represented powerful positive values, and the suburban kitchen, living room, garage, and lawn were presented countless times each night in the television advertisements seen by millions. Food ads conveyed feelings of warmth, security, and safety as the family gathered around an Ethan Allen dining-room set. Dad was portrayed in his workshop, teaching Junior the skills of do-it-yourself projects. Friends captured fun on the lawn and in the backyard with Kodak 8-mm movie cameras. Mothers not only washed clothes for cleaniness' sake, but sent the children off to school scrubbed and dolled up as mini-advertisements for domestic competence. Marketing researcher Burleigh

Gardner noted that, "Mrs. Middle Majority" built her whole life around the home and was the "darling of the American advertiser." No wonder, then, that TV commercials were so dominated by images that were family-based, home-focused, suburban, and almost always white.[7]

Convenience was also a central value marketed to the target households. Instant gratification became a watchword: instant cameras with "pack-film" (brought out by Polaroid in 1963), instant coffee with "freeze-dried" technology (Tasters Choice and Maxim, circa 1966), instant oatmeal (marketed by Quaker Oats in 1966). Anacin promised "fast-fast-fast" relief for headaches. Miles Laboratories promoted "Speedy Alka-Seltzer." Most members of the Boomer generations could easily make a list of their favorites in this genre of products.

The drive toward pre-packaged convenience, unsurprisingly, extended to the marketing of suburban real estate. Miramar, Florida (a Dade County suburban community about 20 miles northeast of Miami and, despite its name, some 15 miles from the nearest ocean beach), was established in the 1950s and marketed as "the world's most perfect community." The concept was to offer "packaged homes in a packaged community."

What does it mean to buy a "packaged" home in a "packaged" community? For many (but apparently not all) of the Miramar families, it means they simply had to bring their suitcases, nothing more. No fuss with moving vans, or shopping for food, or waiting for your new neighbors to make friendly overtures. The homes are completely furnished, even down to linens, china, silver, and a refrigerator of food. And you pay for it all, even the refrigerator full of food, on the installment plan.

Perhaps the most novel and portentous service available at Miramar – and all for the one packaged price – is that it may also package your social life for you. As Mr. Gordon (Robert W. Gordon, the developer of the project) put it: "Anyone can move into one of the homes with nothing but their personal possessions, and start living as a part of the community five minutes later." Where else could you be playing bridge with your new neighbors the same night you move in! In short, friendship is being merchandised along with real estate, all in one glossy package. *Tide* (a marketing magazine) described this aspect of its town of tomorrow in these words:

> To make Miramar as homey and congenial as possible, the builders have established what might be called "regimented recreation." As soon as a family moves in the lady of the house will get an invitation to join any number of activities ranging from bridge games to literary teas. Her husband will be introduced, by Miramar, to local groups interested in anything from fish breeding to water skiing.[8]

If this sounds more than a bit Orwellian today, it might be worth reflecting just how prescient the novel *1984* has proven to be!

Redefining the "dream"

Kenneth Jackson, in *Crabgrass Frontier*,[9] notes that, "the dream house" is a unique feature on the world stage because it bore the utopian ideals of American society, a role that the city or the nation played in other cultures and in other times. As a mechanism of self-definition, there was an element of exclusivity – a "defining against" sense – that conflated suburban growth with the phenomenon of "white flight." Racial and economic homogeneity, it cannot be questioned, were a major part of the appeal of the suburbs. Zoning helped control this, with lot size restrictions, architectural standards, and yard-and-setback requirements enforcing segregation by income that acts as a proxy for segregation by race.

The effectiveness of using income measures to act as a race filter can be seen by looking at the earnings gap between white and black men during the 1960s. Regions differed, of course, but only in relative scale. In 1960, for instance, black men earned 40 percent less than their white counterparts in the North, and 59 percent less in the South, a differential that influenced the "Great Migration." By 1970, the racial gap had narrowed in both regions, but was still very wide, with blacks lagging whites by 36 percent in earnings in the North and by about 45 percent in the South (Figure 3.1).[10] The impact of the civil rights movement helped noticeably, but one effect was to create a social backlash that propelled whites toward the suburbs. Following the *Brown* v. *Board of Education* ruling in 1954, and President Lyndon Johnson's landmark Civil

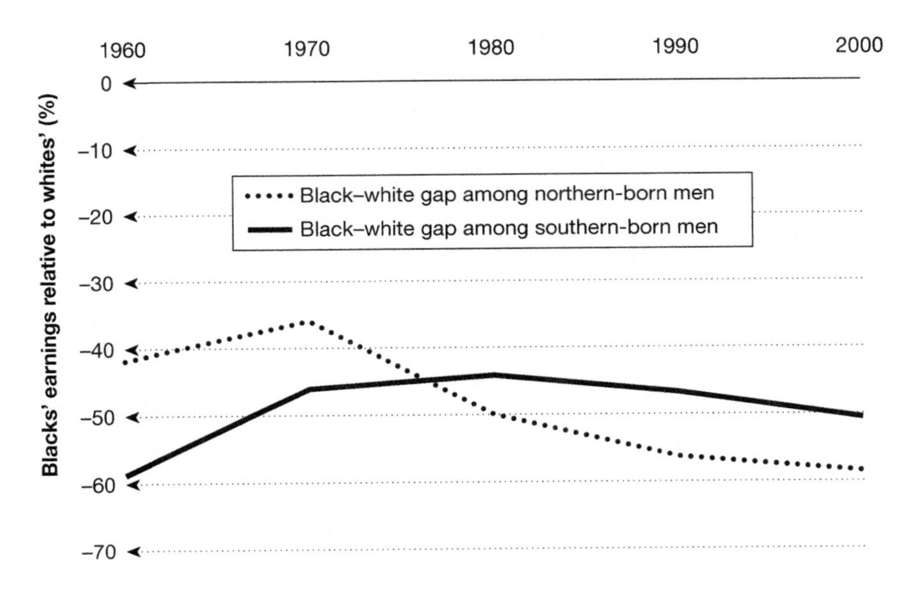

Figure 3.1 Relation among race, geography, and earnings: 1960–2000

Source: "Changes in the Racial Earnings Gap since 1960," Cannon and Marifan, *The Regional Economist*, July 2013

Rights Act of 1964, which outlawed discrimination in the workplace and in all places of public accommodation, whites in significant numbers sought to use residential location itself as a bastion against racial integration and, as Jackson notes, its related phenomena of racial violence on the one hand and racial intermarriage on the other.[11]

It is important not to cast suburbanization as entirely an expression of negative motivation, however. The popular and even scholarly literature about cities and suburbs is quite polarized, with paeans and polemics fairly common and balanced discussion fairly rare. Even in 2014, it is surprising to see a clear treatment of the pros and cons of city and suburban living. But the National Association of Realtors website (realtor.com) does provide such a discussion, and what it has to say in the second decade of the 21st century could well have been said 50 years ago. Here is the NAR summary about the positive aspects of suburban home ownership:

> Suburban living is for you if you love peace and quiet, and you are looking to escape from your work and the stresses of city life. The suburbs usually offer more than urban areas: more room for your home, more greenery and more opportunities to develop friendships with neighbors. A suburban home will cost you less than one of comparable size in the city, and suburban homes are often larger and have more amenities.

Can we doubt that, whatever the public policy influences tilting the urban/suburban choice toward the periphery, households embraced the concepts of "peace and quiet," "escape from stress," "more room, more greenery, more amenities, greater size, and opportunities for friendship"?[12]

In fact, these attributes were intensively marketed by the suburban developers of the 1950s and 1960s, and their appeals clearly resonated with millions of households. It is useful to look at some of the marketing strategies adopted by homebuilders of the time, as studied in Vance Packard's contemporary account:

> A large community development near Chicago faced the problem of selling a thousand houses quickly. To expedite the seemingly formidable task it retained a depth-oriented ad agency in Chicago. The agency called in several psychiatrists for counsel, and a depth study was made to find the triggers of action that would propel prospects into a home-buying mood.
>
> The task of selling the houses was complicated, the probers found, by the fact that men saw home in quite a different light from women. Man sees home as a symbolic Mother, a calm place of refuge for him after he has spent an abrasive day in the competitive outside world, often taking directions from a boss. He hopes wistfully to find in his idealized home the kind of solace and comfort he used to find as a child when at his mother's side.
>
> Women on the other hand see home as something quite different since they already are symbolic Mothers. A woman sees home as an expression

> of herself and often literally as an extension of her own personality. In a new home she can plant herself and grow, recreate herself, express herself freely.[13]

Advertisers were among the first social observers to fully appreciate how influential women were as decision-makers, even in areas where male dominance was generally assumed. For example, automobile repair and maintenance were widely considered a male domain, but it was the woman of the house who was most likely to know the local automobile mechanic, as she was the one tasked with bringing the car in for its inspections and routine work and who spent the most time behind the wheel ferrying children to school and to recreation.[14] Television's presentation of the suburban housewife as the pragmatic manager of family life, often better in touch with the realities of day-to-day functions than the somewhat oblivious husband, was grounded in the domestic experience of many viewers.

It is altogether likely that the home purchase – the most significant investment decision for most families – did have significant input from the male "head of household." Certainly, the amount of comparison shopping that would be typical of experienced female consumers is notably absent in the statistics on home searches by 1950s families. On average, couples looked at fewer than six houses before making this major investment. About 19 percent of homes were selected after only two possibilities were considered. And 10 percent of homebuyers took the very first house they visited.[15]

Major economic decisions such as home purchase might well have reflected the gender gap in the workforce during the quarter-century after World War II. In 1950, women were just under 30 percent of the labor force, a share that rose only slowly to 33.4 percent in 1960 and 38.1 percent in 1970.[16] A large wage gap yawned between the earnings of men and women, too, with women's annual earnings only 60 percent to 65 percent of men's income over the period – a range that persisted well into the 1980s (Figure 3.2).

Veterans' benefits made the largely male cadre of returning GIs the funding conduit for federally sponsored mortgage programs, as mentioned at the start of Chapter 2. The understandable pride that the veterans felt after their service period, and the sense of power that the military victories against the Axis catalyzed, made patriotism a potent trait in the American psyche. And the rapid shift into the mentality of the Cold War era, which began as early as 1946, reinforced the power of "Americanism" as a motivational force.

The real estate industry was quick to seize on this motivation in marketing homes. Indeed, the catchphrase for home ownership was "the American Dream." Although most historians describe the attributes of the American Dream as those enunciated in the Declaration of Independence, namely the "unalienable rights" of life, liberty, and the pursuit of happiness, with their correlates of opportunities for prosperity, success, and upward social mobility, the privately owned home became a tangible proxy for those more-broadly defined values.

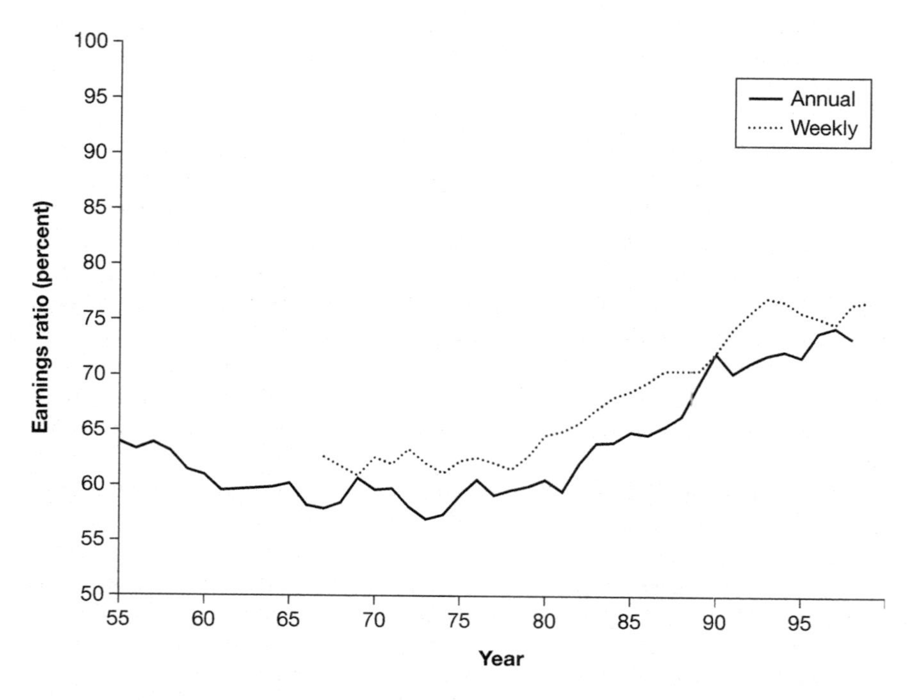

Figure 3.2 Female-to-male earnings ratios of full-time workers: 1955–1999

Source: Bureau of the Census, Population Reports, Series P-60, various issues; Employment and Earnings, various issues; and Census Bureau and Bureau of Labor Statistics webistes. Annual earnings are for year-round, full-time workers

The legendary developer of the suburban tract homes known as "Levittowns," William J. Levitt, said in 1947, "No man who owns his own house and lot can be a Communist." Senator Joseph McCarthy argued that public housing was socialistic and communistic, and that public housing would eventually lead to compulsory racial integration. McCarthy was joined in opposition to the 1949 Housing Act by real estate trade groups, including the National Association of Real Estate Boards and the National Association of Home Builders.[17] In 1957, the National Father's Day Committee promoted the patriotic motif, "Liberty Starts in the Home." And, in 1959, Vice-President (and presidential candidate) Richard Nixon squared off with Soviet Premier Nikita Khrushchev in a suburban model house purposely built for the American National Exhibition in Moscow. The house was represented as being affordable to all Americans (perhaps an understandable propaganda overstatement) and packed with modern gadgetry. Nixon made the point that such a home demonstrated the superiority of capitalism over the communist ideology.[18] Thus, the quest for a home and private lot, virtually always in the growing suburbs, was wrapped in the flag and sold in the marketplace of ideas, as well as in the real estate market.

Desiderata: Better schools and less crime

Households moving to the suburbs were likely to cite personal objectives more than patriotism as the reason for their relocation choices. Better education for the kids and safer neighborhoods led the list of "must haves" for couples raising the Boomer generation. Even if this was often a veiled reference to race-based concerns, the motivations were a search for positive attributes as much as a conscious exercise in prejudice.

As with most social behavior, the education and crime issues are not easy to classify simply. Residential segregation by race intensified. A study of 1967 data by the Census Bureau found that, whereas half of the white population of US metropolitan areas lived in suburbs, 80 percent of the black population was concentrated in the center cities.[19] "Disentangling" the topics of education and crime from the impact of the racial divide makes a certain analytical sense, but may be rightly considered artificial, given their conflation in the actual city–suburban dynamic of the volatile years of mid-20th-century America.

Nevertheless, education and safety gaps should surely be looked at. Measures of comparative educational quality found suburban schools had better-qualified teachers, newer facilities, and higher educational standards. Suburban schools could focus more time and attention on direct educational programming, whereas urban school systems were used to provide a variety of government services, including nutrition programs, social work interventions, and community outreach. Suburban school systems had financial advantages as well, ranging from higher per-pupil grant funding from the states to a higher percentage of local fiscal resources devoted to education, when compared with urban school districts.[20]

A tradition of local control over education has long characterized American culture. Consequently, new school districts grew up as suburban growth sprouted. In and around Philadelphia, for example, the 1950s and 1960s found enormous population growth in its Pennsylvania (Bucks and Montgomery Counties) and New Jersey (Camden, Burlington, and Gloucester Counties) suburban ring. Levitt & Sons spurred this growth by planned communities in both Bucks and Burlington Counties. Local control made the concept of a "metropolitan school district" a non-starter politically. As homebuyers specified that a local 4-year high school was one of the highest priority features desired in a community, suburban governments responded vigorously to their expanding constituencies. The financial resources to accomplish this were partly supplied by local taxation, but partly from statewide funds. The latter resource was a zero-sum game, though, with more money directed to support growth in the suburbs meaning less money available for maintaining and improving urban systems.[21]

Similarly, in the Chicago metropolitan area, the suburban school districts were widely acclaimed, even as the inner-city schools were described as in bleak disrepair, with "broken windows, torn window shades, broken desktops, and badly lighted, worn central hallways." Overcrowded Chicago School District facilities were compared with suburban districts funded by vigorous local economies and strong tax bases willing to invest in new schools and new educational programs. Under such disparate conditions, the concept of the "neighborhood school" exacerbated inequality and divided the metro area into a winner–loser choice favoring the suburbs, as far as residential choice was concerned.[22]

In the face of such stark differences, and with a growing sense of 1960s political activism, pressure arose to use the tool of school busing – transporting students beyond their neighborhoods to the schools with the better educational profiles – to ameliorate the quality of educational outcomes. In practice, this exacerbated the divide rather than healing it.

Although dozens of cities had similar experiences, perhaps it is the city of Boston that most vividly epitomized what was widely called "the busing controversy." The role of schools in social desegregation had been canonized in the 1954 *Brown* decision. In 1966, the Metropolitan Council for Economic Opportunity (METCO) sought to enhance access to integrated education by establishing a voluntary program whereby students from the largely black urban neighborhoods of Roxbury and Dorchester could attend superior schools in seven suburban districts, with a guarantee of continuing through high-school graduation. METCO presented this as a "win–win" proposition. African–American students would be afforded access to the superior facilities of the suburban schools, and suburban students would benefit from a more diverse group of classmates, an experience essential to success in a future, more multiethnic and multicultural social economy.

Though successful in a limited way, the METCO initiative did not establish a model whereby intra-metropolitan school disparities could be addressed. As it was a voluntary program, both the urban students and the suburban districts taking part were self-selecting. On the one hand, that led urban activists to fear that METCO was "cherry-picking" the best of the black students, leaving the local Boston schools with the more-difficult-to-educate youngsters. On the other hand, the suburban school districts represented just a few of the towns surrounding Boston, and even in the participating districts there were mixed reviews about the limited integration experiment in busing. When a broader, mandated busing program was initiated in 1974 in response to a court-ordered desegregation plan, the result was heated conflict that escalated into incidents of violence – and a massive shift of the white student population into private schools to avoid the busing program entirely. Needless to say, the disparity in educational outcomes between urban and suburban schools widened over time.[23]

Kenneth Jackson convincingly summarizes the educational motivation for household migration to the suburbs:

> Millions of families moved out of the city "for the kids" and especially for the educational . . . and social . . . superiority of smaller and more homogeneous school systems. The sprawling, single-story public schools of outlying towns, surrounded by playing fields and parking lots and offering

> superb facilities, new laboratories, and well-paid teachers, became familiar symbols of suburban life and educational manifestations for tract developments. Unlike the locked doors and grated windows of city institutions, they reflected an openness to nature. More importantly, the suburban school promised some relief from the pervasive fear of racial integration . . .[24]

Fear of crime was also a powerful motivator, and one that was all too easily conflated with racial stereotyping to justify labeling suburban growth as "white flight." Later sociological literature tells a more nuanced story, however.

The President's Commission on Law Enforcement (1967) attributes increased crime to density factors in "large impersonal cities."[25] Victor Flango and Edgar Sherbenou, in the journal *Criminology*, hypothesized that property crimes are associated with urbanization.[26] Violent crime, by contrast, was considered to be associated with poverty, following literature that enjoyed wide popular as well as professional readership in the 1960s. These sources included Michael Harrington's *The Other America: Poverty in the United States* (1963), Oscar Lewis' *La Vida: A Puerto Rican Family in the Culture of Poverty* (1966), James Q. Wilson's *The Metropolitan Enigma* (1967), and Daniel Patrick Moynahan's *Toward a National Urban Policy* (1970).[27] Flango and Sherbenou tested the hypothesized relationships and found better fits for the degree of urbanization and poverty than for alternative explanations of the city/suburb safety disparity such as degrees of affluence, stages in life cycle, economic specialization, and welfare expenditure policies. Taken together, the link between the causal factors of poverty and urbanization and the dependent variable of crime presumably provided a powerful motive for those seeking to avoid crime to leave cities, especially cities with substantial concentrations of the poor.

This rudimentary level of correlation was quickly amplified by others. W. Norton Grubb's 1982 *Journal of Urban Economics* paper, "The Flight to the Suburbs of Population and Employment, 1960–1970," concluded that upper-income residents do in fact tend to flee cities with a concentration of poor people. Grubb makes the economic argument that such concentrations raise required levels of public expenditures, and that the movement of households with high marginal tax rates is an attempt to shield themselves from public levies.

Addressing the common conflation of factors, though, Grubb finds no confirmation of a racial motivation (so-called "white flight"). The implication would be that minority households with similar incomes would have similar motivations, and would act on them similarly given equal opportunity. He describes this as a self-perpetuating pattern, whereby each relocation by an upper-income household to the suburbs increases the relative concentration of poverty in the city. Recognizing the difficulty of his prescription, Grubb nevertheless urges cities to concentrate on (unspecified) positive measures to retain their upper-income households, as he observes that attempts to enhance the mobility of the poor show limited, if any, success. Unlike Flango and Sherbenou, Grubb paints a picture whereby poverty and urban concentration become increasingly synonymous.[28]

One thing that certainly could be said for the crime factor is that there was plenty of rational motivation to be concerned about this issue in the 1960s, a decade that (as Figure 3.3 illustrates) saw an explosive rise in rates for both crimes against property and crimes against persons. Those fearing a continuation of this trend would have been absolutely justified, as crime rates continued to rise in the 1970s and, in the case of violent crime, again in the 1980s. FBI statistics confirm a strong positive correlation between the size of a place (city, town, village) and the number of crimes per 100,000 persons. Center cities are in the category of largest places in the FBI classification, and suburban police districts are typically smaller in total population. Those seeking to avoid crime would tend to move from the larger to the smaller places, based upon the data.[29]

Suburban boosters were not shy about pointing out such facts. Kenneth Jackson again provides a pertinent citation in *Crabgrass Frontier*:

> As early as the 1950s, suburban real estate advertisements were harping on the themes of race, crime, drugs, congestion, and filth. The well-to-do could avoid the costs of urban old age by simply stepping over the border, leaving the poor to support the poor. "Escape to Scarborough Manor [a Westchester County suburb of New York City]. Escape from cities too big, too polluted, too crowded, too strident to call home."[30]

Who could resist such an appeal?

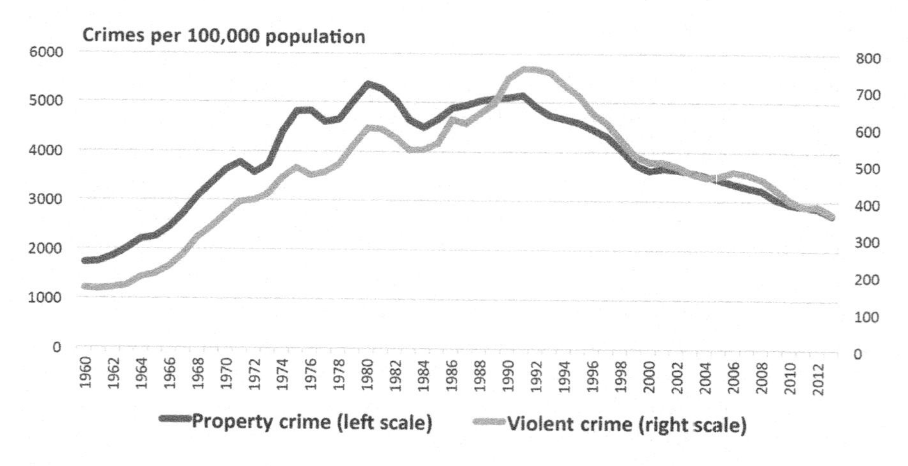

Figure 3.3 Crime in the United States peaked in early 1990s

Source: FBI "Crime in the United States" reports (1960–2013)

The argument from land economics

With all of the anecdotal information from history and the insights provided by sociology and motivational psychology, perhaps the most persuasive and theoretically satisfying treatment of cities and suburbs flows from a spatial analysis developed by three urban economists during the 1960s. William Alonso published "A Theory of the Urban Land Market" in *Regional Science* (1960), followed by his book *Location and Land Use*, in 1964. Edwin Mills contributed "An Aggregative Model of Resource Allocation in a Metropolitan Area" to the *American Economic Review* (1967) and published a book-length treatment of the subject, *Studies in the Structure of the Urban Economy*, in 1972. Alonso and Mills' work was complemented by Richard Muth's *Cities and Housing: The Spatial Pattern of Urban Residential Land Use* (1969), which looked at residential dispersion in metropolitan areas, as a detailed application of his 1966 paper, "Urban Residential Land and Housing Markets" for the Institute for Urban and Regional Studies. Taken together, the thoughts of this trio of scholars set the foundation for the Alonso–Mills–Muth model (A–M–M), which for decades was the basic intellectual framework for the understanding of how and why metropolitan areas grow, and what the economic rationale for land use choices in metro areas might be.[31]

A–M–M is a sophisticated econometric equilibrium model of urban form.[32] Like all models, it is a simplification of the complicated variables that go into shaping the establishment, functioning, and later development of cities. The language is technical, and the mathematics is quite advanced. A–M–M proposes that density/distance gradients descend outward from the center. What is a "density/distance gradient"? In plain English, this concept simply tells us that the greater the "density" – or the number of users seeking to occupy a particular piece of urban space – the greater the "rent" or market price of the space will be. So that is a fairly obvious application of the law of supply and demand that every college student learns in Economics 101.

The next, and equally obvious, step is to note a basic feature of Euclidean geometry (now we *are* talking about the great Greek Euclid, not the village of Euclid of zoning law fame). That feature is the formula for the area of a circle: $A = \pi r^2$. The amount of space available within a circle (think of the areas around a city center) is an exponential function of distance from the center of the circle. Why does this matter? Well, for cities, it is assumed that the center of the circle is the place with the highest economic value (or "land rent"), as that is the point with the more efficient access to all elements of the urban economic marketplace (the greatest density function). Value declines (the "gradient") as distance from the center increases.

The gradient is not "linear" – prices are not cut in half when distance from the center doubles – but "exponential", because of Euclid's law governing the area of the circle. Assuming the amount of economic activity remains constant (models always begin their analytical process with such assumptions), then the greater the radius (distance from the urban center to the suburban edge), the more area there is to share that amount of economic "rent."

To make it simple, let's say that two urban areas exist with a billion dollars of potential economic rent. One metro (A) has a radius of 5 miles, the other's (B) is 7 miles. The circle describing City A is about 78.5 square miles, meaning that each square mile would capture $12.7 million of land rent (on average), but that the perimeter would be much less than the average, because of its distance from the most efficient point, in the center. City B, however, encompasses an area of almost 144 square miles, and therefore the average land rent would be just $7.0 million per square mile, and again the perimeter would be much less than average because of its greater distance.

Euclid also teaches us that the circumference of the circle is π times the diameter ($C = \pi d$). So, the circumference of an urban area with a radius of 5 miles is roughly 31.4 miles, and a city with a 7-mile radius has a circumference of about 44 miles. Imagine how many 1-acre residential lots can be created at the perimeter of each. The smaller metro area would support 795 such housing lots, whereas the larger would have 1,115 homebuilding sites, a 40 percent increase in supply. Again, assuming economic demand remains the same, the wider metro area would see much lower average prices at its edge. This results from a combination of lower average prices in City B, divided by the greater number of lots that can be created in the larger city. (For those following the back-of-the-envelope math to estimate the number of available building lots, there are 640 acres to the square mile, so, at the edge of the circle, the number of acres are calculated as the square root of 640, or 25.3, times the number of miles along the circumference.[33])

Sprawl has its economic price, and perhaps this simple math illustrates for non-econometricians (i.e., most of us!) why prices decline exponentially with distance from the center in the A–M–M model.

The next step in understanding the logic of the model is that different land uses compete for location on the basis of their ability to pay rent. Ultimately, it is the profits from productive activity that pay the rent for developed real estate on the land. (Or even for agricultural land, which is valued as a function of its yield per acre.) The greater the profitability of a property, the closer it can be to the optimal location in the economic center. That is one reason why, during the industrial age, manufacturing facilities could cluster close to the urban center – they were the source of greatest profit. In the post-industrial city, office buildings that are the work space of investment bankers doing multi-billion dollar M&A deals, lawyers billing $800 an hour, and technology firms managing the flow of valuable information out-compete factories and warehouses for center-city space. The measure of a land use's competitive place in the urban area is somewhere along the density–distance gradient and is described as the bid-rent curve, generically illustrated in Figure 3.4.

Theoretically, this is all very clean and neat. So what makes cities messy? (This messiness is not a criticism of the model, by the way. Models are *supposed to simplify*, so that we can understand the basics better. Otherwise, we could just take photos or draw maps of actual cities – closer to reality, certainly, but not as useful for the understanding.)

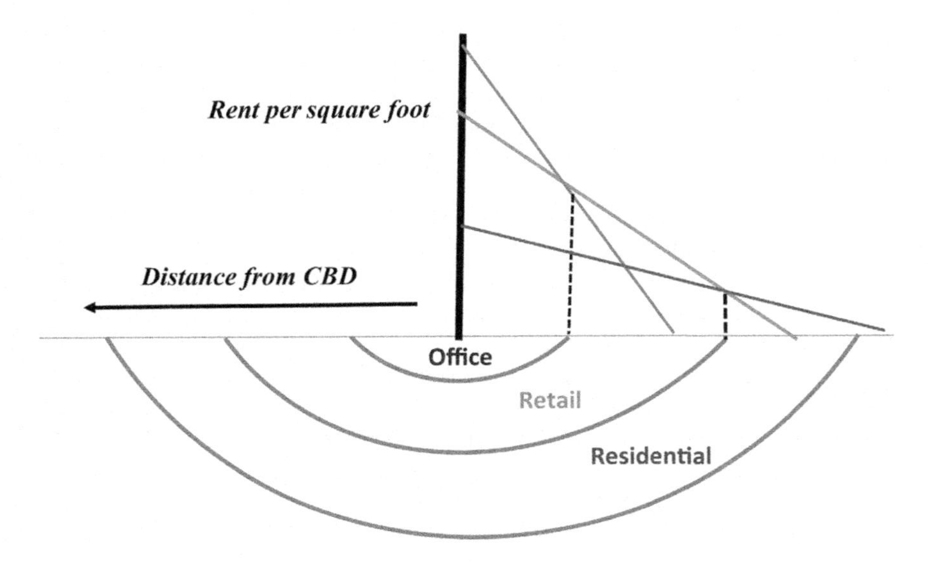

Figure 3.4 Bid-rent curve

There is actually a fundamental principle from real estate economics that contributes to making actual cities more complex than the model (as explained up to this point) suggests. That principle is called "highest and best use." Property is valued according to its best use, as governed by four characteristics. That use is the one that is physically possible, financially feasible, legally permissible, and maximally productive.[34]

Here's how "highest and best use" impinges on the nice, orderly world of land use models such as A–M–M. The model does not distinguish between parcels of different physical characteristics, such as soil conditions, slope of the site, the presence of water, and other topographic considerations. Next, the cost to develop various kinds of vertical real estate can vary tremendously; for instance, the same volume of space will vary in cost depending on whether commercial offices, rental apartments, or a hotel are developed. The "best use" of the site depends upon what income each use can generate compared with the cost to build it. And legally permissible uses look to the zoning codes that public agencies have put in place as a matter of policy. It is inappropriate to value a parcel of land for multistory apartments, if the zoning restricts residential use to single-family dwellings. So, the "productivity" of the land is not just a matter of the density–distance gradient and the bid-rent curve, it is the economic productivity possible given all those other constraints.

Let's just add one other wrinkle to the urban fabric at this point. Zoning is a creature of political entities, and metropolitan areas are rarely coterminous with city political boundaries. Most metro areas are comprised of multiple jurisdictions, each of which creates zoning codes according to its own constituent motivations. Brookings Institution scholar Anthony Downs has acidly commented on the balkanized land use patterns that have emerged, as suburban towns and villages have enacted ordinances to prohibit "LULUs" (locally undesirable land uses) within their boundaries. Such uses could be as noxious as landfills, as noisy as airports, or as disturbing as prisons. They may also simply be uses considered incompatible with local tastes or preferences: a mass transit station that will increase commuters' ability to access the locality, or a multifamily complex that will shift the balance of renters to owners in the jurisdiction.

So, the actual shape of a metro area's land uses will be influenced by the economics considered by the A–M–M model, but substantially modified by highest and best use considerations.

Actual cities also have "legacy" real estate to consider. The A–M–M model envisions quite an abstraction: "an open linear city on a plain," where each resident commutes to a job in the CBD, and each resident takes up "one unit of land" that can be anywhere within the urban boundary. However, not only does zoning complicate matters, but land uses already in place mean that existing buildings need to be considered. Even if the object of solving the model is the distribution of housing, and its pricing, several real-world considerations will alter the shape suggested by the model alone. For one thing, the existing housing will have definite (and idiosyncratic) attributes, including age, design, functional utility, and physical condition. For another, nonresidential property (older office buildings or manufacturing lots, for instance) that has reached a certain point of functional or economic obsolescence may be convertible to housing at some cost that is different from the cost of new residential construction.[35]

It is also critical to understand that cities sometimes have the opportunity to expand in a 360° circle, but often cannot do so. Cities are located where they are for historical reason, and the reason often has to do with topography. Port cities such as New York, Boston, Seattle, or Miami, for instance, have an ocean that constrains the direction of potential growth. Los Angeles is hemmed in by mountains, New Orleans by bayous and Lake Pontchartrain, and Phoenix by Native American reservations. However, there are many cities that do have 360° opportunities: Atlanta, Dallas, Minneapolis, Dayton, Kansas City, and others. So, the analytical implications of the A–M–M model need to be reviewed with such topographical variations in mind. Furthermore, the idea of a circular form contains a further assumption that the metropolitan area is "monocentric," that there is an identifiable core business district at the center, a true CBD. As cities have evolved, that is less frequently the case.

Finally, let's talk about distance. In the model, distance is "frictionless." Movements from one point to another are considered to be the same if the mileage is identical. But no one who has tried to navigate through cities would agree with that. There are better and worse routes to take, depending upon modes of transportation, time of day, relative levels of congestion, and so on.[36]

It may seem that, with all these exceptions, the usefulness of A–M–M is very limited. Mills himself has generously acknowledged that the canonical model may not have the wide application that it once had. Nevertheless, it still has very strong explanatory power, and scholars depend upon the model as an analytical tool even more than a half-century since its development. Many cities still appear to grow in ways that the model would predict.[37]

For the present purpose, understanding the economic underpinning of suburban residential growth, the model indeed has practical benefit. Why is it that more affluent households have had a propensity to choose to live at the urban periphery? This is a question that A–M–M particularly considers. The model suggests that, because the time value of upper-income households outweighs out-of-pocket transportation costs for such households, they are early adopters of "faster commuting options" and choose perimeter locations where they can optimize their housing choices on lower-cost land. In the simplest form of this model, lower-income households are left at the center, ironically using higher-cost land, but left with higher-density multifamily housing units and greater dependency on older and slower commutation options.[38]

For all the abstraction of the model, doesn't this correctly describe several decades of housing decisions for thousands of US families? Economic choice is not just willingness to pay, but capacity to pay. Upper-income households have greater effective freedom of choice about where to live. Less-affluent households not only have fewer options, but are often required to make very difficult trade-offs, such as circuitous journey-to-work patterns from low-cost housing to low-paying jobs, because the value of their time weighs less in the balance than that of their more-affluent urban neighbors. Such trade-offs are well understood within the A–M–M framework and suggest that economic, as well as political, sociological, and psychological, explanations account for the growth of suburbs over the second half of the 20th century.

Forms of change: Maturation

One of the most recurrent errors in anticipating change is the tendency to extrapolate historical patterns without regard to limits. The mislabeling of any extended statistical pattern as a "long-term trend" encourages that error. It is important to recognize that both internal and external conditions may alter the momentum of change. Where those limits, over time, affect the ability of any measured variable to reach a "this far, but no farther" point, we may more correctly identify the operative form of change as "maturation."

Sophocles, in *Oedipus the King*, alludes to the "riddle of the Sphinx," which asks, "What goes on four feet in the morning, two feet at noon, and three feet in the evening?" The answer, of course, is a human being going through the normal stages of life: a baby crawling, an adult walking, and then an elder using the help of a cane. At each stage there are capabilities, and at each stage there are limits. Poor Oedipus went through life oblivious to the fatal limits that were constraining his happiness, unaware until tragedy overtook him.

With an awareness of growth and limits to growth, we can see maturation as a critical form of change. While in the midst of the process, many people unfortunately enjoy the growth without keeping the limits in mind, limits that mean that the trajectory of change over time is not a straight line.

Like most concepts in scientific modeling, the math of maturation (if I can call it that) is based on a theory of equilibrium. But here we find *two equilibria* – an initial condition and a final condition – and a transition between them. Between the creeping baby and the "geezer with his cane" (both slow moving), we have the fast-paced period, which takes up a good bit of our lives. Mathematically, that is described by a period of change that is exponential, in the strict sense.

Think of a lovely meadow with lush grass, plenty of water and sunshine, and a few scattered rabbits. At first, the population of rabbits will grow very slowly. The rabbits have to find each other and form the initial rabbit relationships. There is a period of gestation to consider as well (pretty short: rabbit pregnancy lasts just 1 month). That's the initial equilibrium. But rabbits are prolific breeders. Females become fertile at between 3 and 6 months of age, have litters of seven or eight bunnies typically, and can become pregnant again just a day or so after giving birth. They multiply – like rabbits! So, within a short amount of time, there is a rabbit population explosion – exponential growth as the rabbits do what comes naturally. This goes on until the meadow is so full of rabbits that it strains the carrying capacity of the meadow – the vegetation is munched away, water is not replenished quickly enough, the rabbits get aggressive with each other, predators discover the delicacy of rabbit dining, and so on. Some rabbits decide to hop on over the hill to the next lovely meadow. That becomes the second equilibrium, a mature ecological balance. It is described by the logistics curve, or sigmoid curve, introduced in Chapter 1.

That curve, and the form of maturational change, is extremely common in life. Just about every parent charts that as they record the height-and-weight growth of their kids – including the legendary "growth spurt" of early adolescence. Kids just "shoot up," but only up to a point! The "growth trend" does not last forever. There are limits.

Understanding those limits is critical to evaluating the prospects for change and its consequences. In fact, whenever the concept of "growth" is invoked in economics and in policy discussions, there should be a rule that the very next paragraph should discuss the questions, "For how long?" and "Up to what point?"

One of the great social changes of the post-World War II period was the increase in the female labor force participation rate. As Figure 3.5 shows, the first three decades of the 20th century saw a fairly stable equilibrium in which about one in five women over the age of 16 was in the workforce. The pressures of the Great Depression and then of World War II brought more women into the ranks of those working for pay, and, by 1950, the percentage had increased to 33.9 percent. Over the next 30 years, influenced by greater levels of education, the gradual de-industrialization of work, a higher divorce rate, and the end of the child-rearing period when mothers were caring for the Boomer generation, the female labor force participation rate soared, peaking in 2000. A new equilibrium has apparently been established, since the 1990s, in the 55–60 percent range.

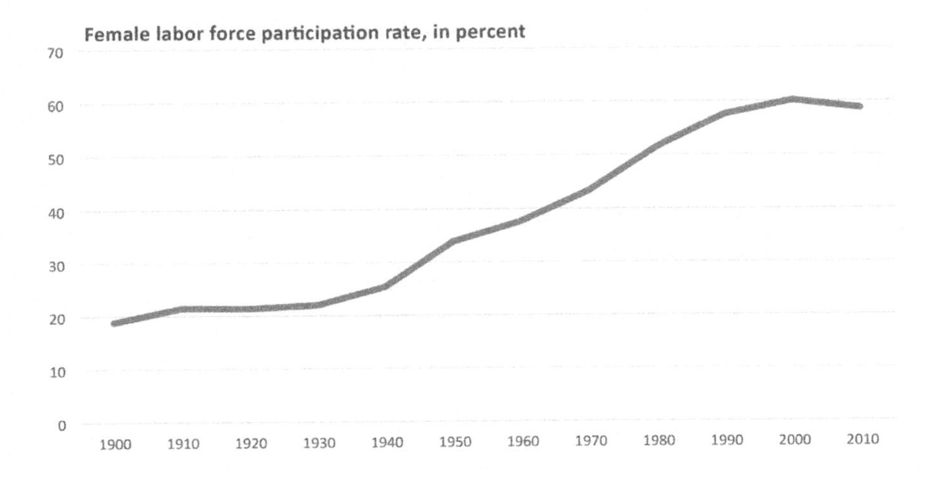

Figure 3.5 After decades of increase, the percentage of women in the labor force has leveled off

Source: US Bureau of Labor Statistics

Home ownership rates have also displayed the sigmoid curve pattern that typifies the form of change that is maturation. Figure 3.6 shows, for the US and for five large states, home ownership percentages from 1900 to 2010. The nation's equilibrium in housing tenure (owner or renter) balanced slightly in favor of renters in the first four decades of the 20th century. But, under the influence of demographic change, a new set of government incentives, and the shift in consumer choices, there was a rapid transition by 1960 in favor of home ownership, and, by 1980, almost two-thirds of US households were homeowners. Despite the impact of the "housing bubble" that afflicted the nation in the first decade of the 21st century, by 2010 the home ownership rate had returned to 66.9 percent from its 2004 peak of 69.0 percent. The economy as a whole seems to be balancing the tenure choice at a narrow range of 64–67 percent owners.

This does vary across the country, of course. Since 1990, populous states such as California, New York, and Texas have seen homeowner rates below the national average. Intriguingly, both California and Texas seem to have achieved their tenure equilibria as early as 1960, whereas New York is still in its transition phase. Florida also found its mature equilibrium rate by 1960, above the US average. Illinois, meanwhile, has moved above the US ownership rate. The highest home ownership states (all 74 percent or higher as of 2012) form an eclectic group, including Maine, Michigan, Mississippi, New Hampshire, and West Virginia. The most renter-heavy states (at 43 percent renter or more) are California, Hawaii, Nevada, and New York, joined by the District of Columbia with a 55 percent renter ratio.

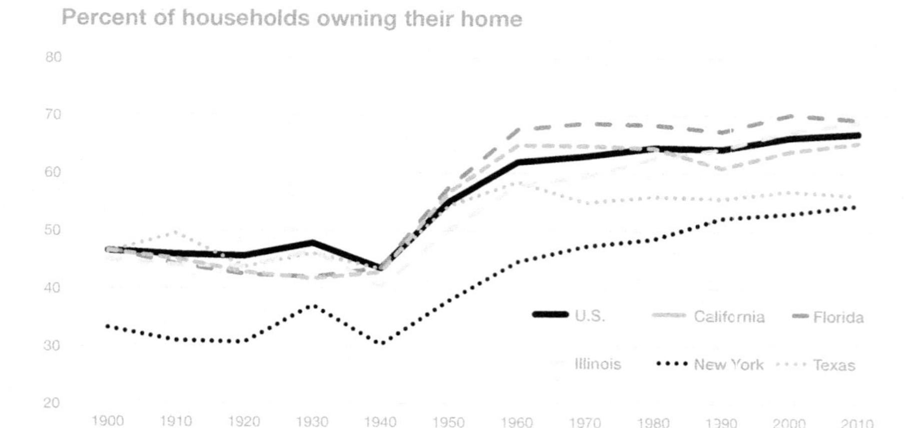

Figure 3.6 Homeownership rate: US and selected states: 1990–2010
Source: US Census Bureau

Growth is important, and the way we measure growth is equally vital to our understanding of how growth indicates change. Both absolute and relative measures need to be examined.

Let's try a trick question. Since 1980, which has been the better growth city, New York or Dallas? Almost universally, Americans would answer "Dallas." They might not know the exact numbers, but embedded in our economic psyche is the "knowledge" that the fast-growing states and cities of the Sunbelt have been outstripping the more sluggish growth of the Northeast and Midwest – the so-called "Rustbelt." The bias in favor of Dallas is not entirely mistaken. Dallas has averaged population growth of about 1.5 percent annually over the past 33 years, whereas New York's growth rate has been just one-third that, at about 0.5 percent.

So, what is the trick to the question? Since 1980, the absolute change in the number of New York City residents has been approximately 1.3 million. The entire population of the City of Dallas is 1.24 million. In other words, over the past 33 years, the five boroughs of New York City have added more population than Dallas, as a city, has accumulated through its entire history. Dallas was founded in 1841 and incorporated as a city in 1856.

To be fair, when most people think of Dallas, they envision the entire sprawling Dallas–Fort Worth Metroplex. That urban conglomeration covers 12 counties and boasts a metropolitan area population of 6.8 million persons. It covers a land area of some 9,000 square miles.

Even so, that regional population is lower than the 8.4 million that New York City compresses into just 309 square miles. And, to compare the New York metropolitan statistical area (MSA) with the Dallas figures, the New York MSA has about 20 million persons living in a 6,720 square-mile land area.

The point here is not to puff up New York by marking some numerical superiority to Dallas. The key item to appreciate is that the standard for evaluating "growth" is much too easily assumed to be percentage change over time, rather than absolute change. Almost universally, stories about "top growth cities" focus simply on percentage change, usually in population and jobs, as the relevant measure. That is an exceptionally misleading perspective, not least because real estate is used by actual people, not abstract percentages.

Since the middle of the 20th century – and for reasons discussed in Chapter 2 – a number of key American cities in the South and West regions of the country have grown both absolutely and in relative terms, as can be seen in Figure 3.7. Los Angeles, Phoenix, and Las Vegas stand out as strong growth cities, no matter how you might care to measure things. And it is obvious that Chicago, Philadelphia, and many other older manufacturing cities have passed their peaks in population and jobs. But, as Figure 3.8 illustrates, an older city such as New York is doing well in its maturity and may be pointed toward an even higher equilibrium, whereas Chicago, Washington, DC, and even Atlanta have lost population in their urban cores (Figure 3.9). What does this mean for land values and the values of developed real estate from city to city? Investors and demographers may (and do) arrive at very different answers to such a question.

For many of the Sunbelt cities, the process of maturation is far from complete. Whereas most of the cities included in Figure 3.7 were well established by the mid-19th century, such is not the case for cities such as Dallas, Miami, and Phoenix, or for Las Vegas, Minneapolis, or Seattle. These are "younger" cities and should be expected to have a greater propensity to grow population in the present era than cities that date from colonial times. Most of these younger cities also have the opportunity to expand in a 360° arc, a differentiation from coastal cities. It is also important to distinguish which cities have been growing

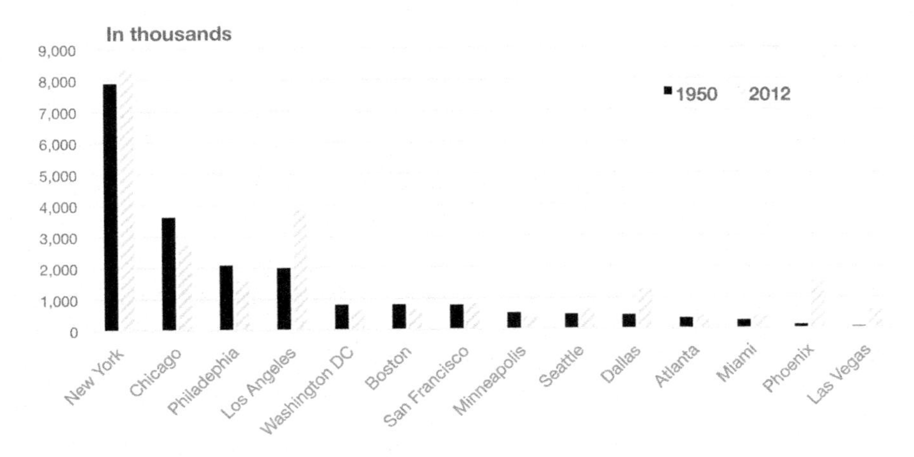

Figure 3.7 Where they stood in 1950 and 2010: Population in selected cities

Source: US Bureau of the Census

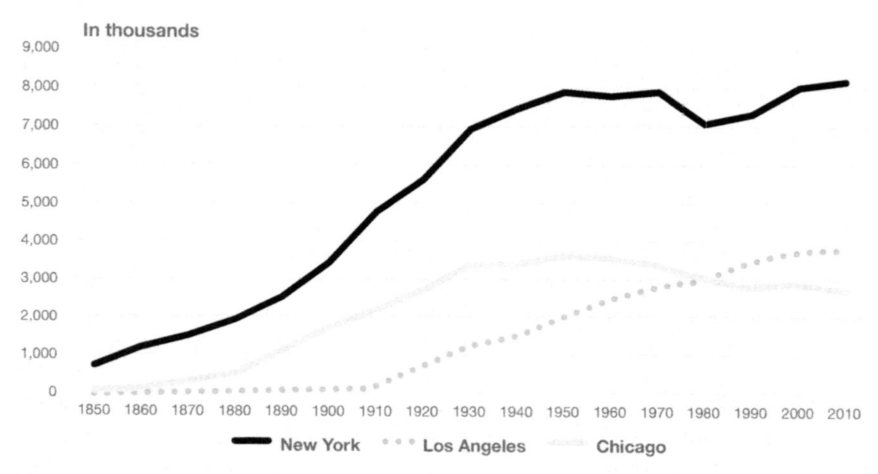

Figure 3.8 Population growth curves for three largest US cities: 1850–2010
Source: US Bureau of the Census, Historical Population of the United States

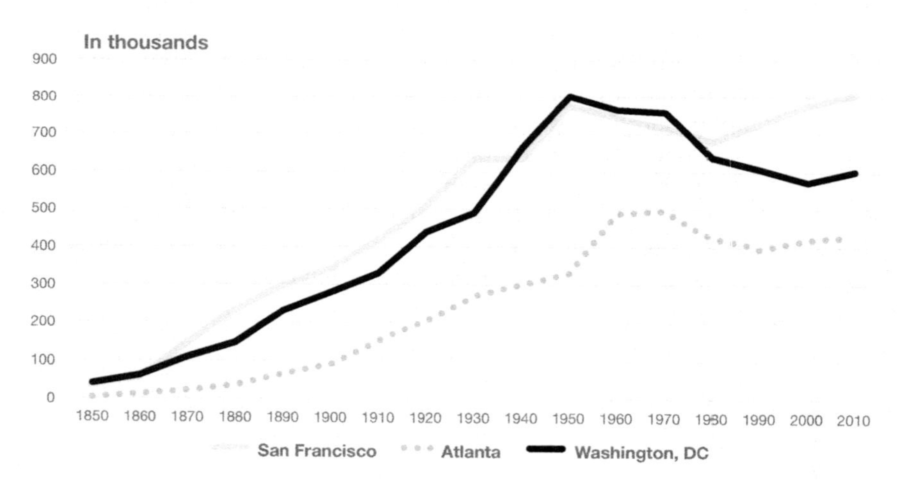

Figure 3.9 Population growth curves of selected US cities: 1850–2010
Source: US Bureau of the Census, Historical Population of the United States

by annexation, rather than just by population fundamentals. Of the nine cities shown in Figures 3.8, 3.9, and 3.1, depicting population change since 1850, the cities of Atlanta, Dallas, and Phoenix owe at least some of their "growth" to the annexation of existing localities in an effort to rationalize political jurisdiction to match the de facto blurring of urban boundaries.

Cities such as Dallas and Phoenix (Figure 3.10), although still in the growth-spurt stage of maturation, will have to confront their individual limitations, limitations that growth-by-annexation only make it more critical to recognize.

For Dallas, the issue will be the point at which the outward expansion of the metro so dilutes its economic energy and clogs its highway arteries that the population and job expansion no longer feed upon themselves. As much as Dallas may see the Red River (the Texas–Oklahoma border, about 75 miles north of downtown) as the only natural limit to expansion, it is highly unlikely that the region can develop the level of gross metro product needed to sustain growth. Dallas will be seeing the leveling off of its growth curve in the decades ahead.

For Phoenix, the issue is likely to be water. As much as Phoenix has enjoyed unbridled expansion, this has come at a thirsty cost. In 1950, the urban land area around Phoenix was just 17 square miles; by 2010, the urbanized area was 475 square miles, and the city's boundaries had been expanded by an aggressive annexation program, bolstering its property and sales tax base. Federal and state infrastructure funding significantly subsidized this growth, as did the provision of water from the Colorado River – a resource that Arizona must share with six other states, including California. After a drought that has lasted 15 years, Lake Mead is at just 39 percent capacity – its lowest level since 1937, shortly after the Hoover Dam was constructed – and the upstream Lake Powell is at 52 percent capacity, with seven of its nine boat launches now unusable. The ripple effects of a drought that scientists see as just the cutting-edge of global warming impacts are broad: not just drinking water, but agricultural production and hydroelectric power depend upon the same water resource that Phoenix has prodigally used to support its sprawling growth. Like other large Sunbelt cities, the dynamic of the flattening maturation curve will be shaping Phoenix's 21st-century future.

Many of the episodes of irrational exuberance in our past have been the result of markets' excitement over sudden and explosive growth, without a sense that the growth would ultimately be reined in by the discipline of limits. John

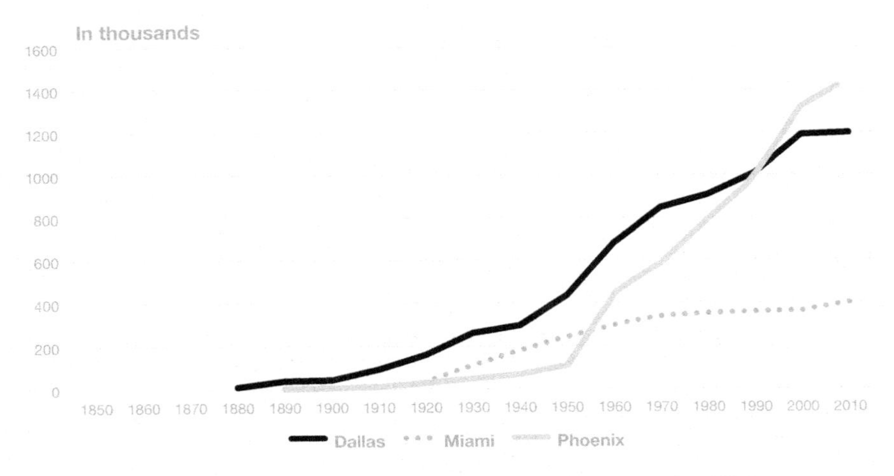

Figure 3.10 Population growth curves of selected US cities: 1850–2010

Source: US Bureau of the Census, Historical Population of the United States

Kenneth Galbraith's slim volume *A Short History of Financial Euphoria*[39] traces our recurrent lapses into the temptation to extrapolate trends into the indefinite future. Those caught up in the real estate syndication phenomenon of the 1980s, the dot-com bubble of the 1990s, and the housing price and financial derivatives debacle of the past decade nicely (if painfully) illustrate how correct Galbraith was when he wrote, "one thing is certain: there will be another of these episodes and yet more beyond." Even so, an appreciation of the form of change represented by maturation can help curb the enthusiasm of speculative euphoria by forcing us to consider what happens when growth encounters finitude, as it must in our bounded world.

Notes

1 Fannie Mae advertised, "We're in the American Dream business." Cited in Dolores Hayden, *Building Suburbia: Green Fields and Urban Growth, 1820–2000*, Vintage Books/Random House (New York, 2003), p. 3.
2 See John Cheever, *The Stories of John Cheever*, Vintage/Reprint Edition (New York, 2000); John Updike, *Rabbit, Run*, Random House Trade Paperbacks/Reissue Edition (New York, 1996); Ira Levin, *The Stepford Wives*, Random House (New York, 1972).
3 Joel Kotkin, *The New Geography: How the Digital Revolution is Transforming the American Landscape*, Random House (New York, 2001), especially Chapter 2, "The Anti-Urban Impulse," pp. 27–51.
4 George Sternlieb and James W. Hughes, *America's New Market Geography: Nation, Region and Metropolis*, Center for Urban Policy Research/Rutgers University (New Brunswick, NJ, 1988).
5 The television shows mentioned are discussed in Ella Taylor, op. cit., pp. 24–31. Greater detail is accessible online at www.classic-tv.com/shows
6 Vance Packard, *The Hidden Persuaders*, Pocket Books/Simon & Schuster (New York, 1957), online at www.keepandshare.com/doc/3542337/the-hidden-persuaders-vance-packard-1957-pdf-february-11–2012–7–09-pm-9–6-meg. See Chapter 8, "Marketing Eight Hidden Needs," pp. 86–94.
7 Maya Montanez Smukler, "Race, Suburbia and the Televised American Dream," ARSC (Association for Recorded Sound Collections), 2008; Burleigh B. Gardner and Sidney J. Levy, "The Product and the Brand," *Harvard Business Review*, March–April 1955; Juliann Sivulk, *Soap, Sex, and Cigarettes: A Cultural History of American Advertising* (2nd ed.), Wadsworth Publishing (Boston, MA, 2012); for an amusing selection of classic commercials from this era, online, see https://archive.org/browse.php?field=subject&mediatype=movies&collection=classic_tv_commercials
8 Miramar is discussed in Packard, op. cit., pp. 215–216, in Chapter 21, "The Packaged Soul?".
9 Kenneth T. Jackson, *Crabgrass Frontier: The Suburbanization of the United States*, Oxford University Press (New York, 1985). See pp. 238–245, "Characteristics of Post-War Suburbs."
10 Maria E. Canon and Elise Marifian, "Changes in the Racial Earnings Gap since 1960," *The Regional Economist*, St. Louis Federal Reserve Bank, accessible online at www.stlouisfed.org/publications/regional-economist/july-2013/changes-in-the-racial-earnings-gap-since-1960
11 Jackson, op. cit., p. 290.

12 Accessed online at realtor.com on July 15, 2014. The post has since been taken down. For a more recent discussion on this site, see www.realtor.com/news/generation-y-prefers-suburban-home-city-condo/
13 Packard, op. cit., pp. 102–103.
14 Ibid., p. 102.
15 Ibid., p. 118.
16 US Bureau of Labor Statistics, "Women in the Labor Force," BLS Report 1052, December 2014; Jing Shen, "Recent Trends in Gender Wage Inequality in the United States," *Journal of Sociological Research 5:1* (2014), pp. 32–49.
17 Robert A. Beauregard, *When America Become Suburban*, University of Minnesota Press (Minneapolis, MN, 2006), pp. 155–157.
18 Ibid., pp. 166–169.
19 US Bureau of the Census, *Current Population Reports P20 – 209*, Table B, p. 5, January 8, 1971.
20 Arthur Adkins, "Inequities Between Suburban and Urban Schools," *Educational Leadership*, Association for Supervision and Curriculum Development, December 1968, pp. 243–245.
21 The classic text on the mismatched spending in education is Jonathan Kozol, *Savage Inequalities: Children in America's Schools*, Harper Perennial (New York, 1988). Philadelphia's situation is discussed specifically in William W. Cutler III and Catherine D'Ignazio, "Public Education: Suburbs," online at http://philadelphia encyclopedia.org/archive/public-education-suburbs/
22 John L. Rury, "Schools and Education," online at www.encyclopedia.chicagohistory.org/pages/1124.html
23 Laura Chanoux, "From the City to the Suburbs: School Integration and Reactions to Boston's METCO Program," Bachelor of Arts Honors thesis, University of Michigan, Department of History, March 30, 2011.
24 Jackson, op. cit., pp. 289–290.
25 This association was not a novel one, but part of the mainstream of social science research at the time. The consensus was influenced by a seminal paper by Louis Wirth, "Urbanism as a way of life," *American Journal of Sociology 44* (July 1938), pp. 1–24. Wirth wrote, nearly three decades before President Lyndon B. Johnson's commission, that large size lends an impersonal flavor to social systems, and, consequently, "personal disorganization, mental breakdown, suicide, delinquency, crime, corruption, and disorder might be expected . . . to be more prevalent in the urban than in the rural community" (p. 23).
26 Victor E. Flango and Edgar L. Sherbenou, "Poverty, Urbanization, and Crime," *Criminology 14:3*, American Society of Criminology (November 1976).
27 Michael Harrington, *The Other America: Poverty in the United States*, Simon & Schuster (New York, 1963); Oscar Lewis, *La Vida: A Puerto Rican Family in the Culture of Poverty*, Random House (New York, 1966); James Q. Wilson (Ed.), *The Metropolitan Enigma*, Anchor Books (New York, 1967); Daniel P. Moynihan, *Toward a National Urban Policy*, Basic Books (New York, 1970).
28 W. Norton Grubb, "The Flight to the Suburbs of Population and Employment, 1960–1970," *Journal of Urban Economics 8* (1982), pp. 348–367.
29 The FBI's statistics on crime in the United States are discussed in Michael J. Hindelang, "The Uniform Crime Reports Revisited," *Journal of Criminal Justice 2* (1974), pp. 1–17.
30 Jackson, op. cit., p. 285.
31 William Alonso, "A Theory of the Urban Land Market," *Regional Science 6:1* (January, 1960), pp. 149–157; William Alonso, *Location and Land Use*, Harvard University Press (Cambridge, MA, 1964); Edwin S. Mills, "An Aggregative Model of Resource Allocation in a Metropolitan Area," *The American Economic Review*

57:2 (May, 1967), pp. 197–210; Edwin S. Mills, *Studies in the Structure of the Urban Economy*, Johns Hopkins Press (Baltimore, MD, 1972); Richard F. Muth, "Urban Residential Land and Housing Markets," Institute for Urban and Regional Studies (Washington University, St. Louis, MO, 1966); Richard F. Muth, *Cities and Housing: The Spatial Pattern of Urban Residential Land Use*, Graduate School of Business, University of Chicago (Chicago, IL, 1969).

32 The model is discussed in Marvin Kraus, "Monocentric Cities," Boston College Working Paper 559 (n.d.), pp. 1–22, online at www.bc.edu/ec-p/wp559.pdf

33 Alternatively, take the square root of 43,560 (the number of square feet in an acre), which is 208.71, and divide that into 5,280 (the number of lineal feet in a mile) for the same multiplier of 25.3.

34 This principle continues to be basic to real estate valuation. It is accepted as a basic analytical foundation for real estate investment as early as Arthur M. Weimer and Homer Hoyt, *Principles of Real Estate*, Ronald Press Company (New York, 1939), pp. 14, 69.

35 Thus, the concept of highest and best use is not static, and the land rent needs to consider all "costs to create" improvement, including (when applicable) the demolition or substantial rehabilitation of existing buildings.

36 Richard Voith, "Changing Capitalization of CBD-Oriented Transportation Systems," *Journal of Urban Economics 33* (1993), pp. 361–376; Richard Voith, "Comment on Baum-Snow's 'Effects of Urban Transportation.'" *Brookings-Wharton Papers on Urban Policy* (Washington, DC, 2005).

37 Mills wrote, "The classical urban model is dramatically contrary to fact" in "Sprawl and Jurisdictional Fragmentation," *Brookings-Wharton Papers on Urban Affairs* (2006), p. 239, although this seems an overstatement in our context, as Mills' subject in this paper was the political balkanization of metropolitan areas. Of more general importance is Mills' observation that, "the easy problems in spatial analysis, specifically the monocentric model, have been solved. Multicentric models are much more realistic but much more difficult." ("A Thematic History of Urban Economic Analysis," *Brookings-Wharton Papers on Urban Affairs* (2000), p. 6. In this 2000 paper, Mills makes specific reference to the A–M–M model as an early model that was fruitful in providing useful and testable insights, but that later models have eclipsed in fidelity to complex detail. The exponential increase in computing power has had much to do with this advance.

38 This remains a thorny issue, especially in America's most prosperous cities, as discussed in Elizabeth Roberto, "Commuting to Opportunity: The Working Poor and Commuting in the United States," Metropolitan Policy Program at Brookings (February 2008).

39 John K. Galbraith, *A Short History of Financial Euphoria*, Viking Penguin (New York, 1993).

4 The hollowing of the center

The push away from the core in the 1960s and 1970s

The blossoming of the suburbs could have been complementary to growth in the central cities, during a period in which the population base of the US was increasing significantly, and its economy was the strongest in the world. There is no real reason to consider the city–suburb relationship a "zero-sum game" in which gains on one side necessarily entail losses on the other. But, for most American metropolitan areas, that was exactly the case between 1950 and 1990. A tenuous balance was maintained for a while, but, after 1970, the expansion of the suburbs ringing the city was fueled by an exodus from the center. This phenomenon became widely known as "hollowing out." Cities, especially the great manufacturing centers, became regarded as artifacts of the Industrial Revolution whose time had passed. The exceptions were the ever-increasing, low-density, automobile-oriented urban areas typical of the South and West regions of the country. A divide was discerned between Sunbelt and Rustbelt.

This chapter examines how and why that happened. Although it is a disturbing study for those who love cities, it is important to learn its lessons. Moreover, it is vital to see how the decline of cities in these decades (shockingly sketched out in advance by Jane Jacobs in her 1961 book, *The Death and Life of Great American Cities*)[1] laid the groundwork for the renewal that has been the principal urban story of the past quarter-century.

As with the suburbs, we find the bards and troubadours of popular music providing the flavor of the times. At 1650 Broadway, at the edge of New York's Times Square, Barry Mann and Cynthia Weill caught an early hint of urban despair, which turned into a major hit record for the British group, The Animals, in 1965.[2] The city was dirty, unhealthful, imprisoning for the aspirational young, and the refrain starkly stated that a better life meant "getting out of this place."

Dirt, pollution, decay: these were not unusual themes for songwriters with a clear-eyed awareness of what was around them. Ed Cobb composed a mock paean to Boston for the California group, The Standells. Boston: dirty water, with lovers, muggers, and thieves down by the banks of the Charles River.

Not that Boston was unique. New York's magnificent Hudson River was polluted, not only by New York City's human and industrial waste, but also by PCBs dumped by the General Electric Corporation, 170 miles upstream near

Schenectady, New York. Such fish as survived in the river were declared inedible by health officials. Meanwhile, New York's air was so smoggy that Manhattan's skyline was often invisible from the Brooklyn–Queens Expressway, just 3 miles away. Mayor John V. Lindsay quipped, "Never trust air that you can't see."

Songs of social commentary and protest gained great traction during the decade of the 1960s, influenced by singer and musicologist Pete Seeger, the transformational lyrics of Bob Dylan, and the powerful influence of urban blues on bands such as the Rolling Stones and the Yardbirds.

Urban violence was amplified on the airwaves. The Stones intoned, "Gimme shelter" from the rape and murder that was "just a shot away." And Marvin Gaye lamented about the "too many" crying mothers and dying brothers in 1971's "What's Going On?"

It wasn't a pretty picture. And that's because urban life wasn't a pretty subject.

As always, though, we found ways to laugh at ourselves. A fictional blue-collar home in the New York borough of Queens was probably the most recognized setting in America.[3] In *All in the Family*, TV producer Norman Lear created a microcosm of the tensions pulling on the urban households of the day. Those tensions strained the fabric of this basic social unit across the dimensions of gender, of education, and of generations. The broadly drawn, bigoted head of household, Archie Bunker, could repeatedly be heard instructing wife Edith to fetch him a beer, and, when she reported on her day, she would be cut off with "Stifle, Edith, stifle." Liberal son-in-law Mike was "Meathead," with his anti-war, pro-civil rights, ecology-friendly ideology subject to Archie's condescending corrections as to "how the world really is." Mike's wife, Archie's daughter Gloria, was just "Little Girl" in Archie's eyes, innocent of the world of work and the decaying city that Archie felt was crumbling around him.[4]

Crumbling it was, indeed.

The Great Society

The hollowing out of the cities was, without question, a complex phenomenon that emerged over many years, spanning at least two decades, and perhaps more. But it is useful to anchor a discussion on two key years, 1968 and 1974, which stand out as historical watersheds. The first is widely regarded as a sociopolitical flashpoint for America, and for its cities particularly. The second date, though less freighted with baggage in the popular imagination, is a pivotal year in socioeconomic terms. A look at the urban trends over time suggests that the two years are closely linked to each other, and to the underlying conditions for the emergence of 24-hour cities.

Events in the US were so tumultuous in 1968 that a sitting American president, Lyndon B. Johnson (LBJ), declined to run for reelection under challenge from leaders within his own political party. The stresses on the nation may be described as a confluence of two powerful forces related to the signatures of the Johnson presidency: his extremely ambitious domestic agenda and the Vietnam War.[5] Undertaken at a time of great national affluence, Johnson's policies were labeled by his critics "guns and butter," a reference to the trade-off that often must be made between growing military spending and funding social programs.[6]

Civil rights were a keystone of LBJ's domestic policy, dubbed "The Great Society" by Johnson himself in a commencement speech at the University of Michigan in May 1964. This was a broad and extremely ambitious program of government action to address issues of poverty, the environment, housing, racism, immigration, and a variety of other contemporary issues. Johnson had been elevated to the presidency upon the assassination of President John F. Kennedy in November 1963, but he established himself with a landslide victory in the national election the following year. With a perceived personal mandate, and with large Democratic majorities in both houses of Congress, LBJ undertook to reshape the face of America. How quickly events overtook this complicated presidency!

Johnson's ambitious programs might be attributed to his enormous ego – and there's no historical question about his sense of self-importance. But they were also, and equally unquestionably, a sincere response to deep-seated American troubles. In an amazingly prodigious effort that shames the current gridlock of US government, the 89th Congress passed nearly 200 bills between January 1965 and January 1967, the most productive output in our history.

Out of all that fevered work, three elements are particularly germane to the subsequent story of American cities. The first is the "War on Poverty." The second is the attempt to transform urban America. The third is the focused effort on civil rights.

It is difficult to recall now, in the early 21st century, but the War on Poverty did not exclusively, or even primarily, focus on America's urban ghettos (Figure 4.1). Poverty was endemic in the rural, elderly, and white populations (in absolute terms, though African–Americans had higher relative concentrations of the poor). The "headline image" of poverty was Appalachia,[7] not Harlem in New York, Watts in Los Angeles, or Liberty City in Miami. Of course, there was plenty of poverty in such areas. But, as Michael Harrington noted in his best-selling *The Other America*, such urban poverty had this in common with its rural cousin – it was isolated from the affluent society to such a degree that it was invisible.[8] Johnson intended to change that invisibility, as a first step to eradicate the condition of poverty in America.

Nevertheless, the public's impression of federal anti-poverty initiatives was conflated with the question of race. In some ways, this was quite understandable. For one thing, the nearly 58 percent poverty rate among African–Americans far outstripped the under 20 percent poverty rate for whites, even though there were three times as many poor white Americans than poor black Americans. The extremely high incidence of rural poverty was a motivating factor in bringing those in the hinterlands to the cities in search of economic

	Total population	% poor	Number of poor
US	179,323,175	22.4	40,168,391
White	158,831,732	19.5	30,972,187
Black	18,871,831	57.8	10,907,918
65+	16,559,580	36.9	6,110,485
Non-metro	66,427,997	47.2	31,486,870
Metro areas	112,895,178	7.7	8,681,520

Figure 4.1 Statistics on poverty in US

Source: US Bureau of the Census, 1960 Census of Population

opportunity.[9] Only 37 percent of the US population lived in non-metro areas in 1960, but such places accounted for 78 percent of the nation's poor. As the War on Poverty took hold, the elderly began to benefit from an increasingly strong safety net, as Figure 4.2, from the American Association of Retired Persons (AARP), illustrates. But at the outset of the LBJ effort, to be an older American was not a "golden years" experience. For many, it was instead a daily struggle of living on the economic edge.[10]

Regardless of their idealistic intentions, LBJ's urban policies had decidedly mixed results. It is far from clear that the good outweighed the bad. A cabinet-

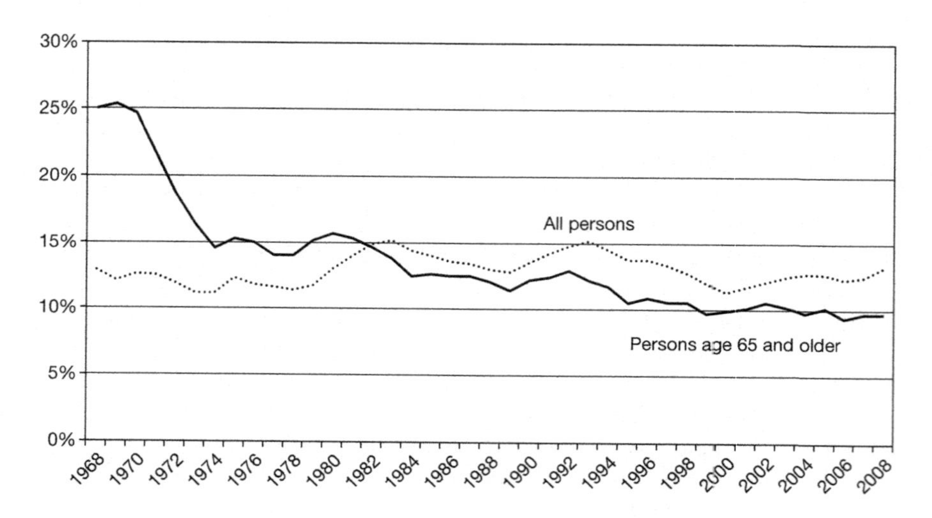

Figure 4.2 Percentage of the population below the official poverty threshold: 1968–2008

Source: AARP Public Policy Institute estimates based on the Current Population Survey, 1969–2009

level Department of Housing and Urban Development (HUD) was established in 1965 and immediately ran into trouble as questions of control arose. Who would "be in charge" of the urban programs? Federal agencies? Local governments? Community groups? Being in charge meant power, certainly; but with power came control of the purse: billions of dollars in funding sourced from US tax coffers.

The urban policies did not enter a vacuum, either. Whether in Washington or at the local level, the prevailing orthodoxy of "urban renewal" was "slum removal" – or, as neighborhood activists dubbed it, "poor removal," or the even more incendiary observation, "black removal." Little heed was paid to the formula for success that had already been articulated in *The Death and Life of Great American Cities*: diversity of use, small block sizes with frequent street intersections, maintenance of a fabric of old and new buildings, and enough population density (daytime and nighttime, workers and residents, and visitors attracted by neighborhood features such as shopping, restaurants, parks, and entertainment) to keep street life active by encouraging lively pedestrian flows for most of the 24-hour cycle.[11] Instead, slum clearance bulldozed sites that were then rebuilt from scratch.

The plans for replacement uses were almost invariably homogeneous: civic districts (often in the style of banal, or even frankly "brutalist," architecture[12]), arts districts that were self-contained and isolated by design from business and residential areas, and, for the dislocated poor, massive housing projects on "superblocks" along the lines of Le Corbusier's "towers in a park" concept, located on cheap land, as far as possible from places of potential employment and supporting neighborhood services.[13]

Worthwhile and promising efforts in education and job training – Head Start pre-schooling, the Job Corps, the Neighborhood Youth Corps – were swallowed up in the political fight over control of federal dollars and had to swim against the tide of physical demolitions and household relocations. The so-called "Model Cities" program was widely recognized as a clunker and was scrapped within a decade of its launch. What was left behind was an urban fabric that was more threadbare, if not already unraveling into tatters. Dick Lee, the mayor of New Haven, Connecticut, one of the Model City showcase programs, said, "If New Haven is a 'model city," God help America's cities."[14]

Although LBJ was himself a product of rural Texas, and the 89th Congress was dominated by conservative Democratic senators and representatives, often from the Deep South, together they found the political will to address problems afflicting cities all across the nation. That some of the programs failed is not surprising, but it would be an error to claim that the whole attempt was folly.

The Public Broadcasting Act, for instance, not only took seriously Newton Minow's challenge to turn television into something better than the commercial stations' "vast wasteland." It brought a vision of collaborative, creative, mixed-population, and mixed-use neighborhood to the attention of several generations.[15] In the midst of everything, a positive (but not problem-free) view of an urban street – dirty and gritty, and even grouchy – was placed before the children of the Baby Boomers (and the Boomers themselves – who could resist?) on a daily basis.

At least as significant was the Immigration and Nationality Act of 1965, which abrogated a xenophobic quota system biased nearly exclusively toward Western Europe. Suddenly, America, as a land of opportunity, could once again welcome Asians, Africans, Latin Americans, and (eventually) Eastern and Central Europeans liberated by the fall of communism in the Soviet bloc. The ambitious, the deprived, the talented, the desperate, the skilled workers, and the laborers, all found a newly opened door. Ethnicity itself, as a lived experience, migrated across America, but principally settled in our cities. The nation that cherished as a symbol the Statue of Liberty was learning once again that "foreign born" need not mean "alien to our ways." More importantly, it remembered that "our ways" could be enriched by the experience, skills, hard work, and brainpower of the foreign born.

Optimism and energy, supplemented by old-fashioned political horse-trading and arm-twisting (often accompanied by LBJ's invitation to "come, let us reason together"), marked the rollout of the War on Poverty and the administration's plans for remaking the cities. Regardless of the mixed outcomes, the nation credited the lofty aims of the programs and extended high approval ratings to the government efforts.[16] But nothing matched the moral fervor of the third prong of the Great Society efforts: redressing racial grievances and establishing in law basic civil rights for all Americans, including African–Americans.

Perhaps the clearest way to encapsulate the tensions, the multiple layers of discord, and the painful march of events is to present a simple timeline of some of the headline occurrences of the 1960s. The end of the era in which legally sanctioned racial discrimination prevailed and the beginning of a period when equal protection under the law became the standard was tantamount to the largest social revolution in a century – equal in importance to the War of Independence itself and the Civil War – and it was fought in the hearts and minds of Americans as much as it was in the streets. For the purposes of this study, the story cannot be told in depth. But it must be acknowledged that the events of this decade shape our cities even today, 15 years into the 21st century.

Race relations timeline: The 1960s

1960 Sit-ins last for 6 months at a Greensboro, North Carolina, Woolworth's lunch counter, to attempt desegregation of public accommodations.

The Students Non-violent Coordinating Committee (SNCC) is established to organize young people around the issues of race and civil rights.

President Eisenhower signs the Civil Rights Act of 1960, providing federal oversight of local voter registration practices to counter anti-black discrimination.

1961 The "Freedom Riders" attempt to desegregate interstate bus travel by booking passage for interracial passenger groups.

President John F. Kennedy (JFK) signs an executive order establishing what would become the Equal Employment Opportunity Commission.

Attorney-General Robert F. Kennedy (RFK) pressures the Interstate Commerce Commission to enforce rules forbidding racial segregation on interstate travel routes.

1962 US Department of Defense officially desegregates American military (with the exception of the National Guard).

JFK signs executive order banning segregation in federally funded housing.

James Meredith enrolls as the first black student at the University of Mississippi, over the objections of segregationist Governor Ross Barnett. A violent riot by white students and others from around the state results in two deaths and hundreds of casualties. JFK and RFK send in 5,000 federal troops and US marshals to suppress the disorder.

1963 George Wallace's inaugural speech as Governor of Alabama proclaims, "Segregation now. Segregation tomorrow. Segregation forever."

Reverend Martin Luther King (MLK) is arrested during a Birmingham, Alabama, demonstration for parading without a permit. He writes the "Letter from the Birmingham Jail," in which he argues a moral duty to carry out forceful but nonviolent public actions against discrimination, including civil disobedience.

During further demonstrations in Birmingham, fire hoses and police dogs are turned on black protesters. Television and photojournalism publish the images, which generate additional support for the civil rights movement.

Governor Wallace attempts to prevent integration of the University of Alabama with his "Stand in the Schoolhouse Door." Federally mobilized troops force him to stand aside. That evening, JFK addresses the nation on television on civil rights, enunciating a principle of "equality of treatment under the law."

Medgar Evers, field secretary of the National Association for the Advancement of Colored People, is killed in Jackson, the capital city of Mississippi.

MLK delivers his "I Have a Dream" speech to more than 200,000 at the March on Washington for Jobs and Freedom.

The bombing of a black church in Birmingham kills four young girls attending Sunday school.

JFK is assassinated in Dallas, Texas. Vice-President LBJ is sworn in to succeed Kennedy.

1964 The 24th Amendment to the US Constitution abolishes the "poll tax," a device used in 11 Southern states to discourage poor blacks from voting.

"Freedom Summer" organizes a voter registration drive in Mississippi to enroll blacks at the polls, and to attempt to challenge the all-white Mississippi delegation to the Democratic National Convention. Three civil rights workers, two white and one black, are murdered as they investigate the burning of a black church. The voter registration effort succeeds. The challenge to the segregated delegation is rebuffed.

LBJ signs the 1964 Civil Rights Act, prohibiting discrimination based on race, religion, color, or national origin, and assigning power of enforcement to the federal government.

1965 The Selma to Montgomery March for voting rights is met by a police blockade leaving Selma. The police attack the marchers with billy clubs, tear gas, and bullwhips, all captured by the news media, with the three national networks breaking into regular programming to broadcast the incident. MLK comes to Selma. A court, petitioned to constrain the police, instead issues a restraining order against further demonstrations. A northern minister, coming to Selma to support the civil rights activists, is murdered by white vigilantes.

Two weeks later, under protection of a federalized National Guard, a 25,000-person march proceeds 50 miles from Selma to Montgomery, petitioning for voting rights.

Five months later, the Voting Rights Act of 1965 is passed by Congress, making illegal any literacy tests, property tests, and other inhibitions to voting.

Race riots erupt in Watts, a black neighborhood of Los Angeles. A 6-day event, this was the largest race riot of the 1960s and stemmed from an incident where a white highway patrolman stopped and arrested a black motorist on suspicion of driving while intoxicated. In the riots, 34 people were killed, over a thousand injured, and $40 million in property damage was inflicted. The Governor of California, Pat Brown, sent in 14,000 National Guardsmen to restore order.

LBJ issues an executive order requiring contractors working on government projects to take "affirmative action" toward greater equality in hiring minority workers.

1966 SNCC links violence against blacks in the US to the bombing of civilian populations in Vietnam and unites in opposition to both.

James Meredith leads a "March Against Fear" from Memphis, Tennessee, to Jackson, Mississippi. Meredith is shot during the march. MLK joins the march as it proceeds. At the end of the march, in Jackson, SNCC leader Stokley Carmichael introduces the theme of "black power" to the movement.

That Fall, the Black Panther Party is established in Oakland, California.

1967 The US Supreme Court, in *Loving* v. *Virginia*, rules that laws in 16 states banning interracial marriage are unconstitutional.

In July, a 5-day riot is sparked in Newark, New Jersey, by charges of police brutality in the arrest of a black taxicab driver. With chants of "black power," demonstrations turn into violent looting, leading to a call-up of the National Guard. In the end, there are 26 deaths, 750 injuries, and more than 1,000 arrests – the most severe civil disorder in New Jersey history.

About a week later, in Detroit, a vice squad raid on an illegal after-hours club triggers a riotous response of bottle throwing and arson. Firemen seeking to control the flames are themselves attacked. The riot spreads to more than 100 city blocks, despite a massive police response and the call-up of the National Guard. LBJ finally orders Army paratroopers to the city to assist in patrolling in tanks and armored personnel carriers. More than 7,000 people are arrested during the 4 days of rioting. A total of 43 are killed. Some 1,700 stores are looted, and nearly 1,400 buildings are burned, causing $50 million in property damage. Some 5,000 people are left homeless.

A watershed year: 1968

Even as frustration, backlash, and confrontation are undoing the initial discipline of the civil rights movement, another stream of protest is gathering momentum as the war in Vietnam escalates. By 1967, demonstrations, largely student-led, become common on college campuses, at public monuments, and along city streets. MLK joins his voice to the anti-war sentiment, and boxing champion Muhammed Ali is indicted for refusing induction into the US armed forces, saying, "I ain't got no quarrel with the Viet Cong. No Viet Cong ever called me 'nigger.'" Civil consensus in America appears to be unraveling. Early in the year, FBI statistics report that crimes of violence have increased nationally by 57 percent since 1960, even in the midst of a long economic expansion.

1968 MLK is assassinated in Memphis, Tennessee, on April 4. He had come to aid in a sanitation workers' strike. LBJ appeals for calm, but riots break out in over 110 American cities within days:

- In Washington, DC, 4 days of turmoil leave twelve dead, 1,097 injured, and more than 6,100 arrested. Additionally, some 1,200 buildings are burned, including more than 900 stores. Damages reach $27 million (more than $125 million in 2014 dollars).

- In Baltimore, Maryland, the disturbances last more than a week. The final toll of the looting, arson, and clashes between authorities and the rioters (including snipers): six dead; 700 injured; 5,500 arrests; 1,050 businesses looted, vandalized, or burned; estimated property damage of $13.5 million (roughly $80 million in 2014 dollars).
- The Chicago neighborhoods of Lawndale, Austin, and Woodlawn go ablaze soon after word of Dr. King's assassination is heard. Before 10 p.m. that evening, 36 major fires are reported, and firefighters battle looters and roaming gangs of vandals, as well as the burning buildings. Mayor Richard J. Daly issues a "shoot to kill" order to police if an arsonist is seen. Approximately 10,500 police are sent in, and, by April 6, more than 6,700 Illinois National Guard troops arrive in Chicago. LBJ also sends 5,000 US Army troops into the city. More than 48 hours of rioting leave 11 Chicago citizens dead, 48 wounded by police gunfire, 90 policemen injured, and 2,150 people arrested. Miles of the city are left in a state of rubble. More than 200 buildings are gutted by fire, with property damage estimated at $10 million (about $59 million in 2014 dollars).

In June, RFK (running for president after LBJ's announcement, in March, that he was not seeking reelection) is assassinated in Los Angeles. The race for the Democratic nomination then becomes a contest between "peace candidate" Eugene McCarthy and LBJ's vice-president, Hubert Humphrey. As fate would have it, the Democratic Convention is held in riot-torn Chicago. Knowing that Humphrey has a nearly assured victory by delegate count, protesters are deliberately provocative, taunting and physically assailing police. The authorities are prepared, massed in force (more than 25,000 police and troops) and retaliate against the provocateurs. The convention leaves the Democrats in disarray. But the "Battle of Chicago" is, in effect, the last of the major riots of the 1960s.[17]

1969 The incidence of urban rioting drops precipitously between 1968 and 1969. There are incidents of racial violence in smaller cities, including York, Pennsylvania, and Camden, New Jersey, with locally destructive effects. But, by and large, the surge of civil unrest due specifically to black–white tensions settles into a simmering cessation of overt hostilities.

However, in Chicago, protests are organized by the Weatherman[18] faction of Students for a Democratic Society. One of the targets of the protests is Judge Julius Hoffman, who presided over the trial of "the

> Chicago Eight" – several leaders of the 1968 protests at the Democratic National Convention.[19] The locus of activist anger has shifted from civil rights to the anti-war movement, but, though the protagonists have changed, the cities are still the battleground of social unrest.

The presidential election of 1968 was hotly contested and wound up being surprisingly close. Richard M. Nixon was the victor, with 43.4 percent of the vote, trailed by Humphrey with 42.7 percent. Alabama's Governor George Wallace had a decisive third-party candidacy, winning 13.5 percent of the vote nationwide – an indication of the potency of the resistance to the reforms pushed by LBJ and his immediate predecessors. Johnson himself understood the cost – to himself and to his party – of his civil rights initiatives, saying upon the occasion of his legislative successes, "We have just lost the South for a generation."[20]

As an unexpected and unwelcome outcome of a civil rights movement launched under the banner of nonviolence, the riots almost immediately became the subject of sociological study. Only later was that research connected to detailed economic causes and effects, but the extension of investigation into the area of jobs, housing, and the effects on urban economies was both logical and inevitable. The hunger for an answer to the question "why?" is one of the deepest of human needs, and the matching of causes to effects closely follows as a motivating force. Although, perhaps, Edmund Burke and George Santayana were overly optimistic when thinking that understanding history can help us avoid falling into its pitfalls,[21] we feel a need to investigate events in the hope that understanding will improve future action.

Gregg Lee Carter[22] investigated what he termed "supra-local causes" of the riots, but contended that the circumstances particular to the cities in which they occurred played a role as well – especially in the relative severity of the violent outbreaks. The supra-local causes included the mechanization of agriculture and the subsequent black migration to cities, the rising expectations among blacks of greater political, social, and economic advancement, and the development of a "black consciousness" that believed that the status and material gap between whites and blacks should be resisted as, in itself, unjust. Carter found that this widely accepted narrative needed to be supplemented by local conditions specific to certain urban areas and claimed to have conducted the first research based on a comprehensive data set for the era of the riots.

The empirical analysis identified a series of conditions positively correlated with the severity of riot violence. Black population size in a given city was found to be important, but not overwhelmingly so. Deep South cities tended to have lower levels of black economic advancement and were less prone to riot, as they had less in the way of "rising expectations" in the face of entrenched structural and cultural racism. The more industrialized cities of the North, by contrast, saw not only rising expectations, but greater personal experience of the gap in material well-being between the races. In fact, the severity of rioting increased in those cities where greater economic equality was perceived to be in reach but yet still denied. Lastly, Carter found that political structures that raised hopes of greater jobs and services, acceding to minority desires, but failed to deliver them after election campaigns were completed, as partisan consensus proved elusive, also served as fuel for the frustrations erupting in violent street disorders.

Many years later, Collins and Margo moved from measurement of causes to an analysis of effects.[23] These authors found that, although not unique in their historical severity (gauged by deaths), the 1960s riots were highly unusual in their frequency, geographic distribution, and level of physical destruction (Figure 4.3). The authors developed a "severity scale" for the riots. Los Angeles, Detroit, Washington, Newark, Baltimore, Chicago, Cleveland, New York, Mobile, and San Francisco were at the high end of the severity scale. The geographic pattern of riots, Collins and Margo argued, shows spillover effects related to media coverage, although this pattern waned quickly over time. But the urban impacts were profound and affected populations – and especially urban labor markets – far beyond the numbers of persons directly involved or the immediate loss of property value.

These authors argued that follow-on effects on physical, financial, and human capital rippled widely: personal income and business revenues fell; local fiscal conditions deteriorated, as the tax base was eroded while the demand for services increased; crime rates rose; and declining property values prompted the migration of the relatively well-off, of all races, exacerbating the concentration of poverty in the inner cities. Such impacts were central to the hollowing-out phenomenon. An additional variable with a measurable effect on riot severity was the relative importance of manufacturing in a city.

Collins and Margo's research shows, moreover, that the riots' effects were not transitory, but persisted for a decade or more, as measured by relative income declines and employment rates (though not in "unemployment," as officially defined, as "discouraged workers" or those exiting the labor force are excluded in the unemployment rate calculations). And, in an indirect measure of white

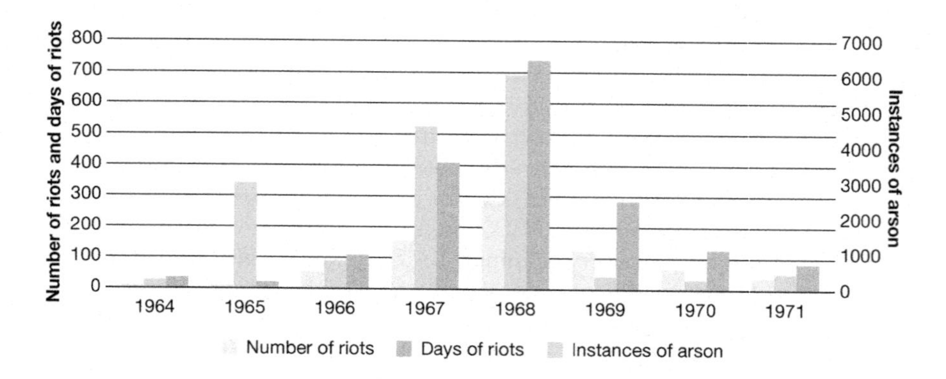

Figure 4.3 1968: A peak year of unrest

Source: Collins and Margo, 2004, using data from Carter, 1986

flight, cities experiencing severe riots had disproportionately high increases in blacks as a percentage of total population.[24] So, in addition to the "pull" of life in the suburbs described in Chapter 3, there was also a powerful "push" of population – largely white, but middle-class blacks as well – away from the troubled core that increasingly was termed "the inner city."[25]

The Nixon years

Federal policies underwent a curious mix of continuities and discontinuities as the Nixon administration took the baton from the Johnson era. Like Johnson, Nixon was a political pragmatist, but Nixon's basic agenda was a sharp departure from his predecessor's. Nixon had served as vice-president under Dwight D. Eisenhower. Eisenhower had come to office after a succession of Democratic presidents and took the approach of consolidating existing programs with an eye to running them more efficiently. That was not Richard Nixon's way.

Ironically, whereas Johnson's presidency sought to build a legacy upon progressive domestic policy, only to be undone by foreign affairs, Nixon wanted to make his mark internationally, but found his résumé tarnished by historic blunders at home. This was not merely the disgrace of Watergate,[26] but also economic and urban policy moves that served narrow, short-term political ends but ultimately contributed to serious national disruptions.

Nixon well understood the meaning of Johnson's remarks about "losing the South for a generation." This was not taken as a guarantee by Nixon, a savvy political operative, but as an opening to shift the balance of power. So a "Southern Strategy" was crafted.[27] The concept was to use the 1968 base of South Carolina (won by Nixon with the endorsement of segregationist Senator Strom Thurmond) and the five states won by George Wallace (Georgia, Alabama, Mississippi, Arkansas, and Louisiana) to establish a reliable electoral base among disaffected whites, not only in the South but across the nation.

One of the pillars of the Southern Strategy was to identify Nixon as the "law and order" choice, a repudiation of the protests of both the civil rights and anti-war elements prominent in Democratic policies. Nixon proposed returning voting rights litigation to state rather than federal courts and opposed busing as a desegregation tactic, expressing faith in "neighborhood schools." He shifted programs from the Office of Employment Opportunity to the larger bureaucracy of the Department of Health, Education, and Welfare and restricted categorical grants (i.e., those with narrowly specified purposes), while consolidating funding into so-called Community Development Block Grants that provided large pools of discretionary funds to state and local governments.[28]

Although the frequency of urban riots decreased under Nixon, the decay of cities, especially those in the heavily populated Northeast and North Central states, accelerated. During the early 1970s, greater attention began to be paid to the substantial differences in living costs and business costs that existed between regions of the United States. The greater densities, more productive economies, and more expansive public services of the states along the northern tier, from the Atlantic Seaboard through Illinois, made them higher-wage and higher-tax venues than the states targeted by the Southern Strategy. Such regional disparities had long been recognized, and equilibrium economics worked in several ways both to sustain some differences and to accommodate change.

States with lower wages had lower cost of living as well, reflected in housing prices, land costs, and tax-supported public services such as education, public transportation, and welfare benefits. There was an equilibrium of low costs reflecting the rough economic reality of "you get what you pay for." The same principle worked, in a complementary way of course, for higher-cost locations.

Inflation has a pervasive impact on price-based differentials and disadvantages economic entities with already high-cost structures.[29] Economic policy under Nixon exacerbated inflation, although his intentions were exactly the opposite. Nixon inherited the guns-and-butter dilemma from Johnson and had exactly the same instinctive political response: spending cuts were unpopular, and failure was even more unpopular. Nixon was determined to exit Vietnam, but felt he could do so victoriously. This would both confound his anti-war opponents and burnish his long-standing anti-communist credentials. But the path to success, in his mind, did not run through the reduction of the military budget but in funding an expansion of the war into neighboring nations, including Cambodia and Laos.

At the same time, Nixon, with a sharp eye on the electorate, sought to influence the Federal Reserve to maintain an easy-money policy to keep unemployment low, and funded large increases in Social Security, Medicare, and Medicaid. Unemployment was just 3.3 percent when Nixon was inaugurated, but was rising at the troublesome rate of 6.1 percent by the end of 1970. Inflation, meanwhile, stood at 3.6 percent in January 1968, rose to 4.7 percent by January 1969 (when Nixon took office) and spiked upward to 6.2 percent by January 1970.[30] The term "stagflation" entered the economic vocabulary, indicating rising inflation in the absence of economic growth.[31]

Nixon's response was to convene an economic summit at Camp David, in August 1971, and announce a sweeping change in economic policy. He instituted a 90-day program of wage-and-price controls, ended the convertibility of the dollar to gold (the linchpin of international currency stability), and imposed a 10 percent tax on imports to limit the impact of now-cheaper foreign goods on US corporations. Although immediately popular both on Main Street and on Wall Street, the so-called Nixon Shocks proved to be only temporary medicine for an economy that was much more gravely ill than was commonly supposed.

The oil-producing nations represented by OPEC[32] were among those negatively affected by the effective devaluation and import tax, as oil revenues were dollar-denominated. It took some time for OPEC to assemble an institutional response, but, when it did, its move made any temporary benefit of the 1971 policies seem puny. In late 1973, OPEC announced a 70 percent increase in the price of a barrel of oil. Rising energy prices rippled through the US economy quickly, and, by December 1973, CPI inflation had risen to

8.9 percent on a year-over-year basis. That was not the end, though. By December 1974, the CPI was up another 12.2 percent – double-digit inflation had arrived on America's shores.

The regional impacts of inflation, however, were not uniformly felt across the country. Although it had a devastating effect on areas of the country that had depended upon inexpensive energy for heating homes and workplaces, for running production lines, and for transportation, the energy-producing areas of the US – most notably Texas, Oklahoma, Louisiana, Colorado, Wyoming, and Alaska – enjoyed boom times. Jobs that were squeezed by soaring energy costs in the Northeast and North Central areas (which began to be called the Frostbelt, or more acerbically the Rustbelt) found themselves being supplanted by the demand for workers in energy exploration, drilling, and petrochemical processing. The old industrial cities of the North now had another reason to shrink, whereas the younger Sunbelt cities saw a new stimulus to growth.

The Great Migration

The second equilibrating force was the mobility of population.

The turmoil that engulfed the cities is often portrayed as a clash between the forces of the status quo and the demands of the increasingly empowered black communities and the rising Boomer generation. But the economy of cities was changing at the same time, and racial or generational frictions need to be understood in the urban economic context. A case can be argued that the very fires that scorched many cities in the 1960s and 1970s set the stage – at least in some place – for later transformations that gave rise to the 24-hour city.

Between 1910 and 1970, one of the great mass population movements in human history reshaped the demography of the United States. Over that span, about 6 million African–Americans moved from the rural South to the great cities of the North.[33] This occurred in two waves: one from 1910 to 1930, a movement called the "First Great Migration," which was arrested by the Great Depression. The onset of World War II and the concomitant surge in production of wartime matériel sparked the "Second Great Migration," which continued for a quarter-century after the war's end in 1945.[34] This crested in the late 1960s, as noted by demographer William Frey:

> At the tail end of the "Great Migration" . . . the 14 states experiencing the greatest black out-migration were all located in the South, led by "Deep South" states of Mississippi, Alabama, and Louisiana. Meanwhile, migration gains during that period [1965–1970] included those in the industrial Midwest and Northeast. These states contained urban industrial centers that, at the time, attracted large numbers of less-skilled black laborers in search of employment.[35]

The attraction was not merely passive. In its first phase, World War I had created a significant labor shortage in the North, as 5 million men, mostly white, entered the military (still segregated until the 1950s). At the same time, immigration from most of Europe was halted. This left steel mills, tanneries, stockyards, railroad companies, and shipyards scrambling for workers. The decline of Southern agriculture as a source of employment, partly a result of mechanization and partly due to the plague of boll weevils that attacked the cotton crop, left black labor as the last large source of available workers. Industry agents traveled south to attract workers, using incentives such as subsidized transportation, better education and housing options, and higher wages to entice workers northward. The African–American populations of New York, Chicago, Detroit, Philadelphia, and Pittsburgh swelled, together with West Coast cities such as Oakland, Los Angeles, and Seattle.[36]

Following World War II, even as the GIs returned, the need for industrial workers continued to expand, and the black migration northward resumed in force. Between 1950 and 1970, for instance, the African–American population in the US grew by 50 percent. Over those two decades, the number of blacks grew in Chicago by 124 percent, in Boston by 161 percent, in New York by 123 percent, in Washington by 91 percent, in Baltimore and Newark by 177 percent, and in Detroit by 120 percent.[37] Over the same time period, manufacturing employment in the US surged from 13 million to more than 18 million.[38] But the steep upward path of industrial jobs halted during the 1970s, just as the Baby Boom generation was entering the job market as a new factor in labor competition. As William Frey notes, the Great Migration then began to show signs of reversing, as the "people follow jobs" dynamic again exerted its equilibrating effects.[39] In part, blacks shared the economic motivations triggered by the evolution of the South's economy (discussed in Chapter 2), but in part they responded to the negative forces of deindustrialization, which brought manufacturing employment in the US down to 15 million by 2001, and down even further to 12 million by 2014.

Deindustrialization and urban decline

It was particularly unfortunate that, just as the rights of African–Americans were clearly codified in law, the level of economic opportunity available in America's industrial cities – for blacks and for whites – began to drop, steeply and quickly.

Though the link was scarcely noted at the time, Wall Street provided some early warning of manufacturing's severe problems. The Dow Jones Industrial Average, the widely followed index of stock prices for 30 iconic US manufacturers, had risen sixfold in value, moving from a level of 180 at the beginning of the 1950s to a peak of 995 in 1966. The Dow first flattened, then shuddered in 1967, and finally collapsed to troughs of 631 in 1971 and again to 570 in 1974. If stock prices are, as presumed in standard financial theory, a reflection of the best available information about future corporate earnings, the market was signaling serious trouble in America's industrial sector.

Indeed, America's manufacturers were not only on the ropes, they were on the run.

In some ways, our companies were victims of our own successes. First and foremost, the Allied victory in World War II had devastated the production capacity of the Axis powers, most notably Germany and Japan. US manufacturing, meanwhile, had been shielded from physical destruction by the oceans and by our counteroffensives. Following the war, America's enlightened policy of foreign assistance reconstituted independent governments in formerly adversary nations and provided material aid through the Marshall Plan[40] (including $2.2 billion to West Germany) and, under a separate program, $2.4 billion to Japan. This magnanimous response was at one level a recognition that excessively punitive measures following World War I had set the stage for the Great Depression in Europe and Hitler's rise to power.[41] It was also an act of self-interest for the US, reestablishing markets for American goods and services as we entered an era of economic hegemony.[42] However, by the 1960s and 1970s, American manufacturing was facing conditions of world trade where nations such as Japan and Germany were producing goods in modern, post-war factories, whereas our own companies were still using plants that were built in the early 20th century.

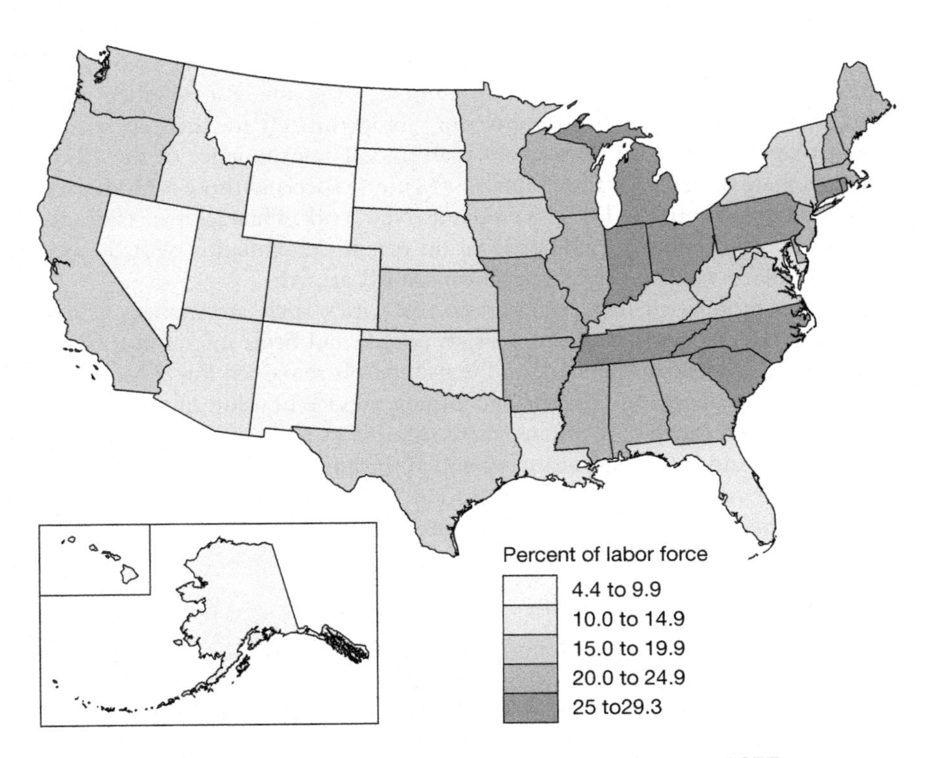

Figure 4.4 Concentration of manufacturing employment, by state: 1977

Source: US Census of Manufacturing

In the face of significant international competition, many American corporations sought to protect their bottom line by reducing costs. State and local economic development corporations were quick to tout advantages of lower taxes, lower wages, and lower business costs, if such were available, as lures to footloose companies. For many, this meant looking south (Figure 4.4). The textile industry that had been a feature of the New England economy since the early Industrial Revolution had migrated, almost *en masse*, to the Carolinas. This was a very early case of the industrialization of the South, dating back to World War I. This trend accelerated over time. However, this early success made Georgia, North Carolina, and South Carolina especially vulnerable to globalization forces later in the 20th century.

Similarly, the automotive industry, which had been centered in the Great Lakes states, found itself locating new assembly plants and supply chains in mid-South states, particularly Kentucky and Tennessee. In many cases, the impetus for change was the existence of so-called "right to work" laws (i.e., laws restricting the power of unions to organize) in the Southern states, a provision of the 1947 Taft–Hartley labor law.

Sadly, however, the manufacturing industries found that they could not cost-cut their way to success. In fact, the low-cost environment of the South turned out to be merely a way station in the migration of production jobs to even lower-cost venues – Mexico, China, India, and Indonesia. Cities whose economic development strategies relied upon being "the low-cost provider" have found themselves in a race to the bottom – unfortunately for their citizenry.

Not that this was at all obvious 40 years ago. The experience of the 1970s was extremely difficult for the "high-cost" cities, especially those with dominant manufacturing bases. For cities such as New York, Philadelphia, Chicago, Detroit, and Baltimore, deindustrialization meant depopulation. For Sunbelt cities, population and business migration spelled growth.

The automobile-friendly configuration of cities such as Atlanta, Dallas, Phoenix, and Los Angeles meant that, as people and firms migrated to these areas, their perimeters expanded, and density levels remained low. One of the imperatives of seeking low cost entailed finding sites for housing and workplaces that were priced cheaply. In an era where suburbs were the preferred locations to live, work, and shop, this meant growth at the edges, and the hollowing out of the center. In America's political system, that also meant that power was shifting from North to South, and from city to suburb.

1974: A second watershed year

In the stream of history, dates are often associated with singular events. The Battle of Tours in 732. The arrival of William of Normandy on England's shore in 1066. The first voyage of Columbus in 1492. The American Declaration of Independence in 1776. The year 1974 does not have such a singular event – although the resignation of Richard M. Nixon as President of the United States, under threat of impeachment, could rise to that level. But it was a watershed year nonetheless, largely because of the sum of events and because of an alteration in trajectory for America's cities.

The year began with the OPEC oil embargo in full force, and the northern cities found Daylight Savings Time pushed back 4 months into January, leaving school children and workers alike beginning their days in darkness and cold. Inflation was at the top of public attention, and those with means were shifting cash into hard assets. Headlines announced record prices for precious metals, and gold, now detached from convertibility to the dollar at $35 per ounce, soared toward $200 per ounce.

Although the era of civil riots had largely abated, racial tensions had not relaxed. Boston saw itself riven by a school busing controversy, as the "neighborhood school issue" proved to be as much a northern sore point as an element of the Southern Strategy.

Depopulation of the older manufacturing cities was at floodtide. Empty blocks of abandoned, often flame-gutted, buildings pocked the landscape of many cities. Crime was on the rise, and fear kept more and more people at home – or packing their bags for the suburbs or Sunbelt. Occasionally, attendance at professional sporting events would provide shocking evidence of the urban decline. In January, a Cleveland Cavaliers basketball game drew just 1,641 patrons. In May, a sparse 4,149 fans took in a baseball game at Philadelphia's Veterans Stadium. Sports facilities with 90 percent of their capacity unfilled told a story more starkly than a shelf of sociology journals ever could.

In the world of finance, the Dow Jones Industrial Average hit a low of 570. Bond houses were scrambling to adjust to changes in Securities and Exchange Commission rules that would abolish fixed commissions, a major shift in their revenue stream. Large exchanges such as New York's were roiled, but smaller exchanges such as Philadelphia's saw the shift as an existential threat. Major US corporations were startled by an anti-trust suit initiated by the US Department of Justice against AT&T, an attempt to "break up the telephone company" and create a competitive and deregulated communications industry. Congress passed ERISA, the Employees Retirement Income Securities Act of 1974, which affirmed that corporate pension funds were proprietary to the beneficiaries, not the corporations. One of the many impacts of this law was a rise in institutional money management, and this led to the enfolding of real estate investment as an asset class into pension fund portfolios. Over time, this made urban real estate an object of institutional investment attention and a recipient of large-scale capital deployment.

Daniel Bell's book *The Coming of Post-Industrial Society*[43] was published. This not only focused attention on manufacturing's decline, but also projected that the economic future would see growth determined by the production of knowledge and the provision of services, rather than the output of hard goods. A shift from blue-collar to white-collar jobs portended a new description of what would drive urban success. Information and creativity took on enhanced importance in the public mind. Education and the arts were understood to be civic assets rather than nice amenities for the few.

At year's end, the magazine *Popular Electronics* offered a hint of how communications, technology, and the information society were going to evolve. The Altair 8800 computer, the first microcomputer available for mass use, was sold by the thousands in its first months on the market. It featured a central processing unit using the Intel 8080 chip and utilized Microsoft's first product, the programming language Altair BASIC.

In retrospect, the year 1974 appears as revolutionary as 1776.

In prospect, however, the future did not appear to be bright for urban America. Depopulation, deindustrialization, civil discord, and negative economic trends brought many cities into fiscal disarray. New York City was surely the most prominent example, as it careered toward what appeared to be certain bankruptcy. An appeal for federal assistance was made, and rejected by the President of the United States, Gerald Ford, prompting a memorable newspaper headline, "FORD TO CITY: DROP DEAD," on the front page of the tabloid *New York Daily News* (October 30, 1975).

If the President of the United States could turn his back on the nation's largest city, what hope could there be for urban America?

Forms of change: Changes of state

Science teaches us that fundamental changes in state represent yet another way in which we find modifications in the world. Such changes differ from temporal alterations such as the cyclical fluctuation around equilibrium, the unfolding of a trend, or the maturation of an organism. In state change, we find an alteration in some object that reflects not only its condition but also its behavior.

For instance, take the molecule H_2O – the chemical designation for water. We learn early in life that water is most commonly found as a liquid, but, depending upon changes in temperature, it can shift its state to a solid (ice, below 32° Fahrenheit or 0° Celsius). The behavior, not only of the molecules, but also of the macro-level condition of water, changes. It no longer "flows" or behaves fluidly, but is both rigid and fairly brittle, as well as being solid. We would never characterize the behavior of the liquid known as water as being either rigid or brittle, but ice certainly can be described that way.

Then, if we heat the H_2O molecules above 212° Fahrenheit or 100° Celsius, the state changes once again. At this point, steam, a gas, is produced. Water and ice are heavier than air and, if released from any container, will obey gravity and move toward the ground. Steam, however, rises; moreover, it does not cohere, but disperses into the air. As the state of this simple physical compound, H_2O, goes through its transformations, we see marvelous alterations in what it does and how it can be used.

A more complex example is familiar to anyone who has spent time in a kitchen. Mix ingredients such as flour, eggs, water, sugar, salt, baking powder, and any manner of flavorings in proper proportions, and a viscous blend known as cake batter is created. Pop that mixture in the oven at 375° Fahrenheit for 30 minutes and let it cook. What comes out looks very different than what went in. It smells and feels different. And, most importantly, it is a very different treat for the mouth. What has happened is known in chemistry as an endothermic change of state, a.k.a., baking.

Sociologically, we know about changes of state, too. Put a crowd of people together, in a sufficient state of stress, and incite them to action. With enough initial stress and the promptings of passions, especially rage or fear or greed, and you have a mob. Interestingly, the individuals in that mob might not imagine themselves engaging as single persons in the behaviors of the mob, and afterward might well return to their normal state the very next day. But together, and under certain circumstances, watch out! (And, lest some sniff that *they* would never indulge in such unseemly and dangerous activity, the psychology of herd behavior at sporting events is not that much different. And then there is the kind of financial euphoria that leads to bubbles of all sorts, from tulipmania in 17th-century Holland to the housing frenzy of early-20th-century America.[44])

Lastly, even in the progress of science, we have experiences of "state changes," or what have been termed "paradigm shifts." Probably the most widely read book in the philosophy of science is Thomas Kuhn's *The Structure of Scientific Revolutions*, which describes the conditions under which Ptolemaic cosmology gave way to the Copernican view of the universe, how the pre-modern theory of "the humors" of the body and environmental "miasmas" was replaced by the germ theory of disease, and the physics of Sir Isaac Newton was changed by early-20th-century physicists including Einstein, Heisenberg, Planck, and Bohr. This kind of intellectual change of state doesn't happen very often, but, when it does, it alters the way we view the world in profound ways.[45]

Historically, we have witnessed the economic impacts of the agricultural revolution that transformed our species from survival of hunter–gatherer societies to division of labor economies that gave rise to cities in the first place. Then we experienced the Industrial Revolution that mechanized agriculture and introduced mass production of goods, expanding cities exponentially and moving society beyond handicrafts to manufacture. Now it appears that a technological revolution of information and communication is again beginning to reshape urban living. That may well turn out to be a similarly profound change of state for our cities.

To give just a few examples of how the knowledge economy is reshaping cities, we might start with Manuel Castells' 1989 analysis.[46] Twenty-two years before the 2011 Occupy Wall Street movement pushed the divide between "the one percent and the ninety-nine percent" into public consciousness and political debate, Castells saw that technical skills generated a cadre of highly-paid jobs, while driving minorities, women, and immigrants disproportionately into an expanding "service jobs" sector with comparatively low wages. Such jobs can be found in healthcare, child-care, hotels, restaurants, personal services, taxi and limousine services, and other fields, often occupations depended upon by the affluent knowledge workers. Castells described the result as a "dual city" that is characterized by a "spatial structure that combines segregation, diversity, and hierarchy."[47]

Mitchell Moss and Hugh O'Neill, in a prescient 1991 paper,[48] saw the largest US city evolving toward a city featuring high demand for office-based jobs, partly in finance but also in what have become known as the TAMI industries (technology, advertising, media, and information). NYU President John Sexton, in a 2007 lecture, called attention to the ICE sector (intellectual, cultural, educational) as a driver for the city, and also as a talent magnet.

Richard Florida extends and elaborates on this concept in *Who's Your City?*[49] Florida discusses "the basics" when it comes down to places with the greatest attractiveness to a mobile population: good schools, good healthcare, a positive diversity of strong centers of spirituality and religion, and cities with high levels of tolerance. Such factors trumped more conventional metrics such as housing affordability, great transportation, or population homogeneity.

In *Smart Cities*, Anthony Townsend discusses elements such as the Internet of things, autonomous vehicles, ubiquitous public WiFi, and the concomitant data-gathering associated with embedded technology.[50] Whatever you might call it, and whatever the acronym *du jour*, it is becoming obvious that a new "state of the city" is emerging. Although this might be termed the drive toward a "happy city," the duality identified by Castells decades ago is intensifying.[51]

Recently, my colleague Rosemary Scanlon and I teamed with Richard Florida and Steven Pedigo to produce a report entitled "New York City: The Great Reset," which examines in detail the reshaped city – resilient and prosperous for many, but with a services class that is both large (1.9 million workers) and growing. The issue is that, in New York's "change of state," this service sector represents 51 percent of all jobs in the city, but captures just 35 percent of wages. We propose a series of steps to address the future of an urban society where the answer to income inequality is the generation of higher-income mobility.[52] Time – and effort – will tell whether our recommendations will be effective. But it is clear that the very issues facing New York are a function of its evolution – the emergence of a reinvented city that has undergone a change of state over the past quarter-century.

Notes

1 Jacobs, op. cit.

2 Bob Stanley, *Yeah! Yeah! Yeah!: The Story of Pop Music from Bill Haley to Beyoncé*, W.W. Norton (New York, 2014), p. 72.

3 The chairs from the living-room set of the Bunkers' home can now be seen at the Smithsonian Institution.

4 James Roman, *From Daytime to Primetime: The History of American Television Programs*, Greenwood Press (Westport, CT, and London, 2005), pp. 102–106.

5 Among the many sources for information on the LBJ presidency, see online http://millercenter.org/president/lbjohnson/essays/biography/; *The Washington Post* published a retrospective, "The Great Society at 50," on May 17, 2014.

6 Known to economists as a "production possibility frontier," the concept accepts that there are finite economic resources available to achieve desired objectives, necessitating often-difficult choices among desired goods.

7 A memorable set of photographs from that era can be found at http://life.time.com/history/war-on-poverty-appalachia-portraits-1964/, some with contemporary captions from *Life* magazine, and others from *Life*'s archives, uncaptioned and previously unpublished. The title of the magazine's article was "War on Poverty: Portraits From an Appalachian Battleground" (January 31, 1964).

8 Michael Harrington, *The Other America: Poverty in the United States*, Simon & Schuster (New York, 1962).

9 A continuing theme in urban migration, as noted in Edward Glaeser's *Triumph of the City: How Our Greatest Invention Makes Us Richer, Smarter, Greener, Healthier, and Happier*, Penguin (New York, 2011). See Chapter 3: "What's Good About Slums."

10 The passage of Medicare (1965), the healthcare program for the elderly, is one of the Great Society programs that has survived a half-century of shifting political preferences in the United States. By shielding senior citizens from the burden of rising medical expenses, it has not only lessened the economic stress on aged households but has contributed to significantly improved lifespans for Americans.

11 Jacobs, op.cit., Parts I and II (Chapters 2–12).

12 The term "brutalist" comes from the French for "rough concrete" (*beton brut*), but brutishness is deliberately suggested by the architectural designation for the heavy and coarse material and blockishness of the genre. Le Corbusier is considered the originator of the style, the American architectural and development firm of John Portman & Associates its best-known practitioner, especially in downtown Atlanta's Peachtree Center. See "Brutalism," *Encyclopedia of 20th Century Architecture, Volume I*, edited by R. Stephen Sennott, Fitzroy Dearborn (New York, 2004).

13 See Jacobs, op.cit., p. 4.

14 Cited in *The Washington Post*, art. cit., May 17, 2014.

15 Yes: *Sesame Street*, first aired in November 1969. The program showed not only that races and religions could live in harmony, but that even monsters of various pedigrees could be lovable.

16 In November 1964, 77 percent of the public felt that government could be trusted "just about always" or "most of the time." That declined somewhat to 65 percent by January 1967, and 61 percent by November 1968. See the Pew Research website, online at www.people-press.org/2014/11/13/public-trust-in-government/. Such figures can be compared with the approval ratings of recent years, which have ranged between 10 percent and 25 percent.

17 Sources for the chronology include Bernard Grun, *The Timetables of History*, Simon & Schuster (New York, 1963); *Baltimore Magazine*, "100 Years: The Riots of 1968," July 2007; and the International Civil Rights Center and Museum, online at https://sitinmovement.org/history/america-civil-rights-timeline.asp

18 The group took its name from a line in Bob Dylan's song, "Subterranean Homesick Blues," that noted, "You don't need a weatherman to tell which way the wind blows."

19 The Chicago Eight were found "not guilty" of the charge of conspiracy to incite a riot, though five were convicted of a charge of crossing state lines with the intent to riot. Those convictions were later vacated by an appellate court, citing Judge Hoffman's bias against the defendants.

20 This statement was directed privately to LBJ's aide Bill Moyers and has recently been under attack as inaccurate and undocumented by Republican advocates. Moyers, however, in a November 5, 2008 interview with Terry Gross on National Public Radio, affirms Johnson's observation. A recording of that interview can be found

online at www.npr.org/player/v2/mediaPlayer.html?action=1&t=1&islist=false&id=96648963&m=96650934

21 Many have made the point that those who don't know history are doomed to repeat it. But others take a dimmer view of our ability to learn from experience. Kurt Vonnegut (in *Slapstick*) noted, "History is a series of surprises. It can only prepare us to be surprised yet again." Economist John Kenneth Galbraith takes the gloomy position (in *A Short History of Financial Euphoria*) that, "Recurrent descent into insanity is a not wholly attractive feature of capitalism. The human cost is not negligible, nor is the economic and social effect." Historian Barbara Tuchman's *The March of Folly* considers the tendency of leaders, since at least the Trojan War, to repeat actions contrary to self-interest.

22 "The 1960s Black Riots Revisited: City Level Explanations of Their Severity," *Sociological Inquiry 56:2* (1986), 210–228.

23 William J. Collins and Robert A. Margo, "The Labor Market Effects of the 1960s Riots," NBER Working Paper 10243, National Bureau of Economic Research (Cambridge, MA, January 2004).

24 Collins and Margo candidly acknowledge, though, that their findings are suggestive rather than definitive. Precise mechanisms tracing the cause–effect relationships cannot be measured in detail. And there are surely significant "omitted variables" (such as education, detailed industry and occupational distributions, and non-governmental social structures) that could conceivably have had significant parts to play. Nevertheless, these authors persuasively argue that their analysis captures the reality that the 1960s riots made a lasting imprint on American cities. For our purposes, we can note that many of the worst affected cities somehow became the places that evolved into "24-hour urban centers" within a generation.

25 The Merriam-Webster Dictionary Online indicates that the first known citation of this term to describe impoverished minority neighborhoods in American cities dates from 1961.

26 "Watergate" denotes the break-in at the Watergate offices of the Democratic National Committee headquarters by operatives of Nixon's "Committee to Re-Elect the President" (unbelievably, known by the acronym CREEP). The subsequent attempt to cover up the crime, orchestrated from the White House itself, ultimately led to the President's resignation under threat of impeachment in 1974.

27 Notoriously, the Republican campaign strategist Lee Atwater revealed the cynical premise behind the Southern Strategy in a 1981 interview, with the proviso it not be published for attribution until after his death. *New York Times* columnist Bob Herbert quoted Atwater in a 2005 article:

> You start out in 1954 by saying, "Nigger, nigger, nigger." By 1968 you can't say "nigger" – that hurts you. Backfires. So you say stuff like forced busing, states' rights and all that stuff. You're getting so abstract now [that] you're talking about cutting taxes, and all these things you're talking about are totally economic things and a byproduct of them is [that] blacks get hurt worse than whites.
>
> (*New York Times*, October 6, 2005)

28 Dubbed "the New Federalism," the devolution of both funding and planning to more local government reduced the leverage of the national government to pursue Johnson-era Great Society programs without explicitly voiding their rationale.

29 For instance, let's say that income-based price differences keep the price of a basket of groceries to $50 in one location, whereas the same basket of goods costs $70 elsewhere. If the price of food jumps 10 percent over time, Location A sees the grocery bill rising to $55, but at Location B it increases to $77 . . . expanding the cost differential to $22 from $20 between the two places. Unless incomes increase proportionately, the standard of living in Location B declines relative to Location A.

30 Inflation figures are the year-over-year change in the consumer price index (CPI), retrieved from the Bureau of Labor Statistics website, online at http://data.bls.gov/pdq/SurveyOutputServlet
31 The real GDP of the US economy (in chained 2009 dollars) was $4.72 trillion in 1969, $4.71 trillion in 1970, and $4.91 trillion in 1971, according to the US Bureau of Economic Analysis.
32 The Organization of Petroleum Exporting Countries.
33 For a lucid discussion of the Great Migration in the context of cultural history, see online, http://kenanmalik.wordpress.com/2014/12/21/jacob-lawrence-and-the-great-migration/
34 See William H. Frey and Alden Spears Jr., *Regional and Metropolitan Growth and Decline in the United States*, Russell Sage Foundation (New York, 1988).
35 Frey, "The New Great Migration: Black Americans' Return to the South 1965–2000," The Living Cities Census Series, The Brookings Institution (Washington, DC: May 2004), pp. 3–4.
36 http://afroamhistory.about.com/od/segregation/p/Causes-Of-The-Great-Migration-Searching-For-The-Promised-Land.htm
37 Population figures for cities, broken down by race and ethnicity, can be found at the Census Bureau's website, online at www.census.gov/popest/data/historical/index.html
38 Long-term changes in manufacturing employment can be tracked on the Bureau of Labor Statistics website, online at http://data.bls.gov/timeseries/CES3000000001
39 Frey, art. cit.:

> Southern metropolitan areas, particularly Atlanta, led the way in attracting black migrants in the late 1990s. In contrast, the major metropolitan areas of New York, Chicago, Los Angeles, and San Francisco experienced the greatest out-migration of blacks during the same period.

40 Termed by Winston Churchill "the most unsordid act in human history."
41 See John Maynard Keynes, *The Economic Consequences of the Peace*, Harcourt Brace (New York, 1920), now widely available on the Internet in PDF form.
42 In 1941, *Time* magazine publisher Henry Luce editorialized that America needed to forego isolationism and use its power to spread democracy as a matter of national policy. After the war, Luce and others saw the state of the world as a "Pax Americana" in the democratic/capitalist sphere as much as a "cold war" with adversaries across the Iron Curtain.
43 Daniel Bell, *The Coming of Post-Industrial Society: A Venture in Social Forecasting*, Basic Books (New York, 1974). Bell's seminal insights have since prompted a point–counterpoint of considerable interest. Steven S. Cohen and John Zysman argued, in *Manufacturing Matters* (Basic Books: New York, 1987), that, although industrial employment has declined, output has increased. Jeffrey Mayer, of the US Department of Commerce's office of Economic Statistics and Assessment (1995), noted that US technology production had altered the composition of American manufacturing, illustrating some of the points made by Cohen and Zysman. In 2013, the Partnership for a New American Economy – an initiative spearheaded by New York Mayor Michael Bloomberg – studied the positive correlation between immigration trends and manufacturing-job growth in a number of large US cities ("Immigration and the Revival of American Cities," online at www.as-coa.org/articles/immigration-and-revival-american-cities).
44 See John Kenneth Galbreath's *A Short History of Financial Euphoria*, Penguin Books (New York, 1991) for a brief, lucid, and entertaining account of speculative bubbles through the centuries.

45 Thomas Kuhn, *The Structure of Scientific Revolutions*, University of Chicago Press (Chicago, IL, 1962).
46 Manuel Castells, *The Information City: Information Technology, Economic Restructuring, and the Urban Regional Process*, Blackwell (Oxford, UK, 1989).
47 Ibid., pp. 187, 224–228.
48 Mitchell Moss and Hugh O'Neill, "Reinventing New York," Working Paper at the Taub Urban Research Center of New York University (New York, 1991).
49 Richard Florida, *Who's Your City? How the Creative Economy is Making Where You Live the Most Important Decision of your Life*, Basic Books (New York, 2008). See especially Chapter 7, "Job-Shift," and Chapter 10, "Beyond Maslow's City."
50 Anthony Townsend, *Smart Cities: Big Data, Civic Hackers, and the Quest for a New Utopia*, W.W. Norton (New York, 2013).
51 See Charles Montgomery, *Happy City: Transforming our Lives through Urban Design*, Farrar, Straus & Giroux (New York, 2013), and Jan Gehl, *Cities for People*, Island Press (Washington, DC, 2010).
52 Ricard Florida, Hugh F. Kelly, Steven Pedigo, and Rosemary Scanlon, "New York City: The Great Reset," a White Paper from the School of Professional Studies/ Schack Institute of Real Estate at New York University (New York, 2015). Accessible online at www.pageturnpro.com/New-York-University/67081-The-Great-Reset/ index.html#1

5 Central tendencies versus the hollow-core story

A half-century of push and pull in America's cities

A word, "yuppie," is probably as clear a signal of a shift in the fortune of cities as we are likely to find. It's an acronym, really, and stands for "young urban professional." Both the Merriam-Webster Online dictionary[1] and the Oxford Dictionary[2] identify its coinage as arising in the 1980s and define the word this way: "a young college-educated adult who is employed in a well-paying profession and who lives and works in or near a large city" and "A fashionable young middle-class person with a well-paid job."

Although there has always been a hint of derision in the use of the term, its emergence to describe a large enough population segment in America's cities to require an identifier tells us that suburban growth and the hollowing out of the center city are far from telling the entire story. A key subplot was taking shape as early as 35 years ago. Its narrative was strengthening as the final quarter of the 20th century proceeded.

As usual, popular culture gives us broad hints about the changes. Blue-collar and suburban settings were the norm in the "golden age" of television. Idyllic depictions of small-town or rural life, such as *The Andy Griffith Show* (1960–1968) and *The Waltons* (1971–1981), provided sentimental nostalgia for an idealized community life. Shows set in actual big cities tended to be focused on crime (*Dragnet* [1952–1959, and 1967–1970], *Columbo* [1971–1978], *Kojak* [1973–1978], *Baretta* [1973–1978], *Miami Vice* [1984–1989]) or on the chaos of big city hospitals (*Medical Center* [1969–1976], *St. Elsewhere* [1982–1988], *Chicago Hope* [1994–2000], *ER* [1994–2009]). Like the popular music cited in Chapters 3 and 4, TV both captured and validated the outward thrust of population from the city centers.

But a different kind of show began to celebrate young urban professionals and the lifestyle that cities provided them. *The Mary Tyler Moore Show* (1970–1977) was set in a Minneapolis TV newsroom and featured the workplace as its setting, with single men and women across an age spectrum depicting a "new normal," contrasted to the home-and-family sitcom. *Murphy Brown* (1989–1998) celebrated its unmarried eponymous heroine, a Washington, DC-based investigative journalist. A Boston bar attracted accountants, postal carriers, and psychologists in *Cheers* (1982–1993), its patrons served by a staff of ex-athletes, doe-eyed liberal arts majors, and sharp-witted housewives trying to make ends meet. In New York, *Seinfeld* (1989–1998) and *Friends* (1994–2004) introduced America to the world of hip Manhattan apartment living, shared by entertainers and show-biz wannabes, office workers, anthropologists, aspiring chefs (and, again, a postal worker). The glamor and the glitz were ratcheted up by *Sex and the City* (1994–2004). The celebration of sophisticated city life appeared coast to coast with *Frasier* (1993–2004), set in Seattle, and *The Big Bang Theory* (2007–present) in Los Angeles.

Black America also saw its aspirations depicted as positive urban possibilities. *All in the Family* (1971–1979) spun off *The Jeffersons* (1975–1985), where Archie Bunker's African–American neighbors found themselves "moving on up to the East Side, to a deluxe apartment in the sky." And in brownstone Brooklyn's gentrifying neighborhoods, we could see, on *The Cosby Show* (1984–1992), a black family with a doctor as father and lawyer as mother, raising a crew of mischevious youngsters, with typical childhood and adolescent issues, but who were youngsters who were clearly destined for middle-class and even beyond-middle-class affluence.

White (collar) flight: The corporate headquarters exodus

The difficulties facing the urban workforce as cities hollowed out went way beyond the loss of production jobs, at whatever skill level.[3] The Industrial Revolution not only took advantage of concentrated urban populations for production work, it made the nation's big cities the centers of command-and-control functions for business. In 1917, as Figure 5.1 shows, 62 percent of the headquarters of the largest US corporations were located in just ten cities, all situated in the northeast quadrant of the country. That meant that the cities were the locus of the white-collar job base as well. Executives and managers, clerks and secretaries, billing agents and bookkeepers, receptionists and switchboard operators, and armies of other workers – typically housed in downtown office buildings – were directly employed in the corporate sector. Moreover, a network of service providers in the financial, advertising, legal, insurance, and accounting industries were linked both by function and by proximity to corporate headquarters.

Into the 1950s, the big cities retained their role as centers of big business. In 1957, the same ten cities still accounted for 60 percent of large corporate headquarters. Just 17 years later, though, those cities had 65 fewer headquarters operations, and their share of the national total had dropped under 48 percent. Large urban centers were hemorrhaging white-collar jobs as well as blue-collar jobs, at least in the kinds of company represented by the Fortune 500 top industrial corporations.

An insightful 1977 study[4] by Columbia University's Conservation of Human Resources project acknowledged that the mere decline in numbers could be somewhat misleading as an indicator. The Fortune 500 list is fluid for quite a number of reasons: companies are added and subtracted from the list based upon annual revenue, and firms can be reclassified owing to factors such as mergers. The investigators nevertheless recognized that business mobility was a reality, and that "an increasing number [of large corporations] no longer feel compelled to remain in one of the nation's large metropolitan areas."[5] They cited improved and less-costly air transportation, highway transportation, telephone communications, and the greater numbers of married women in the labor force as factors contributing to the national trends.

Drilling down further into the data, the investigators found that, between 1957 and 1974, the 25 largest metropolitan areas had seen a net decline in headquarters of 33 firms; the 25 next-largest metros had gained 14 headquarters; and "all other [smaller] places" had gained 19 corporate headquarters. Whether jobs followed people, or people followed jobs, the trends were not good news for the big cities.

Because the Columbia study was directly focused on New York City, the investigative team was able to access corporate executives for both firms that had relocated away from the city and those that had chosen to remain. Interviews revealed significant differences in viewpoints between the two sets of decision-makers.

	Number of headquarters		
Metropolitan area	**1917**	**1957**	**1974**
New York	150	144	107
Chicago	32	54	48
Philadelphia	26	21	15
Detroit	18	17	12
Boston	29	6	8
Pittsburgh	27	24	15
Baltimore	6	1	2
St. Louis	8	13	10
Cleveland	13	16	17
Buffalo	7	6	3
Total of 10 metro areas	316	302	237
All other places	184	198	263
Grand total	500	500	500

Figure 5.1 Headquarters location of Fortune 500 firms by metropolitan area: 1917–1974

Source: Fortune 500 directories for 1957 and 1974; data for 1917 from *Moody's Industrials* and annual reports[6]

Firms moving operations out of New York cited New York State's high personal income tax as a key factor. They added improved commuting for senior and middle managers, greater ease of assigning middle managers to headquarters, and the ability to consolidate corporate functions into a single location. Tellingly, they also alluded to an "escape from the negative aspects of the urban environment."[7]

Executives of firms keeping their headquarters operations in New York (the majority of firms, in fact) had another perspective. Although not denying the difficulties cited by their peers, these corporate leaders felt that the deal-making environment for both suppliers and customer groups was superior in the central city. Agglomeration benefits the network of corporate services firms – the "ability to call meetings . . . with marketing and advertising, legal and accounting, and banking, on short notice." This was considered a major efficiency. Moreover, it was noted that, "New York remains an important drawing card for young talented people that want to be part of a fast-paced environment."[8] To acknowledge this in a report issued in 1977, the nadir of New York City's employment decline, was significant indeed.

Generational change and the preference for cities

Long-term shrinkage

Oceans of printers' ink have been put to paper, tracking the population decline of the deindustrializing cities of the United States. As can be seen in Figure 5.2, there are half-a-dozen representative cities that have seen unremitting loss of residents in the half-century from 1960 to 2010. These are the prime cases of hollowing out – cities that now have a 50-year history of demographic decline during an era when the national population increased by 72.3 percent, or nearly 130 million people.

These are just the "big city" examples. Smaller cities that were once significant places on the map of urban America have experienced similar fates. For instance, New Orleans, Cincinnati, and Milwaukee are all listed by Holli[10] among the

City[9]	Population decline (1960–2010)	50-year population loss (%)
Detroit	956,367	57.3
Cleveland	479,235	54.7
St. Louis	430,732	57.4
Baltimore	318,063	33.9
Pittsburgh	298,638	49.4
Buffalo	271,449	51.0

Figure 5.2 Large cities with persistent population loss in post-industrial shift

15 most historically significant US cities. Milwaukee ranked 11th in population among US cities in 1960 – just missing the Top Ten. But, by 2010, its size rank had dropped to 31st on the basis of a 146,491-person drop in population, representing a 19.8 percent decrease. New Orleans ranked 15th in city size in 1960, but had lost 286,696 residents (45.2 percent) by 2010 and ranked just 51st. Similarly, Cincinnati's size rank fell from 21st to 65th, as its population loss tallied 205,607, or 40.9 percent.

Wholesale population loss sapped the urban strength of cities across northern New York State (Rochester, Syracuse, and Utica), Ohio (Akron, Dayton, Toledo, and Youngstown), and Indiana (most notably in Gary, whose 98,026-person decline in population (55.0 percent) dropped it from 70th to 396th place in the national rank order of city size. Nor was the phenomenon restricted to the Rustbelt. Birmingham, Alabama, for example, shrunk its residential base by 128,650 persons (37.7 percent) over the 1960–2010 period, falling from 36th place to 101st.

Turnaround stories

The foregoing is the hollowing-out story so often told, depicting America's older cities as mere artifacts of the Industrial Revolution. Obviously, though, decline has not been universal. The successful renaissance of another set of cities is now widely recognized.[11] A close look at population patterns across cities reveals a group that posted losses between 1960 and 1980, only to shift into population growth thereafter. Those cities include New York, San Francisco, Boston, and Seattle. Their common attribute: the ability to retain and grow population between the ages of 25 and 34, the young-worker cohort that contains the yuppie.

Any comparative look at age groups over time has to account for the variation in generations. The Boomer generation, born in the two decades following World War II, was, until recently, the largest single demographic cohort in American history. However, the generation now moving into the workforce (variously known as Generation Y, or the Millennials) is even larger in size than the Boomers (roughly 77 million Boomers and 83 million Millennials). Therefore, it is helpful to look at the 25–34 years age group as a percentage of total population. This is shown for the US and for the four cities that saw a post-1980 (positive) reversal in their total population trends in Figure 5.3.

The data tell quite congruent stories for the West Coast cities of San Francisco and Seattle.[12] Both cities lost about the same number of inhabitants between 1960 and 1980: San Francisco, 61,340, and Seattle, 61,223. But the Boomers surged into both cities in the 1980s and 1990s, and, by 2010, the 30-year population gain for San Francisco was 126,261 and for Seattle it was 114,796. That's a pretty remarkable turnaround for established cities.

The year 1980 was the year that that first half of the Boomer generation hit this particular age range. Those 25–34-year-olds had been born between 1946 and 1955. The second half of the Boomer cohort, those born from 1956 to

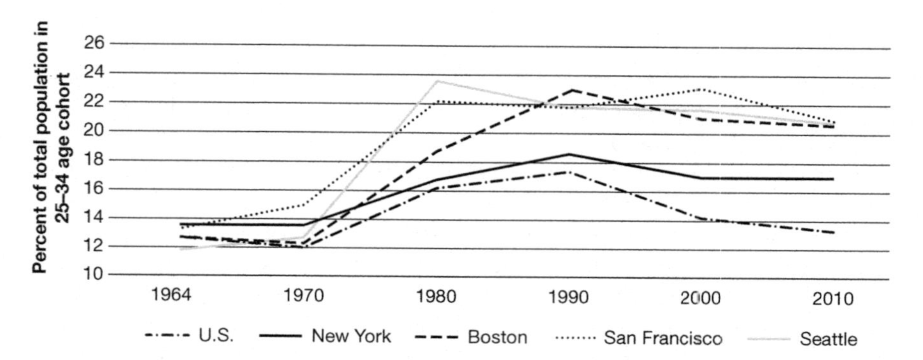

Figure 5.3 Several cities have attracted the key generational cohorts in their early career years

Source: Data from US Census Bureau decennial counts; analysis by Hugh F. Kelly Real Estate Economics

1965, formed the "young worker" age range in 1990. As Figure 5.3 shows clearly, both San Francisco and Seattle soared above the national norm for this age group in the 1980 census, and they have maintained sharply higher proportions of young workers ever since. What the graph doesn't show is that both cities enjoyed strong growth in this cohort during the 1960–1980 period, when total population was in decline. For those two decades, San Francisco registered an excellent gain of 52,996 residents aged between 25 and 34, and Seattle posted a solid increase of 40,318 young workers. So, even before the Boomers hit the job market in force, San Francisco and Seattle were demonstrating their attractiveness as places to live and work for those beginning their careers – a demographic and economic feature that is hidden beneath the fact of overall population loss in those cities. And, in the three decades between 1980 and 2010, that strength has persisted, albeit at a slower pace. The numbers are as follows: over those 30 years, San Francisco added 16,898 young workers, and Seattle added 20,021.

More strikingly, since 1960 – over the entire half-century covered by these data – the expansion of the numbers of 25–34-year-olds in each of the cities *exceeds the total population change experienced by those cities.* Not only have the young workers helped reverse a tide of population loss, they have reshaped – and continue to reshape – the socioeconomic profiles of San Francisco and Seattle profoundly.

Boston, with its amazing collection of colleges and universities, is a logical candidate for attracting the young-worker cohort. The narrative arc of Boston's age-demographic change follows the pattern of San Francisco and Seattle, though perhaps more intensely. Boston also lost population between 1960 and 1980 (134,203) and then gained residents in each of the next three decades (a total of 54,600). And, while this was happening, its young-worker cohort rose by 18,420 between 1960 and 1980, and by an additional 20,773 in the three decades leading up to 2010. What is different, of course, is that Boston's post-1980 rebound still leaves the city down in total population by 79,603 compared with a half-century ago, even with the additional 39,193 in its 2010 young-worker cohort. In this way, Boston differs from the San Francisco and Seattle experience and could be said to be even more dependent upon its attractiveness to the 25–34 age group.

The researchers for the Conservation of Human Resources Project[13] were insightful in their comment that New York remained a mecca for the young, talented, and ambitious. With the city in the midst of a cataclysmic population outflow – and without the benefit of data that would emerge in the 1980 Census of Population – it took intellectual courage as well as observational acuity to make such a call. Maybe the huge numbers could already be felt, even if they had not been officially counted. The facts are these: even though New York's total population in 1980 was 710,345 fewer than it was in 1960, young people in the 25–34 age group jumped up by 146,769 over the two decades. It might safely be said that no one would have had the chutzpah[14] to predict the increase in New York's overall population that actually occurred between 1980 and 2010 (a gain of 1,103,484) and the concurrent surge in the young-worker cohort of 189,728. The change in New York's population after experiencing this demographic roller-coaster was a net gain of 393,149 over a 50-year span, largely accounted for by the increase in its young-worker cohort of 336,047. To put these figures into perspective, New York City's total net gain was equivalent to the entire population of Minneapolis, Minnesota, and its young-worker-cohort growth was more than the total population of Anaheim, California.

In *Edge City: Life on the New American Frontier*, Joel Garreau made this sweeping statement: "Every single American city that is growing, is growing in the fashion of Los Angeles, with multiple urban cores."[15] Although this statement was originally made in 1991, Garreau kept it as a lead sentence on his website in 2015. As the foregoing look at the East Coast cities of New York and Boston and the West Coast cities of San Francisco and Seattle indicates, Garreau was only marginally correct at the time of his book publication. And he is sadly out of date now.

America is not inexorably moving toward the model of the multi-nodal city as the sole urban growth form. Period. Full stop.

I am not going to argue that the 24-hour city is the only growth option: I will strongly assert that we should be wary of any "one size fits all" interpretation of American urban dynamics. The nation is heterogeneous, in fact and almost by definition. The "melting pot" metaphor of the past, with its implied assimilation and leveling of differences, has long since fallen out of favor in the sociological literature.[16] Diversity is a plus that is increasingly recognized – and celebrated.

Persistently growing cities

That said, Garreau did put his finger on a set of cities that share the common characteristics of sustained growth, attractiveness to the young-worker cohort, automobile dependence, and a sprawling tendency to disperse economic activity into a network of subcenters rather than in a dominant downtown.[17] That set of cities includes Los Angeles, Phoenix, Las Vegas, Dallas, and Miami. All five of these cities have posted population growth in each of the decades between 1960 and 2010 and exceeded the US average share of population in the 25–34 age group. As argued above in Chapter 3, this set of cities is earlier in the sigmoid curve of maturational change, with some noticeable differences between them in terms of the best decades for growth and how recent experience has affected their trajectories.

At first glance, this could be construed as a simple Sunbelt story, influenced by all the factors mentioned in the earlier discussion of the expansion of sports locations. Plus, of course, the particular surges of the Baby Boom and Millennials. Overlaid, however, are several city-specific elements that add important detail to the overall trends.

Take Los Angeles as a case in point. During this half-century, LA supplanted Chicago as America's "second city" in terms of population, as its resident count grew from 2,479,015 in 1960 to 3,789,621 in 2010. The decades of the most significant growth were the 1960s and the 1980s, as shown in Figure 5.4. And growth has decelerated extremely since then. The 1990s had less than half the growth of the 1980s, and then the first decade of this century saw less than half the growth of the 1990s. The local demographic change failed to keep up with the national population trend. Since 1980, LA's population increase, although very significant at 847,435 persons, has seen it lose ground to the "first city," New York, which added more than 1.1 million people over the same period.

In many ways, Los Angeles was able to accommodate new Angelinos by filling in empty space. As a city, LA spans 469 square miles in land area (compared,

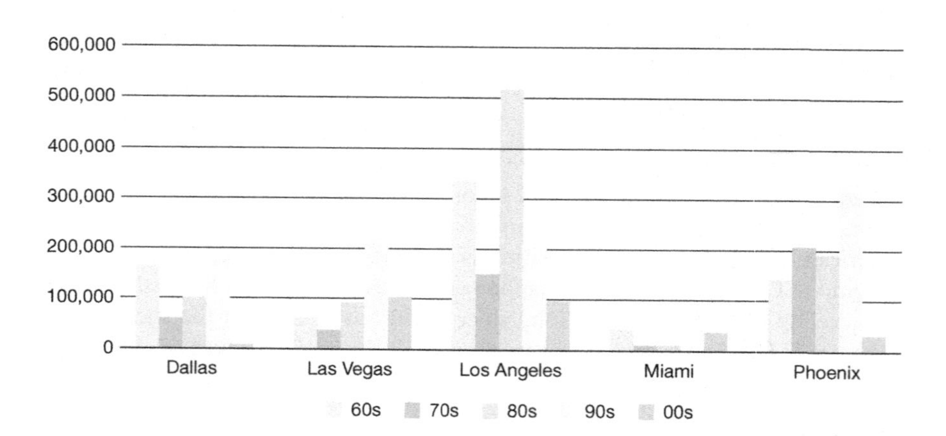

Figure 5.4 Population change by decade: Five consistently growing US cities

Source: US Bureau of the Census decennial counts

say, with New York's 304 square miles). So, with 54 percent more land area but 68 percent fewer residents (as of 1960), LA had substantial physical capacity to spare. Not only did it have a much lower population density, its built environment was largely shaped by its freeway patterns, which encouraged settlement along routes such as the 405, 10, 101, and 110 highways. In 1960, there was still abundant agricultural land within the city limits, especially in the San Fernando Valley, and this was the locus of LA's most significant population expansion.[18] Homeowners and residential real estate developers were a powerful political force, favoring housing over commercial and industrial concentrations, and, hence, income-producing properties tended to cluster based upon automobile access to population clusters. This could be seen as a contrast to more traditional city patterns, where housing development was influenced by proximity to workplace.

Whatever the interaction between commercial and residential forces, though, it left Los Angeles in the form of a sprawling, deconcentrated metropolis by 1980. I had the opportunity in the early 1980s to tour the city by helicopter, in the company of an East Coast office developer, at the behest of the LA City Council. The council wanted to encourage the creation of a modern skyline for downtown. My initial, amazed reaction on a street-level inspection of the central city was that it was instantly recognizable from memory of the 1952–1957 television series *Dragnet.* After our aerial tour, the developer (my consulting client) remarked to me privately, "This is not a city. This is nine Dubuques[19] laid end to end. It's 100 miles wide and two stories high!"

Since then, LA has done a notable job of altering its approach. It now has a world-class skyline downtown. It has invested in a modern (if limited) mass transit system. And someone walking its downtown streets during the evening now would see an evolving office, multifamily residential, entertainment–culture complex and retail–restaurant choices that Sgt. Joe Friday of *Dragnet* would not recognize. If LA's maturation curve seems to be flattening, that is not entirely to be suggestive of the end of population growth for this city. It just means that vertical rather than horizontal capacity will be the key to the future here. It is telling that Angelinos talk about the "*Manhattanization*" of their city.[20] That is a 180° turn from Joel Garreau's expectation in *Edge Cities.*

California and Texas are America's two most populous states. If Los Angeles is the signal example of West Coast brio, Dallas may be the epitome of Texas swagger. In the 1956 musical comedy *The Most Happy Fella*, lyricist Frank Loesser captured its spirit in the song "Big D," describing "Dallas, where every home's a palace/'cause the settlers settle for no less." Population size partially accounts for the growing political power of the two states and their large cities. California and Texas have given America five of its ten presidents since 1960, and four of those held the White House for 31 of the past 54 years (through 2014).

With that power has come considerable largesse. In 1964, a federal agency, the Civil Aeronautics Board, ordered the cities of Dallas and Fort Worth to come up with a plan and a site for a new regional airport. In the Texas tradition of "thinking big," the two cities presented a design for an airport whose size – 17,500 acres – was larger than the island of Manhattan. When the new airport opened in 1974, it made Dallas/Fort Worth a natural hub location, approximately midway between the two coasts. And, midway between the two cities, the creation of the airport spurred massive highway construction projects to provide access, programs that were, of course, 90 percent federally funded.

The 1960s and the 1990s were the most significant growth periods for Dallas between 1960 and 2010. Over time, though, Dallas has consistently enjoyed high percentages of young workers, peaking at 22.3 percent in 1990 (when the US norm was 17.4 percent) and still a relatively high 18.4 percent in 2010 (when the nation's standard was 13.3 percent). Dallas has stressed modernity and flash – witness its neon-highlighted skyline – and fostered entrepreneurialism as a civic and business trait. Most growth in the Dallas area, however, has occurred outside the city itself. There is powerful expansion, particularly in its northern suburbs, including Frisco, Plano, and Denton, as north–south highway routes are connected by new circumferentials such as the President George H.W. Bush Tollway.

During the 2000–2010 decade, the City of Dallas population growth was a meager 9,236 persons. This has prompted the city to refocus efforts on its downtown growth as a mixed-use urban center.[21] The challenge is stated as follows:

> Downtown Dallas certainly has tremendous resources, advantages and recent successes – but it can and must do more in the coming years. So how do we continue to make the vision a reality? Achieving future success will take targeted efforts to overcome key challenges and bolster Downtown's overall livability, competitiveness and attractiveness. Downtown Dallas 360's over-arching strategies must therefore be bold and "transformative."
>
> They must change the conventional wisdom in Dallas for how to "get things done." They must be viable in a new 21st century economic paradigm. And they must work together to shape a premier urban environment that attracts the best and the brightest from around the region, country and world.

So Dallas, like Los Angeles, is embracing an urban vision that says its future will not look like the half-century just past. Ironically, the post-World War II suburban growth model is now being understood as inadequate, a kind of "dated modernism" that is both unsustainable environmentally and undesirable economically.[22] Although this model has supported Dallas's growth for 50 years, a new approach warrants adoption. That "new" approach is actually an old one: it is the model of New York, Boston, and San Francisco, cities that leapfrogged Dallas's population growth in the early 21st century.

The first time I was in Phoenix, around 1980, I stayed at the Hyatt Hotel in its quiet downtown district. I decided to see some of the area early one morning and set out along Cave Creek Road before dawn. After less than 15 minutes of driving, I found myself watching the sunrise over the Arizona desert. From the center of town to desert isolation – 15 minutes. Just 2 years later, I made that same drive along Cave Creek Road. But, by then, the desert landscape had been filled with strip malls, fast-food joints, auto sales lots, and the whole panoply of suburban America at its most banal. That's one consequence of the growth that has seen Phoenix's city population triple over the course of 50 years.

Like Dallas, Phoenix saw that growth brake to a sudden halt between 2000 and 2010. Between the 1990s, when Phoenix welcomed 337,642 net new residents, and the last decade, the growth rate dropped by about 90 percent. The net change for the 10 years ending 2010 was just 31,974. Even more critically, the absolute number in the young-worker cohort fell by 16,597 (and the number of Phoenicians aged 20–24 dropped as well, by 21,912). Unless this pattern changes soon, there is trouble brewing for Sun City.

It remains to be seen whether Phoenix can resume its previous explosive growth. In many ways, it might be said that this city's economy grows on the basis of growth. That is, it is the continued influx of population that drives the other key elements of its economy: construction first and foremost, but also retailing, personal and professional services, and the allied industries of transportation, utilities, and banking. Most of Phoenix's economy is "local," that is, its businesses provide goods and services to those living nearby, rather than "exporting" to the rest of the nation and to the world. So. if population inflow ceases, that ripples out across the economy right away.

This is why the housing bubble and its subsequent collapse affected Phoenix so powerfully. The Case–Shiller Home Price Index[23] registered a price decline of 56 percent for homes in the Phoenix area after 2005, with widespread foreclosures locally. The Great Recession that spread over the nation (and the world) as subprime mortgage derivatives brought down the banking system stopped the Phoenix economy in its tracks. How it will restart, and the degree to which it regains its prior vigor, are critical questions.

I returned to the Phoenix area most recently in the spring of 2013. I took the new light rail tram from Tempe, where Arizona State University is located, directly to downtown Phoenix at about 5 p.m. on a gorgeous afternoon, with brilliant sunshine and moderate temperatures in the upper 70s F. Walking through the downtown, I noticed enormous change from my first visit 30 years earlier. There was a baseball stadium and a basketball arena. There was a new convention center, with associated hotels. There was a performing arts center and an attractive museum. What there was not was people. The sidewalks were virtually empty, and, although I had intended to find a nice restaurant for dinner, I was prompted to get back on the tram to Tempe.

What this shows is that successfully growing a downtown is not merely a list of economic development ingredients. It is a recipe, with the proper mix of live–work–play elements and the chemistry or interaction between them. As the change of state discussion at the end of Chapter 4 stresses, the ingredients themselves don't create a satisfying dish. Phoenix's chances of bringing its young-worker cohort back into the city, with the positive economic energy that it generates, are going to depend upon serious attention to creating a new recipe. Perhaps its leaders have only to look a streetcar ride away, to Tempe, for some hints as to what that recipe is.

Another of the "sand states" cities, Las Vegas, might seem to resemble Phoenix in growing explosively until the housing crisis and then finding itself in the headlines for all the wrong reasons. But Las Vegas, which I count among the 24-hour cities, despite its sharp differences from its bigger city companions, did not see quite the demographic deceleration that Phoenix did. And Las Vegas remained on the positive side of the 2000–2010 ledger in the young-worker cohort (though it did see a drop in its 20–24 years age group).

What are the differences, other than that Las Vegas is pretty much a one-of-a-kind city?

That thought actually begs the question, because it is Las Vegas' *sui generis* character that is at the heart of the differences of the so-called "Sin City."[24] An isolated desert location was transformed, initially, by government activity. The creation of Hoover Dam in the 1930s brought thousands of construction workers to the area in the short run and provided the entire region with a reliable water source at Lake Mead. Hard as it is to imagine today, the Nevada Test Site for atomic weaponry, 65 miles north of Las Vegas, was a source of tourism business for local hotels, where the seismic rumble, blast, and mushroom cloud could be viewed as a kind of "super fireworks display."[25] In the 1950s, the Test Site was Nevada's second largest employer – after mining and well ahead of the still-nascent casino industry. But it is suggestive that the primary economic impact on Las Vegas was not the employment and income effects at the facility, but in the city's ability to exploit the Test Site (and, to some degree, the Hoover Dam) as a tourist lure.

Tourism is a major reason why Las Vegas has grown so spectacularly, with the 2010 Census counting 583,756 residents, nine times the number (64,405) living in this city[26] in 1960. It also helps explain why Las Vegas's economy did not contract as much as Phoenix's did in the post-2007 housing collapse. Where Phoenix depended inordinately on construction activity, Las Vegas had revenue sources that were more broadly based.

One way of comparatively measuring local economies is with an economic technique called the location quotient (LQ). This compares the share of an industry in the local economy with the share that industry has in the national economy. There are several uses for LQs. They tell us in more subtle ways than simply raw numbers the relative importance any industry might have in two metro area economies. An LQ of 1.00 means that any industry – say, the information industry – has exactly the same share of local employment as it does at the national level. Baltimore, for instance, has an information industry LQ of 0.65, whereas Los Angeles has an information industry LQ of 1.89. That means this industry is underrepresented in Baltimore (and, therefore, Baltimore is probably buying information services from other places), whereas it is overrepresented in LA (suggesting that Los Angeles is likely to be selling such services on a net basis).

As of mid-2014, Phoenix had its highest LQs in financial services (1.64), real estate and leasing (1.49), and professional and business services. These industries are by and large based upon activities within the local economic base. Las Vegas, on the other hand, has one enormously high LQ industry, leisure and hospitality (3.01), and one other strong LQ that is locally based, namely real estate and leasing (1.55).[27] Leisure and hospitality is very much an "export industry." Las Vegas is selling a service in its hotels, restaurants, and casinos that is being paid for by incomes earned elsewhere, but spent in Vegas.[28]

Thus, a key difference between the two cities is in how linked each one is to the national and global economies. Every city is linked to the broader economy, of course, but the degree of linkage varies. Las Vegas, for good or ill, is more tightly connected than Phoenix; Phoenix depends more on local economic feedback loops, and so, when housing construction fell in the subprime collapse, most sectors of the Phoenix economy dropped as a result.

Miami presents a kind of contrast to Las Vegas and Phoenix. Its LQs are moderately strong – around 1.20 – for a whole series of industries: travel and tourism, professional and business services, personal services, trade, transportation, and utilities. As such, Miami has a more diversified economic base than its sand states counterparts. Yet it was afflicted by the subprime crisis just as seriously – particularly in its condominium housing market. Because the US mortgage problems had worldwide ramifications, Miami saw its inflow of South American, Middle Eastern, and northern European investment dry up for a period of years. So even diversified economies are prone to cyclical downturns.

What is notable in Miami, as seen in Figure 5.4, is that its 1960–2010 population increase is consistent, but at a much more modest pace than the other four cities shown on the chart. This is because Miami is hemmed in, both in a physical and in a political sense. Physically, Miami has the Atlantic Ocean to its east and the Everglades to its west. Dallas, Phoenix, and Las Vegas have the potential to grow in a 360° arc, and Los Angeles has, until recently, had substantial undeveloped land tracts within its boundaries. Politically, Miami is located in Dade County, Florida, which has no fewer than 19 cities and 15 other "incorporated places." Miami has not been able to grow by annexation, as have other cities. All told, then, Miami's population growth potential (within its city limits) has been rather more constrained than the other persistently growing cities.

The generational effect has made a significant difference for Miami as well. During the 1980s and 1990s, the young-worker cohort was somewhat smaller in Miami than the national average. But, in the first decade of the 21st century, Miami had 16.5 percent of its population in the 25–34 age group, compared with the US average of 13.3 percent. This was also a noticeable point of distinction compared with Las Vegas (14.2 percent in the young-worker group) and Phoenix (15.6 percent) in the 2010 Census figures.

Young and old, Miami's population has seen an increasing Latin American influence. By the 1980s, Miami had become the "main economic gateway" to South America and the Caribbean. It had become a basing point in the geographic organization of trade, production, and markets. Miami also became a concentration point for international capital, and a locus of command and control for global enterprises. It has also been a prime port of entry for refugees from Haiti and Cuba, a source of labor across the income and occupational spectrum, especially in the small-business and entrepreneurial segments of the economy. And Latin American money – some clearly legitimate, some manifestly less so – flooded the city as early as the 1970s and continues to do so.[29]

Of the five persistently growing cities I've just discussed, two (Las Vegas and Miami) I have classified as 24-hour cities in the course of my research, and two (Dallas and Phoenix) are 9-to-5 cities, both in their history over five decades and even today. Los Angeles, whose downtown has been a 9-to-5 location until recently, is showing visible signs of developing as a 24-hour locale. We observe that cities can and do change, although, once imprinted, their urban character is difficult – and expensive – to change. One thing is clear, though: sheer growth is not enough to create 24-hour live–work–play vitality. The recipe is at once more complicated and more elusive than that.

Cities with stop-and-go growth

Impressionistically, Figure 5.5 is bound to raise some eyebrows – especially for those accustomed to thinking about urban growth as a metro area phenomenon. But this chapter is focusing on the central cities themselves, the areas that suffered the hollowing out of decline and that have had such mixed experiences in seeing the return of populations. Speaking of a recent era of "urban renaissance" is not enough. The details are wildly different, from city to city. Consistent growth, consistent decline, or a manifest turnaround are three storylines. But the five cities examined in Figure 5.5 present a fourth, more complicated, and perhaps more puzzling, narrative.

Atlanta is so often held up as an avatar of growth in the New South that finding it lost population in the 1970s and again in the 1980s seems counterintuitive. Then, realizing that the net change in the city's population since 1960 has been *negative* 67,452 just intensifies the shock. All of Atlanta's growth over the past half-century has been suburban, and, unlike (say) Boston, it has not been able string together decades of central city growth to reverse the hollowing-out trend. So, even though there's a facile assumption that Atlanta is a good example of cities "that are growing in the manner of Los Angeles," that assumption is contradicted by the census numbers. LA has been growing its central city as well as its periphery. Atlanta has not yet shown it can accomplish this.

Such an observation is not new. The Urban Land Institute noted, in the mid-1980s, that, although the metro area population growth had proceeded at a

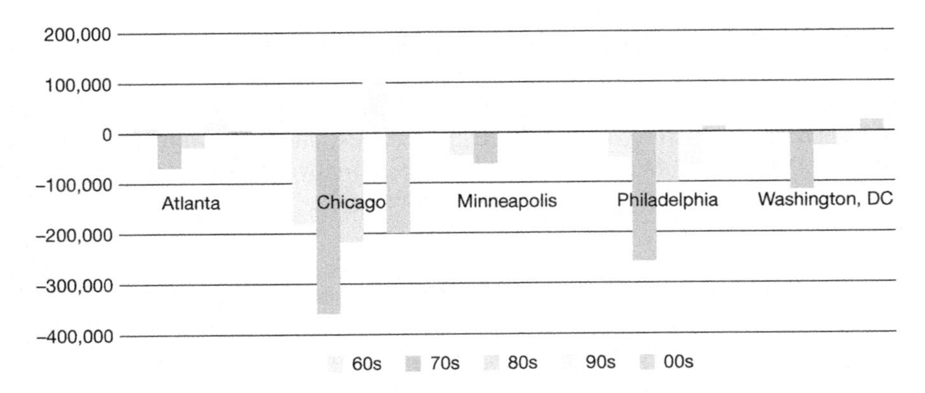

Figure 5.5 Population change by decade: Five cities with stop–start growth

Source: US Bureau of the Census decennial counts

2.3 percent per year rate, "population within the city limits of Atlanta has continued to decline."[30] Nevertheless, the city of Atlanta has been able to sustain its cadre of young workers at above-national-average levels throughout the 1960–2010 period, a testimony to its continued attraction to those beginning their careers. What, then, has kept the city itself from overall population growth and, especially, from the dynamic interactions that spark downtown vibrancy?

The *New Georgia Encyclopedia* identifies "three dominant forces" affecting Atlanta's historical development: transportation, race relations, and an intangible element it calls "the Atlanta spirit." [31]

Transportation will be considered among the characteristics of 24-hour cities in Chapter 7 of this book. But three key features of the Atlanta transportation picture should be noted here. First is the tremendous confluence of the interstate highway system, with I-75 and I-85 converging to form a single north–south superhighway through the city and that superhighway intersecting with the east–west I-20 at a point proximate to Atlanta's downtown. Add to these the circumferential I-285 and the state route Georgia 400 to the northern suburbs, and we can see the commitment Atlanta made to the automobile as the primary transport means for its citizens.

The highway system not only serves to move people but is an important physical delineator of neighborhoods, an effective marker of boundaries.[32] Atlanta, south of I-20 and west of I-75/85, has become predominantly black, as the city's population moved from 38.3 percent African–American in 1960 to 67.1 percent by 1990. The north and east areas in the city, and into the adjacent suburbs, remained (and remain) predominantly white.[33]

Atlanta's MARTA system (Metropolitan Atlanta Rapid Transit Authority) began to put its rail system in place in the 1970s, using about $800 million of federal funds. During the 1980s and afterward, the system was extended to Hartsfield International Airport and on to two spurs in the north central area (to North Springs, just beyond I-285) and in the northeast (to Doraville, just inside I-285). The influence of racial factors on the extent and location of MARTA have long been the subject of public discussion. The Brookings Institute noted as much in a 2000 study,[34] acknowledging that MARTA rail has been limited to Fulton and DeKalb Counties, and expansion into other metro counties has been resisted, sometimes on overtly racial grounds. A researcher from the Virginia Policy Review writes:

> Having worked for two summers in Atlanta, my most prominent memory was the long commute between the white suburbs north of Atlanta, where jobs are plentiful, to its impoverished black neighborhoods downtown. I would have used public transportation, however the MARTA suspiciously ends right before white suburbia begins.
>
> . . . The lack of transportation infrastructure in these counties has prevented minorities in the city from accessing areas with high job growth including Gwinnett, Cobb, and Clayton County. MARTA has not had any transit route development since 2000.[35]

The keepers of the Atlanta spirit have doggedly sought to present the city as a progressive and inclusive place. There is evidence that this effort has borne fruit. For instance, Atlanta is now known as a city that is quite hospitable to the lesbian, gay, bisexual, and transgender (LGBT) community.[36] The city successfully promoted its progressiveness in its successful bid for the 1996 Summer Olympic Games, and its leading corporation, Coca-Cola, has long promoted racial, ethnic, and gender tolerance as a feature of its advertising. Atlanta has a major cultural complex at the Woodruff Arts Center, in its Midtown neighborhood, also an area of high-rise offices, hotels, and residences. So the city is far from standing pat but, as we will see, it still faces significant impediments to achieving 24-hour-city status.

By contrast, Chicago is generally accorded recognition in the "24-hour city club."[37] Unlike Atlanta, Chicago has a large, vital, attractive downtown that functions as the core of its city and region. In the Loop and across the Chicago River along North Michigan Avenue, you can find everything that defines a world-class urban center: corporate office buildings, high-end residential towers, an expansive park that is well used by the citizenry, great shopping, wonderful museums, a skyline of architectural diversity, and a lakefront that is both beautiful and accessible.

But, despite all this, Chicago saw its citywide population drop by 854,886 between 1960 and 2010. Worse, for most of that time, Chicago was barely above the national average for the percentage of the young-worker cohort in its population – even as virtually all other large cities remained attractive to those looking for great starts for their careers. That did change after 1990, and, by 2010, Chicago found 19.1 percent of its resident population in the 25–34 years age group, significantly above the US average of 13.3 percent. Arguably, it is the Millennial generation and its preference for the urban lifestyle that gild

Chicago's 24-hour-city reputation, as much as its undeniable physical assets. But it is likely truer to note that it is the combination of physical and human capital assets – the relationship between them – that accounts for both the vibrancy of Chicago and for its determined, sustained attention to enhance its downtown core.

I first visited Chicago in 1969 and have spent a good deal of working time there over the past 35 years. One thing is for sure: the City of Chicago is a lot more than its downtown. A second thing is also sure, and surely evident to anyone venturing into the city's neighborhoods: this city is still deeply scarred by the economic and social turmoil of 50 years ago. Worse: the politics that marked that era a half-century ago remain a structural feature of Chicago governance. The "Chicago machine" still has the power to get things done – and to prevent things from getting done. And, on a net basis, the evidence is that many people are detached from the spectacular downtown live–work–play environment – and have been voting with their feet to locate somewhere else. The result has been that Chicago has been decisively supplanted by Los Angeles as America's "second city."

Why should this be so?

For one thing, the "image" of Chicago was always more differentiated than that of the manufacturing cities dominated by a single industry – such as Detroit or Pittsburgh or Akron, or even Houston and Las Vegas. But the reality was that Chicago was an incredibly powerful center of the manufacturing economy and the dominant city in the section of the country that Joel Garreau termed "The Foundry" in *The Nine Nations of North America*. The steel industry, the stockyards, shipbuilding, and automobile production helped define Chicago's South Side for decades.[38] Zenith, Motorola, and Western Electric manufactured electronic and communications equipment for consumers and businesses. In the pre-video game and Internet era, Bally pinball machines and Wurlitzer jukeboxes provided entertainment to generations.[39] Chicago, like so many other cities, was unprepared for the destructive impacts of America's rapid deindustrialization in the second half of the 20th century.

Mayor Richard J. Daley (in office from 1955 to 1976) grew up in one of Chicago's working-class neighborhoods and, while a night-school law student at DePaul University, held a job in the stockyards. Working his way up through Chicago's Democratic Party organization, upon his election, the "first Mayor Daley" (his son Richard M. Daley held this same office from 1989 to 2011) focused on the revitalization of the historic downtown area, "the Loop."[40] The Loop had seen no new construction since the 1920s, until Daley established the city's first Planning Department to reorient the city toward future growth. By 1964, 20 million square feet of new office space had been constructed downtown, and, by the end of Daley's incumbency, the size of the downtown office district had doubled. Daley is credited with creating an environment of liberal tax benefits, flexible zoning, targeted city services, and (importantly) the cooperation of the municipal bureaucracy on big real estate projects that transformed the city center.[41]

Those same resources, sadly, were not made available throughout the city, especially in its poorer, working-class residential neighborhoods, in a city that had been a major beneficiary of the Great Migration from the American South. Michael Gecan of the Industrial Areas Foundation tells the story this way, contrasting Cook County (home of Chicago) with the immediately adjacent suburban DuPage County:

> In the 1950s, there was no way of knowing that we were living at the city's high point. The massive political, economic, civic, and religious institutions had seemed as solid and stable as glaciers to those living with them or in their shadows. From the second story of our double-brick corner house we could see the tavern that we once owned, the then-modern building that housed Newark Electronics, where my father and I would someday work, and the row of houses obscuring the view of the Tootsietoy Company where my mother would be employed . . . By the mid-1980s it was all declining . . .
>
> Major corporations had fled Chicago and moved to DuPage's open spaces and tax-friendly towns. Working-class homeowners on the West and Southwest Sides of the city sold their bungalows and bought ranches, Cape Cods, and new townhomes in Wheaton, Naperville, and Downers Grove. Families troubled by the city's public schools happily sent their children to shining new facilities and well-equipped classrooms . . .
>
> The leaders of older cities like Chicago and counties like Cook have shown how to focus on the few . . . Many of the paths of laudable pursuits have been closed or semi-privatized – walled, gated, guarded – for some time . . . The old bungalow bedrock of the city – blue collar and tax-paying – has disappeared.[42]

The disproportionate focus of Chicago, especially a succession of political leaders including both Mayors Daley, on its world-visible downtown at the expense of its ethnic and minority neighborhoods is a major reason for the disconnect between the Windy City's reputation as a 24-hour city for commercial real estate development and investment, and its troubling statistics on a wider range of sociological metrics, beginning with continued population shrinkage. The need for Chicago to improve public–private cooperation in addressing its core problems is recognized in the 2015 edition of *Emerging Trends in Real Estate*, which ranks the Chicago areas (city and suburbs) no better than 18th in its 2015 rankings.[43]

It is just 400 miles from Chicago to Minneapolis – about a 6-hour drive and less than an hour's flight time. They are two very different cities, located in two separate "nations," as described by Garreau: The Foundry for Chicago, and The Breadbasket for Minneapolis.[44] They are certainly distinct in size, in population density, in age distribution, and in physical/topographical characteristics. But they are the same in this respect: as cities, they have seen population ebb and flow over the past half-century, with the result of net population loss since 1960 (20.8 percent in the case of Minneapolis; 24.1 percent in the case of Chicago).

The City of Minneapolis had a population of 382,007 registered in the 2010 US Census, or about 13.4 percent of the metro area population of 2.9 million.[45] Despite its northern location, Minneapolis sprawls much like many Sunbelt metros, and its core is a weaker center of gravity than those of 24-hour cities such as New York, Boston, San Francisco, or even Chicago. Minneapolis proper is a rather compact city, measuring only 54 square miles in land area, giving it a population density of 7,134 per square mile. Still, this is fairly low density by urban standards. Mass transit thrives in direct relation to density of population, and so it is not surprising that it is only recently that Minneapolis has been investing significantly in alternatives to the automobile. Daily ridership on Metro Transit – which serves the entire Minneapolis–Saint Paul region – is under 300,000 per day, and 86 percent of this is via bus.[46]

Harsh Minnesota winters have prompted the development of a skywalk system connecting the office buildings of the city's CBD. This system was inaugurated in 1962, as a means of "revitalizing downtown" after General Mills moved to the suburbs (1955) and the area's first suburban shopping mall opened in Edina (1956). The "Skyway System" of enclosed bridges now connects 80 blocks of downtown Minneapolis and is the world's largest such network. Like Houston's CBD tunnel system, the Skyway is a response to uncomfortable weather conditions and has become a kind of "alternative streetscape." It is an effective alternative, and that is somewhat problematic in siphoning foot traffic away from downtown retailers in both cities.[47]

Notwithstanding the pattern of overall population change, Minneapolis has been able to outstrip the nation in the concentration of the young-worker cohort, which has been above 20 percent of total city population since 1980. The occupations that draw young people here are diverse, but it is telling to examine the occupations that have the highest location quotients: arts, media, and design (LQ – 2.47); legal (LQ – 2.02); engineering, computers, science (LQ – 1.48); and food preparation and service (LQ – 1.36). The profile of what Richard Florida has termed "the creative class" is clearly well represented here, with spin-off employment generation into the "services class" economy that complements the lifestyles of the "creatives."[48]

Its attraction to young workers, however, has not positioned Minneapolis on the list of 24-hour cities. Over the years, *Emerging Trends in Real Estate* has hedged its positive comments because of this: "Low unemployment and limited in-migration threaten longer-term growth prospects of the Twin Cities and other Midwestern markets that aren't primary immigration magnets or 24-hour powers,"[49] and this: "Neither a 24-hour power nor a Sunbelt boom-town, Minneapolis must be careful not to follow St. Louis and Detroit off investors' radar screens."[50]

That cautious outlook has persisted, and the 2015 edition of *Emerging Trends* ranked Minneapolis and its region 30th out of 75 US markets in its investment and development prospects.[51] The investors' opinions and the net population growth data seem to be in accord for this city.

Many Philadelphians chafe at the continued designation of their city as a 9-to-5 center, asking "what more do we need to have?" to don the 24-hour-city mantle. They point to the modern skyline of the city, finally rising "above Billy Penn's hat," after a long-standing "gentlemen's agreement" to build no higher than the top of the statue of William Penn on City Hall was abrogated by the construction of One Liberty Place in 1986. Since then, a half-dozen skyscrapers have risen above that once sacrosanct limit.

Jane Jacobs had effusively praised Philadelphia's Rittenhouse Square neighborhood as far back as 1961, though she noted that it was the only one of the city's squares that was successfully revitalizing its surroundings.[52] Urbanist Eugenie Birch of the University of Pennsylvania studied downtown residential neighborhoods across the US and pronounced Philadelphia's center "fully developed" as a downtown housing market.[53]

There is a magnificent array of museums on Ben Franklin Parkway, world-class music on Broad Street with the Philadelphia Orchestra (just down the block from the city's hockey arena), good shopping along Market Street, anchored by Macy's (in the original Wanamaker's department store building), and strong tourist traffic drawn by Independence Hall and the Liberty Bell. The University of Pennsylvania is a short ride from downtown, as is Temple University and Drexel University. Mass transit – both within the city and by commuter rail from the suburbs – distinguishes Philadelphia from auto-dependent Atlanta, Dallas, and Phoenix. How could such a city not be recognized as the equal of a Boston or a San Francisco, if not New York?

But, as in Phoenix, the ingredients have not yet combined in a successful recipe. With its stops and starts, Philadelphia finds itself with 476,306 fewer residents than it had in 1960. And between 1960 and 2000, its cadre of young workers barely kept up with the national average as a percentage of total population — which means it has actually been shrinking its young-worker labor force: the age 25–34 cohort numbered 253,894 in 1960 and fell to 224,864 in 2000. Since then, there has been a rebound in this age group, and it is now marginally above the US norm. But even so, at 246,062, there were fewer young workers in Philadelphia in 2010 than in 1960.

Partly, it appears a question of critical mass: has Philadelphia assembled enough resources in its downtown to generate 24-hour energy? Partly, it is a question of agglomeration: do the industries and occupations that comprise this urban economy generate enough synergy to make the whole greater than its parts? LQs are less than 1.00 for many key occupations here: management, business, and financial (LQ = 0.80); engineering, computer, and science (LQ = 0.83); even in education, the LQ is just 1.00. The LQ for back-office jobs (sales, office, administrative support) is also about 1.00. Community and social services, however, has a robust 1.63 LQ; healthcare support jobs are about the same (LQ = 1.62), and firefighters and police are even higher at an LQ of 1.83. This all leads to a local median household income that is 30.2 percent below the national median. It is any wonder that young workers are not rushing into the City of Brotherly Love?

The last of the "start and stop" cities is, perhaps, another surprise – the nation's capital city, Washington, DC. Government centers are frequently touted as "recession-resistant," less prone to ups and downs compared with places more directly exposed to the business cycle. The impression has become widespread that, as the size of government has grown, and the federal government above all has become an ever-more-larded bureaucracy, Washington as the seat of government must also be attracting more and more people as a consequence.

The data belie this, of course. Figure 5.5 describes this, showing that Washington as a city has lost population in four of the last five decennial censuses. Washington has dipped in population by 161,233 residents since 1960, a 21.1 percent decline that is approximately the same as the shrinkage in Minneapolis and Philadelphia (in percentage terms). As Chapter 4 discussed, Washington was traumatized by the racial unrest of the 1960s (and was the scene of several massive anti-Vietnam War protests as well). But there was more than that afflicting the city.

There is a structural issue in government that rankles many citizens of the District of Columbia: although Washington elects a representative to Congress, that representative has no vote. To express the local resentment about this anomaly, license plates issued by the District's Department of Motor Vehicles bear the motto "Taxation Without Representation." The issue of self-governance ties in with the checkered history of home rule for Washington. Until 1973, the city did not have its own elected mayor or city council, or even school boards and health departments, but was administered by federally appointed commissioners.

Sadly, the first generation of local officials was, to understate the case, less than exemplary. Marion Barry was elected the second Mayor of Washington in 1978 and served three terms until 1990. By 1989, *The Washington Monthly* would run a story headlined, "The Worst City Government in America."[54] While in office, Barry was imprisoned for 6 months on drug charges and did not run for reelection. But he did run again in 1994 and was returned to office.

The city effectively lost home rule in 1995, as it moved to the edge of bankruptcy. Congress established, and President Bill Clinton quickly appointed, a Financial Control Board to oversee city budgets and operations for an 8-year period.[55] Against such a local background, the extended population decline in Washington becomes far more understandable.

While the city population was declining, the metro area was growing nicely. Metro District of Columbia population jumped from 1.9 million in 1960 to 4.7 million in 2010, as suburbs in nearby Maryland and northern Virginia flourished. The first decade of this century, however, saw a glimmer of light shining on central Washington's fortunes. The city's population rose by 22,664 residents, and its attraction for the young-worker cohort intensified, with the

25–34 cohort actually outstripping the total population gain. In fact, by 2010, there were more Washingtonians (124,745) in this age range than there were in 1960 (109,451), despite the overall shrinkage of 118,177 over the half-century.

With more youth have come greater vitality and the redevelopment of neighborhoods that had been virtually abandoned – including some within a short walk of the US Capitol. So, like Chicago, Washington has been able to fashion a 24-hour-city reputation on its young singles, on a strong residential market, and nightlife in neighborhoods such as Georgetown, Dupont Circle, Mount Pleasant, and Capitol Hill. One Census tally is not enough to include DC among the "turnaround" cities – but it is certainly a change that is a positive sign for this very visible urban center, where every change is closely observed by the elected officials of national government.

Forms of change: Cycles

In my experience, the most frequent topic real estate economists are asked to address, from the podium, in occasional articles, or in informal conversation, is this: Where are we in the cycle? Early in their careers, real estate practitioners (like most business people) become aware that events are often shaped by cyclical behaviors. Veterans tell novices, "This is a cyclical business." Recessions and recoveries are couched in terms of being phases of the business cycle, and real estate booms and busts are seen as extreme cases of the cycle.[56] The link between the real estate cycle and the larger business cycle is often the volume and price of credit. Economists are prone to return to the "financial instability hypothesis" of Hyman Minsky that posited five successive moments in the credit markets: displacement, boom, euphoria, profit taking, and panic.[57]

Displacement happens when some exciting new product comes into play. Boom follows when investors, often fueled by low-cost debt, bid up the price of that product – often causing market participants to blur any distinction between price and value. A rush to "get in while the getting is good" occurs, as euphoria (also known as speculative fever) pervades the marketplace, with the business press and general media fanning the flames with breathless stories of escalating wealth. As retail investors flood into the market, some astute individuals sense the presence of "dumb money" and begin to be sellers rather than buyers, taking profits while they can. Then, prices flatten and start to decline, if speculation has created a spread between current price and sustainable future economic performance. As more and more participants realize this, the balance shifts from a surfeit of buyers to a surfeit of sellers. Panic ensues, and prices plummet.[58]

Although awareness of cycles is clearly embedded in the popular, professional, and academic mind, and the term trips easily from the lips, clear thinking about cycles requires quite a bit of effort. Fortunately, there is a very rich literature on the subject, which need not be reviewed here.[59] Let me instead focus on a few fundamental concepts.

Cycles can be defined as periodic fluctuations around equilibrium. Periodic, in this sense, implies a certain recurrence of patterns, and this is considered a hallmark of cycles. Fluctuation underscores the idea that cycles are a species of change, harkening back to the pre-Socratic philosopher of flux, Heraclitus, and his famous dictum that, "you cannot step into the same river twice."[60] And, finally, this brief definition notes that equilibrium is the central tendency around which cycles oscillate.

There is something comforting about the notion of cycles. For one thing, they suggest a degree of predictability. The perceived existence of a repeating pattern gives rise to the expectation that our historical experience gives us a reasonable capacity to anticipate the unfolding of the future. Then, too, we can expect – especially in difficult periods – that there will be a fundamental tendency, mathematically assured, that a return to balance or "reversion to the mean" lies ahead, and the cycle simply needs to be waited out.[61] Even more soothing is the very idea of equilibrium, which implies stability and order at the center of the welter of events.

Against this, we have a classic warning from Gottfried von Leibnitz, who, with Newton, is one of the developers of calculus: "Nature has established patterns originating in the return of events, but only for the most part." [62] Even in a cyclical system, there are surprises to be found because of the unceasing flow of the river of events. We never step into the same river twice. Cyclical oscillation in markets is not deterministic in the pure sense suggested by the sine curve shown in Chapter 1. Other forms of change are at work simultaneously with the procession of cyclical phases, so where we come out of a cycle is a different point from where we entered it. And, as it turns out, equilibrium is not actually a point at which we find rest, but just an idealized construct of a point that events blow through as they pass from one unstable excess to another.

But though pure predictability is not possible in complex systems such as the economy, or even in the narrower realm of real estate markets, the study of cycles can and does give us important information. Let us consider, for instance, the key parameters of amplitude and frequency of cycles.[63]

So, in Figure 5.6, we see the cyclical pattern of Manhattan office rents, construction volumes, and vacancy rates over a period of time beginning in 1960. The rents are expressed in constant dollars, to filter out the effects of monetary inflation over time. Although rents are clearly responsive to the supply/demand variables of new development and overall occupancy, the question still remains as to why the amplitude of rents – peaks and troughs – remains so consistent over time. I often ask students if they can explain that regularity, and if they think that knowing "how high is up and how low is down" is useful information to have as real estate market participants. Although they agree immediately that it is useful to know how much rent can potentially rise or fall as cycles change, it takes some effort to recognize why this is the case. And, without knowing the "why," it is impossible to explain the "what" – that is, the level that peaks and troughs will likely hit the next time around.

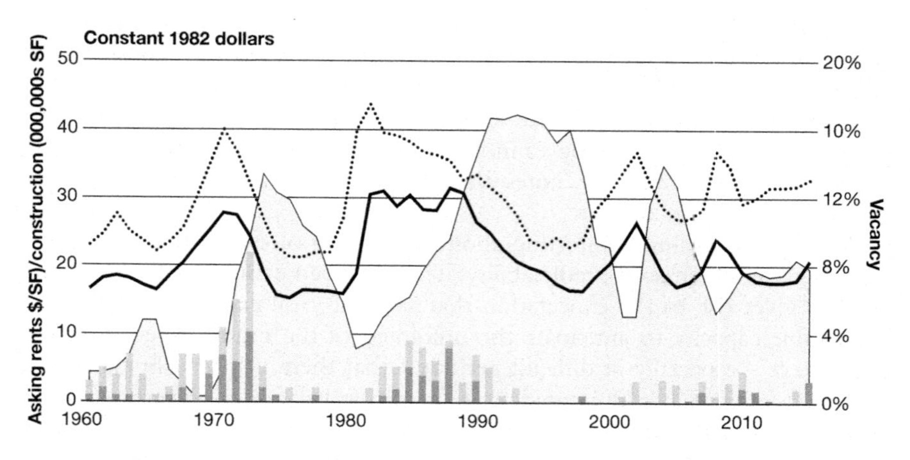

Figure 5.6 Manhattan primary office rental trends

The ebbs and flows, peaks and valleys in the Manhattan office market are vividly displayed in this admittedly too-complicated graphic. Let's focus only on the rent lines. A close correspondence will be noted in the "real" (that is, inflation-adjusted) rents over the long period from 1960 to 2014. That correspondence is closer during the time when rent is low, but still evident at the peak of rents, although less exact. That is because building owners must cover their operating expenses by the rents they charge, and those expenses are well proxied by the CPI, used to deflate rents over time. At peak, however, rents are a function of the cost of construction: new development cannot be financed until rents have risen to a level sufficient to satisfy lenders that the value of the completed project will exceed the amount lent to build it, with profit to the builder. That rent is also influenced by inflation, but by a measure of inflation called the building cost index (BCI), which differs from consumer prices in that the BCI specifically tracks labor and materials costs in the construction sector.

Despite the inexactness of the correspondence at peak and trough, it can be seen that rents do fluctuate within a discrete range. That range helps the careful analyst consider the question of amplitude, "How high is up, and how low is down?"

The graph also gives us some insight about frequency, but not in the way many would like to consider the answer. The periods from peak to peak and trough to trough are irregular in duration. Some cycles are long, at least as far as the Manhattan office market is concerned, and some are much shorter. And they are asymmetric: the adjustment from trough to peak seems to be more sudden than the decline from peak to trough. That is owing to the interactions between the inflationary factors affecting rent and the other factors, such as the volume of new construction entering the market and the level of vacancy rate experienced over time. It is useful to know that rent cycles do not come with reliable regularity, even if the basic pattern of cycles can be anticipated.

We do not have to discover a lockstep determinism in cycles for an understanding of this form of change to be helpful.

Similar lessons can be drawn from Figure 5.7, which displays the percentage change in US employment from January 1960 to January 2015. For roughly 25 years (1960–1985), employment gains peaked between 5 and 6 percent, but, in the 30 years since, peaks have been more in the 2–3 percent range. We might suspect that age demographics are at work in this downshifting of peaks, but we can also detect the influence of globalization. It is useful to remember, too, that the denominator of the fraction – the absolute size of the labor market – was increasing over time, and this makes it harder to continue relative growth. More jobs, in an absolute sense, have to be added to achieve the same level of percentage change as the total number of workers rises.

At the bottom of the cycle, economic recessions typically cost the US economy 1–2.5 percent employment loss, until the catastrophe of the 2007–2009 financial crisis hit. With this important exception, the amplitude of cycles was reasonably regular. And, as in the case of the Manhattan office market, the frequency was decidedly irregular, and asymmetric on the side of growth.

Again, not lockstep determinism, but the study of cycles in employment does yield much useful information.

We can also examine cycles at the level of local employment. Figure 5.8 drops down from the national level seen in Figure 5.7 to view job change in three metro areas: New York, Houston, and San Jose. The graphic shows how the amplitude and frequency of employment change differ from place to place. Each of these metro economies, of course, differs in the composition of its job base. New York is now largely a finance and advanced business services metro. Houston, although diversifying, still relies upon the energy industry. And San Jose relies upon its technology firms as its economic basis.

There is not an especially high correlation between the employment cycles of the three metros. Houston's correlation with New York is just 0.217, and the San Jose–Houston correlation is 0.412. New York and San Jose are

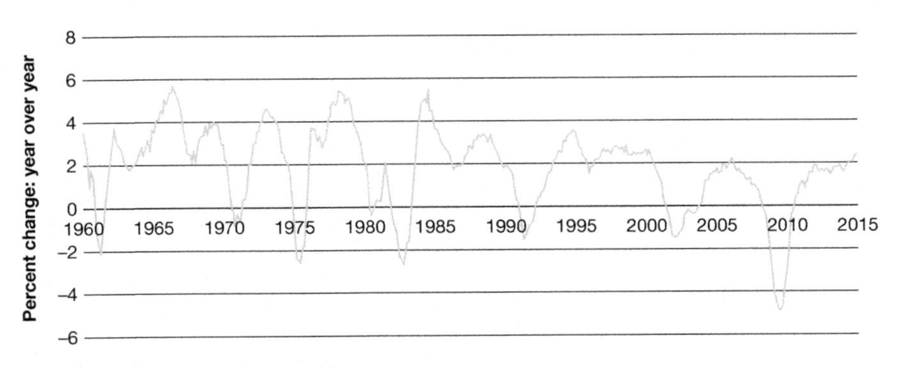

Figure 5.7 Cyclical change in US employment: 1960–2015

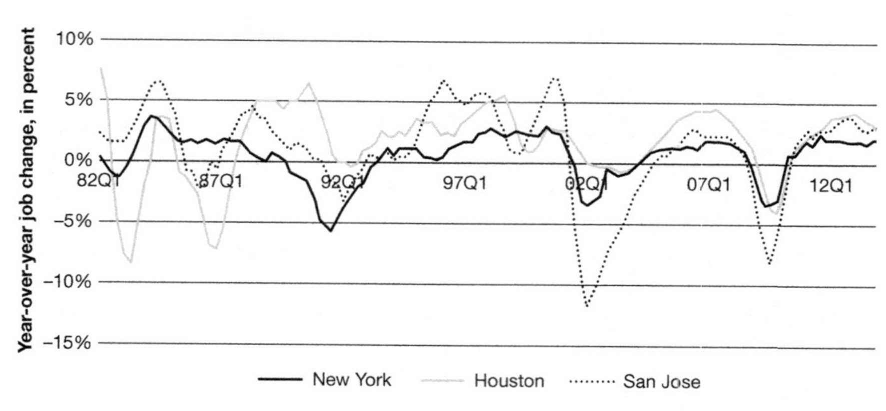

Figure 5.8 Volatility of cycles since 1982 for New York, Houston, and San Jose

somewhat more correlated at 0.685. All the correlations are positive, though, indicating that they are generally responding together to a broader influence, namely the US business cycle of expansion/recession. Still, the imperfect correlations attract investors who seek to gain some diversification benefit from placing money in local economies, so that each hedges the others' risk.

Houston and San Jose each have significantly higher mean growth rates over time, with 1.6 percent on average for Houston, and 1.1 percent for San Jose, versus just 0.5 percent for the larger and more mature New York economy. But the amplitude of the cycles is manifestly much wider for Houston and San Jose, as seen in the peaks and troughs in Figure 5.8. Statistically, the standard deviation of change for New York is just 1.9 percent, whereas it is 3.0 percent for Houston and a highly volatile 3.7 percent for San Jose. So, on average, a growth-oriented person might prefer Houston or San Jose, but a risk-averse individual might like the relatively more stable performance of New York over the long haul.

Cycles, then, present us with a fourth basic form of change, to go along with the trends, maturation, and change of state previously considered. It is becoming clearer how complex our understanding of urban changes must be, if we are to grasp our subject as a whole.

Notes

1 www.merriam-webster.com/dictionary/yuppie
2 www.oxforddictionaries.com/definition/english/yuppie
3 It should be decisively noted that blue-collar jobs are at various skill levels and represent a broad range of income levels. High-rise construction workers, elevator repair workers, electricians, plumbers, subway operators, surveyors, boilermakers, and aircraft maintenance workers are among the blue-collar trades making above-average incomes, according to Bureau of Labor Statistics surveys.
4 The Corporate Headquarters Complex in New York City. Eli Ginzberg of Columbia was the director of the project, with Matthew Drennan of NYU the principal

investigator, assisted by research associate Robert Cohen at Columbia, Thomas Stanback at NYU, and Charles Brecher of the New School for Social Research.

5 Ibid., p. xvii.

6 Ibid. Table 5.1 appears on page 38 of the Conservation of Human Resources report.

7 Ibid., p. 42.

8 Ibid., pp. 43–44.

9 All six of the cities shown in Figure 5.2 are included among the 15 most historically important American urban centers in the 1820–1980 period, as part of the Project for the Study of American Mayors. See Melvin G. Holli, *The American Mayor: The Best and the Worst of Big-City Leaders*, Pennsylvania State University Press (University Park, PA, 1999).

10 Ibid.

11 See Edward Glaser, *The Triumph of the City: How Our Greatest Invention Makes Us Richer, Smarter, Greener, Healthier, and Happier*, Penguin (New York, 2011).

12 These two cities are grouped together in Joel Garreau's marvelously readable book *The Nine Nations of North America*. Published in 1982 by Avon Books, the regional divisions described in the text have stayed remarkably up to date. San Francisco and Seattle are both in a narrow band of geography hugging the Pacific Coast that Garreau labels "Ecotopia."

13 Op. Cit.

14 A Yiddish term well known to New Yorkers and thought by some to be almost a definition of the New York character. It means "shameless audacity."

15 Garreau, *Frontier*, p. xxx, and online at www.garreau.com/main.cfm?action=chapters &id=2

16 See, for instance, Albert Bisin and Thierry Verdier, "Beyond the Melting Pot: Cultural Transmission, Marriage, and the Evolution of Ethnic and Religious Traits," *Quarterly Journal of Economics* (August 2000), and Brent Nelson, "The Melting Pot – Then and Now: Tracking the Idea through the 20th Century," *The Social Contract* (Spring 1996). The pivot point in the reconsideration of the melting-pot theory is arguably the publication of Nathan Glazer and Daniel Patrick Moynihan's *Beyond the Melting Pot: The Negroes, Puerto Ricans, Jews, Italians and Irish of New York City*, MIT Press (Cambridge, MA, 1970).

17 Interestingly, though, the Appendix of *Frontier* that lists edge cities shows that cities with strong downtowns support larger numbers of edge cities in their respective metropolitan areas. This suggests positive ramifying effects for strong urban cores and a possible dilutive effect for metro areas that fail to concentrate their economic energy. Such topics will be discussed at greater length in Chapter 7 of this book. Automobile commutation patterns are an obviously related variable that will also require later discussion, as will population density across cities, as well as within each city.

18 See the discuss online at www.newgeography.com/content/002372-the-evolving-urban-form-los-angeles

19 Dubuque, Iowa, is a small city along the Mississippi River.

20 Predictably, Joel Kotkin is negative about this trend, as seen in his column "Why the Rush to Manhattanize L.A.," *Los Angeles Times*, August 12, 2007. Others see such a shift as the next phase of Los Angeles' growth, as indicated in this blogpost from a local radio station, online at http://blogs.kcrw.com/whichwayla/2013/05/la-grows-up-the-manhattanization-of-los-angeles

21 The City of Dallas's downtown plan, in fact, has explicitly redirected efforts toward developing a city with 24-hour urban characteristics. See online at www.downtown dallas360.com/Content/10000/ThePlan.html

22 See Peter Calthorpe and William Fulton, *The Regional City*, Island Press (Washington, DC, 2001).

23 The most widely followed benchmark for housing price performance in the US. Accessible through Standard & Poors, online at http://us.spindices.com/indices/real-estate/sp-case-shiller-us-national-home-price-index

24 Las Vegas is coy about its naughty reputation. Although the opprobrium surrounding gambling and alcohol has faded considerably from America's moral code in recent decades, Las Vegas's reputation as a hard-drinking casino town gave rise to the "Sin City" soubriquet. Prostitution is legal in the state of Nevada, but not in Las Vegas itself – though the practice is clearly widespread there. Since 2003, the city has promoted itself with an ad campaign with the tag line, "What happens here, stays here." This followed a previous, less-successful campaign that promoted the city as a family-friendly venue, with theme-park-like rides and attractions.

25 Mary Manning, "Atomic Testing Burned Its Mark," *Las Vegas Sun Times*, May 15, 2008.

26 The growth in the entire Las Vegas urbanized area has been equally rapid, from about 89,000 in 1960 to nearly 1.9 million in 2010; see online at www.demographia.com/db-uza2000.htm

27 Location quotients accessed online at www.brookings.edu/research/interactives/metromonitor#/M29820

28 This fact is recognized in the long-used nickname for the city, "Lost Wages."

29 Jan Nijman, *Miami: Mistress of the Americas*, University of Pennsylvania Press (Philadelphia, PA, 2011). On the "shadow economy," Nijman notes that, "willing participants included old-time mobsters, ex-CIA agents, some *marielitos* [Cuban boatlift refugees], bent bankers, opportunitistic lawyers, corrupt police officers, small airplane owners, petty criminals, and an ambitious new crowd of South American gangsters" (p. 63). The television program *Miami Vice*, which came on the air in 1984, captured much of this *noir* ambience, but somehow enhanced Miami's appeal (Nijman, pp. 66–67).

30 Urban Land Institute, *Development Review and Outlook 1984–1985*, "Atlanta Metropolitan Area Market" [prepared by Hammer, Siler, George Associates], ULI Publications (Washington, DC, 1984), pp. 145–146.

31 Andy Ambrose, "Atlanta," *New Georgia Encyclopedia* (June 5, 2014), online at www.georgiaencyclopedia.org/articles/counties-cities-neighborhoods/atlanta

32 Kevin Lynch speaks about the importance of "edges" in understanding urban location, in his *The Image of the City*, MIT Press (Cambridge, MA, 1960), pp. 62–66.

33 *Georgia Encyclopedia*, art. cit. The most recent census data do indicate some change underway in the 21st century's early years, with an increasing white population in the city and greater numbers of African–Americans in suburban Atlanta. This incipient trend, however, has not made a major impact on the racial majority composition of either the city or its suburbs.

34 Margaret Pugh (principal author), "Moving Beyond Sprawl: The Challenge for Metropolitan Atlanta," Brookings Institute Center on Urban and Metropolitan Policy (Washington, DC, 2000).

35 Sarah Colliers, online at http://virginiapolicyreview.com/2014/02/12/marta-offers-equality-a-seat-on-the-bus/

36 The Advocate.com survey places Atlanta fifth in the nation in its 2014 ranking of LGBT-friendly cities. See *Atlanta Journal–Constitution* article in its Business Section, January 9, 2015. During my summer 2014 visit to Atlanta, a local cab driver pointed out a condo building where, he said, Elton John lived, and the driver spontaneously mentioned this LGBT openness as an Atlanta hallmark.

37 See, for instance, *Emerging Trends in Real Estate 1995*, pp. 39–40, *Emerging Trends in Real Estate 1996*, p. 4, and *Emerging Trends in Real Estate 1997*, pp. 5, 24–29.

38 Rod Sellers, *Chicago's Southeast Side Industrial History*, Southeast Historical Society (Chicago, IL, 2006).

39 Mike Matejka, "The Historical Development of Industry and Manufacturing in Illinois," online at www.lib.niu.edu/1999/iht639941.html
40 The Loop specifically refers to the downtown area encircled by the lines of the Chicago Transit Authority's elevated train lines.
41 Holli, op. cit., pp. 109–111.
42 Michael Gecan, *After America's Midlife Crisis*, *The Boston Review*/MIT Press (Cambridge, MA, 2009), pp. 28, 30, 36–37. In the passage immediately following the one quoted, Gecan compares Chicago with New York City, which much more successfully fought what he terms "a virulent and advanced form of civic cancer" (p. 39).
Disclosure: I worked with Mike Gecan and others during the 1980s putting together the Nehemiah Housing Program in East Brooklyn, an impoverished (indeed, desolated) neighborhood, with the East Brooklyn Congregations organization. Some 3,500 Nehemiah houses – single-family brick row houses – have been built in the community as a result.
43 Urban Land Institute and PricewaterhouseCoopers, op. cit., p. 52.
44 Garreau's "Nine Nations," with their archtypal cities, are as follows: New England (Boston); The Foundry (Detroit); Dixie (Atlanta); The [Caribbean] Islands (Miami); The Breadbasket (Kansas City); The Empty Quarter (Denver); Mexamerica (Los Angeles); Ecotopia (San Francisco); and Quebec (Quebec City). Note: Garreau considers New York City one of a kind, and not emblematic of any of the regions.
45 St. Paul, Minneapolis' "twin city," has about 100,000 fewer residents. St. Paul is, however, the seat of Minnesota's state government.
46 Data are from 2014, accessed on Metro Transit's website, online at www.metrotransit.org/
47 Here's what Houston's official website says about the subject: "In a city of more than two million people, you would think our downtown streets would be filled with pedestrians during the lunch hour. Well, that's true, but most of them are walking underground, in the downtown Tunnel System." See online at www.houstontx.gov/abouthouston/Downtown-Tunnel-System
48 Florida rates Minneapolis a "best buy" city for young singles/college graduates in *Who's Your City?*, Basic Books (New York, 2008), pp. 237–239.
49 *Emerging Trends 2001*, p. 40.
50 *Emerging Trends 2002*, p. 42.
51 *Emerging Trends 2015*, p. 32.
52 Jacobs, *The Death and Life of Great American Cities*, p. 92ff. and 104ff.
53 Eugenie L. Birch, "Who Lives Downtown?" Living Cities Census Series, Metropolitan Policy Program: Brookings Institution (Washington, DC, 2005).
54 Jason DeParle, January 1, 1989.
55 Michael Janofsky, "Congress Creates Board to Oversee Washington, DC," *New York Times*, April 8, 1995.
56 "Historically, real estate has exhibited the most severe cycles of any asset class . . . Economists of every stripe recognize that real estate plays an important role in recurring business cycles." From Susan M. Wachter and Anthony W. Orlando, "Booms and Busts in Real Estate," *Wharton Real Estate Review 15:1* (Spring 2011).
57 John Cassidy, "The Minsky Moment," *The New Yorker*, February 4, 2008.
58 It is interesting to note that Minsky is often invoked in the aftermath of bursting real estate bubbles. See Benjamin M. Friedman, "The Minsky Cycle in Action: But Why?" *FRBNY Quarterly Review*, Spring 1992–1993, and Janet M. Yellen, "A Minsky Meltdown: Lessons for Central Bankers," President and CEO, Federal Reserve Bank of San Francisco, speech delivered at Bard College on April 16, 2009.
59 Here are just a few of the resources that focus specifically on the real estate cycle. The first three cited contain excellent source lists for exploring the literature in greater

depth: Fred E. Foldvary, "Real Estate and Business Cycles: Henry George's Theory of the Trade Cycle," paper presented at Lafayette College (June 13, 1991), online at www.folvary.net/works/rcbc.htlm; William C. Wheaton, "Real Estate 'Cycles': Some Fundamentals," *Real Estate Economics 27:2* (Summer 1999); Stephen A. Pyhrr, Stephen E. Roulac, and Waldo L. Born, "Real Estate Cycles and their Strategic Implications for Investors and Portfolio Managers in the Global Economy," *Journal of Real Estate Research 18:1* (1999); Glenn R. Mueller, "Predicting Long-Term Trends and Market Cycles in Commercial Real Estate," Working Paper 388, funded by NAR Transact Program, October 24, 2001; Eric Goldschein, "The Complete History of US Real Estate Bubbles since 1800," January 10, 2012, online at www.businessinsider.com/the-economic-crash-repeated-every-generation-1800-2012-1

60 Fragment 217, cited in G.S. Kirk and J.E. Raven, *The Presocratic Philosophers*, Cambridge University Press (Cambridge, UK, 1957), p. 196.

61 A notion that underlies the admonition that policy makers should not attempt to intervene in market cycles, which are held to be self-correcting.

62 This famous line, from a letter from Leibnitz to the mathematician Jacob Bernouilli, is quoted with approval in Peter L. Bernstein's excellent, *Against the Gods: The Remarkable Story of Risk*, John Wiley (New York, 1996), p. 4. John Maynard Keynes used this quote in his 1921 *Treatise on Probability*.

63 It gives me pause to realize that most contemporary economics students do not have the awareness that the "AM" and "FM" settings on their automobile radios stand for "amplitude modulation" and "frequency modulation," and that it is by varying amplitude and frequency that radio signals convey "information," in the case of the car radio, the music or talk being transmitted.

6 Trends behind the trends

Thomas P. "Tip" O'Neill of Massachusetts was Speaker of the US House of Representatives from 1977 to 1987. He is closely identified with the phrase "All politics is local," a lesson he learned when he suffered his one and only electoral defeat in running for the Cambridge City Council in 1935.[1] If politics is local, how much more so is real estate, whose cliché is that the three most important considerations are "location, location, location"? European languages fix this with their term for real estate, *immobilier* in French, and cognate nouns in German, Italian, Portuguese, and Spanish.

Yet politics and real estate have both been yanked into much wider contexts in recent decades. Globalization, technology, and environmental forces that respect no artificial boundaries make the movement of money, people, ideas, and (literally) the winds of change inescapable concerns of politicians down to the level of the ward or community district. So too for real estate, as trends in design, planning, building systems, and the financing of property spread across the globe at warp speed.

Cities are pre-eminently the place where these forces – what we might call the "trends behind the trends" – work themselves out in concrete form.

The world is flat – or is it?

New York Times columnist Tom Friedman has forged an international reputation by examining the impacts of globalization, most famously in his 2005 best-selling book, *The World Is Flat: A Brief History of the 21st Century.*[2] Deeply impressed by the impact of globalization in leveling the competitive environment, as footloose businesses roamed the world in search of the most economical places for production in an era when technology and the knowledge it embedded were ever more portable, Friedman recounts the rise of Bangalore, India, to world-class status as a city that leveraged the opening of a Texas Instruments plant in 1985 into a reputation as "the Silicon Valley of India."

Globalization cannot be overstated as an influence on the deindustrialization of America and the consequent reshaping of US cities. Both "-zations" are usually considered the results of large and impersonal forces sweeping over the world economy and, to a large extent, impervious to human intervention.

Of course, like everything economic, quite the contrary is true. There is nothing economic that is *not* of human creation. Adam Smith's invisible hand may be invisible, but it is a human hand and, perhaps, as in the title of one of Nietzsche's books, "all too human" in causes and in consequences.

In his *Globalization and its Discontents*,[3] Nobel Prize winner Joseph Stiglitz describes and decries policy making at the World Bank and the International Monetary Fund that he contends increased, rather than mitigated, third-world poverty while larding on debt for nations ill equipped to repay it. Thus policy – a human construct – contributed to economic and political instability, when the putative intent was quite the opposite. Over-borrowing and over-lending are two sides of the same coin, and it is a common-sense observation of financial economists that risk increases in direct proportion to leverage, as the downside costs of volatility wipe out borrowers' equity, while leaving the lenders with collateral of compromised value. Globalization contributed to the destabilizing of the economies of Thailand, Argentina, and Russia in the late 1990s. And the troubles came home to Wall Street with the collapse of the largest US hedge fund, Long-Term Capital Management, in 1998.[4]

In the United States, third-party presidential candidate H. Ross Perot colorfully described the domestic effects as a "giant sucking sound"[5] sending US jobs overseas by the millions. This perspective was admittedly grounded in experience, and its impacts on one US industry – furniture making – has more recently been described in detail by Beth Macy in her book, *Factory Man*.[6] But Stiglitz is making another point – that global trade pacts were shaped by governmental and corporate interests that expanded the profits of those with the size and scope to exploit international inefficiencies, while harming the poorest countries in the world in the meanwhile.[7] Only conspiracy theorists would claim that the consequences in the US and abroad stemmed from a Machiavellian cabal. But the results have been exactly the opposite of the principal theme of a flattening world sounded by Friedman. With a few notable exceptions, the gap between the most- and least-prosperous national economies appears to be widening,[8] and, in the United States itself, the level of income inequality in 2014 places us in the company of Bulgaria, the Philippines, Uruguay, and Cameroon.[9]

The expectation that globalization, and macroeconomic forces generally, would lead to a leveling of differences is based upon the Newtonian scientific model discussed earlier in this book, toward the end of Chapter 1. The Second Law of Thermodynamics predicts that systems operate under the law of entropy, the tendency toward uniformity. This concept undergirds predictive theories and drives them toward equilibrium solutions because it is understood to represent the "arrow of time." Increasing entropy rides on time's forward motion, with time understood to be an irreversible process.[10] As a theory, it is elegant in its clarity and simplicity. It has been powerful as a predictive tool as well. But it does turn out not to be a final, all-encompassing answer.[11]

The difficulty is that the theory sees relative advantage being competed away (or "arbitraged," as the financial economists would say). Such a forecast has trouble matching up with the experience of economic geography, especially the tendency of economic activity to concentrate rather than disperse. Since Alfred Marshall more than a hundred years ago, economists have noted the power of agglomeration, and Harvard's Michael Porter has built a solid reputation on his analyses of "clustering" over the course of several decades. As a counterpoint to Friedman's thesis, *Creative Class* researcher Richard Florida published an article in *The Atlantic Monthly* entitled "The World is Spiky."[12] Florida contends that, although there is much substance to Friedman's argument for a flattening world under the influence of globalization, there is a very different economic topography to chart. Looking at variables such as population, light emissions, patent awards, and scientific citations, Florida and his colleagues find that "surprisingly few regions truly matter in today's global economy."

Later on, in Chapter 10, we will look in some detail at some of the reasons why complex systems such as cities produce exceptions to the phenomenon of entropy, which is, after all, supposedly based on an invariant law.[13] But we are all familiar with the tendency of cites to sort themselves out with distinctive differences, rather than all looking alike. At least, I will claim, that is true for an identifiable set of cities, the 24-hour cities.

To be concrete, let us look at a few characteristics that are *not* universally shared by all cities, factors that are, therefore, emblematic of the "spikiness" that Florida and his colleagues adduce from their research. For a start, let's look at so-called "global cities," where we might suspect that "globalization" will be most in evidence.

Global cities

In 1991, I led a consulting team responding to a client who wanted to look at the cities with the best potential for moving up the North American urban hierarchy over the two decades ahead. The client, who must remain confidential, was seeking to understand where the best places for real estate investment and development might be, considering long-term potential. I believed at the time (and still do) that linkages to the global economy would characterize such locations.

Figure 6.1 shows the results of an analysis at that time, factoring in metropolitan area performance over a set of five variables:

- export trade in goods;
- financial flows;
- the presence of US-based multinational corporations;
- the presence of foreign-owned firms doing business in the US;
- international travel access.

Obviously, the first thing to note is that, if Friedman's "flat world" thesis considers worldwide impacts, the choice of a set of US cities to investigate seems to miss the point. But let's accept that, even in a flat world, the exposure

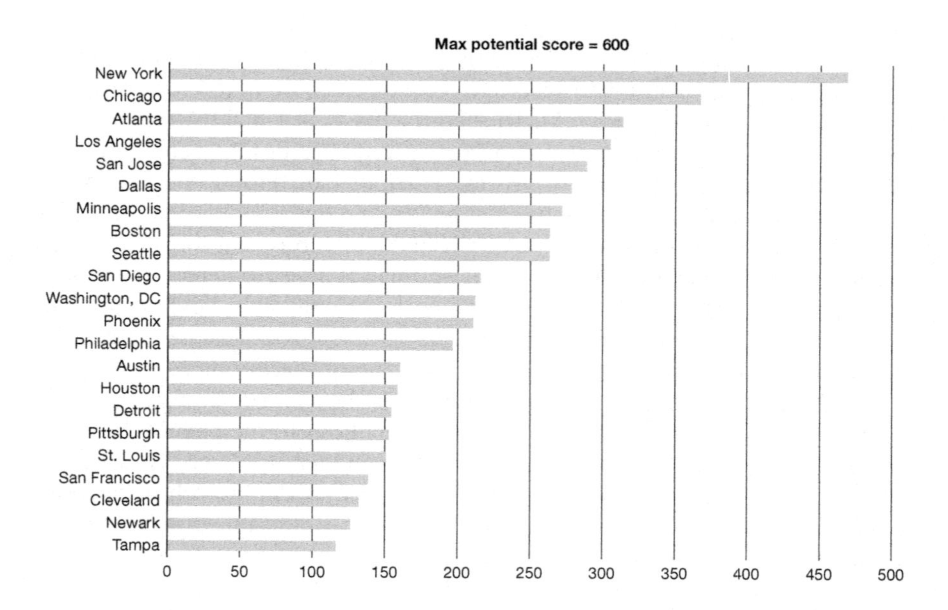

Figure 6.1 Linkages to the global economy: US metros

to international commerce may vary from place to place – a fairly modest allowance – and that this has implications for urban economies.

Relative advantage in exporting is a critical variable, as local economic bases are sustained and ramified by the production of goods and services consumed around the world, indicating success in competitiveness and, importantly, bringing a flow of net revenue from abroad into the domestic economy. Taken together, the presence of US multinational and international firms doing business in the US is not merely a sign of corporate choice of places to do business: it also implicitly recognizes the importance of management and organizational expertise as a factor in production. And a measure of international travel underscores the key role that infrastructure – in this case, airports – plays in supporting economic activity (as well as recognizing that international business travel and tourism function as services exports for many cities).[14]

The idea of linkage is a key concept. Cities do not exist in isolation. In the past, concepts of linkage and their impact on urban economic form were concentrated heavily upon the core city and its hinterlands.[15] The dramatic improvement over time in communications and transportation technology has expanded the linkage concept to more formal studies of urban hierarchies and networks. Globalization has pushed the discussion beyond national borders.[16]

Saskia Sassen claims we may be able to understand the new global economic order only if we recognize why key structures of that economy are ***necessarily*** (her emphasis) based in cities. The first edition of *The Global City*[17] helped focus scholarly attention on the role of a few key cities as hubs of the world economy.

Her key cities – New York, London, and Tokyo – were chosen because of their pivotal roles in the movement of money around the world.

Flows of capital are a defining element of globalization and essential to understanding the role and function of global cities. But they are not the sole descriptors. Indeed, Sassen is at pains to specify a complex of characteristics. In the contemporary economy, few places can be considered isolated from the influence of globalization, as Friedman argued in *The World Is Flat*. David Clark discusses the theme of urban interdependency in the late 20th century, making the point that so-called "world cities" represent critical concentrations of production, finance, and services. He argues that the development of very large conurbations in the developing world is a response to the all-enveloping net of global capitalism.[18]

Sassen makes a more targeted point. As advances in transportation and technology have facilitated the worldwide reach of corporations, the integrity of the enterprises depends upon skilled management processes of planning, reporting, and selection, and orchestration of effort. These skill sets have concentrated in specific geographic locations.

Even giant multinational corporations need to identify and attend to core competencies, and, therefore, firms depend upon an array of outside specialists for both upstream and downstream production functions. Sophisticated "producer services" tend to cluster around major corporations and follow the client to locations around the world. Hence, law firms, accountancies, consultancies, as well as advertising, engineering, architecture, and information technology firms, have also become worldwide enterprises.

Globalization means dispersal, in this sense, but also has manifested itself in concentrations of producer services in key cities, precisely to facilitate access to such services and to make necessary travel more efficient. Naturally, too, all the advantages of agglomeration can be more effectively realized by such an economic configuration.

Global cities have ambiguous relations with the nation-states within which they reside. Global cities simultaneously serve the worldwide interests of transnational corporations and (in general) act as immigration magnets. This can mean that their interests run contrary to the objectives of national governments.

For instance, each country is likely to see its interests as entailing the retention of its existing industry and employment base to the greatest possible degree. But the commercial logic of globalization drives the redistribution of work to its most profitable locale. Xenophobia may be more or less strong as a cultural phenomenon. Japan's long history as a "closed society" may be taken as one case in point, with Tokyo's demographics illustrating the result. The US has built its self-identity as "a nation of immigrants," and this continues to be apparent in cities such as New York, Miami, and Los Angeles. But, even in the United States, a strong strain of nativism remains and is politically powerful.

There is a tendency to regard the nation-state as the primary economic unit, in part because (as Sassen points out) that is the level at which the most comprehensive and frequently reported data can be accessed. By contrast, Jane Jacobs had already stipulated in *Cities and the Wealth of Nations*[19] that the national economy is something of an abstraction, a summation of the work that is accomplished by localities.

McKinsey director Kenichi Ohmae sounds this as a recurrent theme in *The End of the Nation State*[20] and *The Borderless World*.[21] He argues that efficiency is to be found at the urban–regional rather than national level. Such city-regions are the ports of entry into the world economy, and they provide "economies of service." By the latter concept, Ohmae means an infrastructure of communications, transportation, and producer services specialties.

As a management consultant, Ohmae sees the urban question filtered through the organizational imperatives of business enterprises. From this perspective, he judges that the centralized command-and-control business model, where all units "report in" to headquarters, which evaluates performance and redirects operations, is inappropriate to effective transnational business management. At the enterprise level, Ohmae suggests the hierarchical model is less useful than a network model that emphasizes decision-making closer to the customers, the nurturing of human capital, and critical reliance on both formal and informal communications. He stresses that creating the climate for innovation is the key to competitive success, and states, "Few businesses succeed because of control. Most make it because of motivation, entrepreneurship, customer relations, creativity, persistence, and attention to the "softer" aspects of organization, such as values and skills."[22]

The configuration sketched by Ohmae is quite consonant with an anentropic pattern of emergent self-organization. Recognizing the dispersal of production functions around the globe, often, but not always, to "least-cost" locations, it can be observed that the world economy does not present itself as a featureless plain, but one where congregation of activity is more the norm than the exception. And it should be vigorously pointed out that "least cost" is not synonymous with "greatest benefit." Much of the spikiness of the world is created by imperatives that relate to maximization of benefit rather than the lemming-like pursuit of minimization of costs.[23]

Technology and the "death of distance"

Cities have long benefited from advances in transportation and communication technologies. In 1807, Robert Fulton and Robert Livingston launched the first commercial steamboat service between New York City and Albany. The opening of the Erie Canal in 1825 created a shipping lane from the port of New York into the Midwest. New York soon leap-frogged Boston and Philadelphia to become America's largest and economically most important city. Communications technology tied cities closer together, facilitating economic networks. Samuel F.B. Morse, with the technical assistance of NYU's Leonard Gale, solved the problem of sending telegraphic signals over long distances by 1838. Morse's

Magnetic Telegraph Company was established in 1845 to connect New York with Boston, Philadelphia, Baltimore, Buffalo, and points west.

Although the first electric light was invented by Englishman Humphry Davy as early as 1809, it was not until 1879 that Thomas Edison made a bulb that had a long enough life for practical use. The Edison Electric Illuminating Company was incorporated in 1880, and, by 1882, a square mile of New York City was lit by electric light, the harbinger of a city that could imagine 24-hour daily commerce.

Technology advocates and visionary futurists have often leapt from the premise of increasingly convenient and cheap communications to the conclusion of a flatter world (in Friedman's sense). Frank Lloyd Wright, the great American architect, saw technology as a force liberating people from fixity in location.

Historian Robert Fishman extended the argument as he studied the evolution of urban characteristics. As commercial areas began to arise in what were once exclusively residential suburbs, Fishman's narrative suggested that the "locus of growth and innovation" had shifted from city to suburb, or what he termed "technoburbs." As communications and technology spurred decentralization, the need for face-to-face contact in an urban setting was superseded. He credited Frank Lloyd Wright – and H.G. Wells – as being "the prophets of the technoburb."[24]

A widely discussed feature of the rapid changes introduced by telecommunications and the Internet in the past 25 years has been the putative reduction in the importance of place. Joel Kotkin, in *The New Geography: How the Digital Revolution is Reshaping the American Landscape*, puts the case this way:

> The rise of the digital economy is repealing the economic and social geography of contemporary America. . . . Information supplants energy and conventional manufacturing as the critical source of wealth. . . . These changes profoundly alter the very nature of place and its importance by deemphasizing physical factors. . . . Wherever intelligence clusters, in small town or big city, in any geographic location, that is where wealth will accumulate.[25]

Robert Atkinson, of the Progressive Policy Institute, evaluated the impact of technological changes on cities, concluding that technology is replacing formerly critical physical connections, allowing more of the economy to be operated at a distance. In his view, the information superhighway weakens central-city and inner-suburban economies, favoring outer suburbs and exurban locations. The overthrow of the "tyranny of distance and time" in the digital economy reduces the imperative of agglomeration and consequent importance of denser and higher-cost cities.[26]

Atkinson describes phenomena that have been colorfully named "technoburbs" by Fishman and "nerdistans" by Kotkin. In the same number of HUD's *Cityscape* journal, a trio of University of North Carolina scholars assert that knowledge-based jobs are locating with increasing frequency in the relatively newer multinucleate cities of the South and West, and in the suburbs of older MSAs.[27]

The literature contains many articles linking technology, innovation, and competitiveness. The results, as might be expected, are far from uniform.

Jane Pollard and Michael Storper examined employment change in a dozen major US metropolitan areas, focusing on three industry clusters they propose as being "dynamic engines of employment growth." The clusters were identified as:

- intellectual capital industries (including manufacturing in "old technology" fields such as organic chemicals, industrial apparatus, and measuring and controlling devices; also financial services, high-order business services, and healthcare);
- innovation-based industries (including manufacturing industries in biological products, engines, aeronautics, and engineering equipment and instruments); and
- variety-based industries (including a broad range of nondurable manufactured goods; printing and publishing; durable-goods production, including several categories of machinery and equipment; as well as motion picture production and allied services).[28]

Although this amalgamation of industries may appear to overly stress industrial production, these authors are aware of the argument that strong downstream linkages connect manufacturing to services industries, and that economic policy ignores manufacturing at its peril.[29]

Pollard and Storper's results are summarized in Figure 6.2 and make the point that the impacts of technology are heterogeneously spread across America's urban landscape. Interestingly, they do not conclude that multinucleate Sunbelt cities hold a comparative advantage as technology advances.

There is a significant literature suggesting that place continues to have a key role, even in a highly networked economy. Peter Dreier and his colleagues titled their 2001 study *Place Matters*.[30] Ronald Mitchelson and James Wheeler studied the flow of information in the global economy and found that information exchange has the anentropic tendency to cluster activity, a kind of gravitational effect that enhances, rather than erodes, the position of cities high on the urban hierarchy. Equally important, these authors stress that it is precisely the exchange of information that is critical, pointing to a networking function within the system of cities. This can lead to a self-reinforcing emergence of advantage, a common concept in systems theory. Cities with high information-generating characteristics sustain very high growth rates for additional information flows over time.[31]

Edward Malecki's 2002 study of the economic geography of the Internet's infrastructure concludes that the web reinforces, rather than dilutes, agglomeration economies in urban markets. Technological change creates economic

	Single specialty	Multiple specialties	No specialty
High growth	*Dallas–Fort Worth*: innovation-based growth in aircraft and radio–television equipment, with some intellectual capital growth in computer programming and data processing; innovation growth peaked in mid-1980s, though, economy has a growing dependence on US technology trends *San Diego*: innovation sector shows an LQ above 2.0, led by aircraft and parts, electronic components and accessories, engines and turbines, and radio–television equipment; spin-off growth in computers and software, R&D labs, management consulting and other intellectual capital fields *Phoenix*: innovation sector leading very rapid economic expansion, with radio–television equipment and office and computing equipment most prominent; however, LQs declined in the decade following 1977	*Los Angeles*: growth across all three "dynamic" sectors; mainly in innovative industry cluster, but also in all other production sectors, including such routine industries as apparel; high military spending insulates LA from economic shocks *San Francisco*: solid growth in innovative cluster and intellectual capital cluster; broad high-technology, including hardware and software in computers, communications equipment, aero-nautical applications, and precision instruments *Boston*: LQs of 1.4 for both innovation and intellectual capital clusters; educational and scientific research establishments featured; office and computing equipment and software also strong	*Atlanta*: though a very strong-growth city overall, Atlanta did not achieve a LQ above 1.0 for any of the three dynamic clusters; this city looks like it is simply dependent upon regional demographic expansion, servicing the Southeastern states *Minneapolis–St. Paul*: innovation-based industries had an LQ of just over 1.0 in 1977, but this declined thereafter; however, the metro area did post above-national rates of overall employment growth; this may reflect the urbanizing trend across the upper Midwest region
Low growth	*New York*: intellectual capital growth in financial and business services; has not grafted on innovative production and is declining in variety-production and routine industries *Pittsburgh*: overall decline in jobs base, especially due to steel industry decline; intellectual capital growth in banking, healthcare, and engineering/architectural and other business services	*Chicago*: Concentrations in intellectual capital and variety-production, but neither area is strengthening its position – either in the financial services areas such as commodity trading or in the nondurable manufacturing areas such as printing and publishing and pharmaceuticals	*Detroit*: this MSA was found to be in absolute decline, even in its core automotive industry; intellectual capital industries gained relative share during the 1980s, though, and had an LQ approaching 1.0 by the end of the decade, especially in consulting and in computer programming and software

Figure 6.2 Pollard and Storper comparative matrix

disturbances that give rise to new clusters – an anentropic effect that belies the assertion of the irrelevance of place.[32] Owing to the costs of installing bandwidth, the primary measure of Internet accessibility, high-volume information flow spaces are early in the queue of infrastructure development, as such places promise faster payback periods. This becomes self-reinforcing, as greater speed and volume of information flows improve the economic competitiveness of those cities where infrastructure is first introduced.[33] Even with all the hyperbole about the new economy, such "old economy" concepts as the persistence of initial advantage and economies of scope remained fundamental.

In Figure 3.7, earlier in this book, we looked at population growth (1950–2010) in a selected set of cities, the 14 cities that we will later examine in comparing 24-hour cities with 9-to-5 cities. Let us see how those cities compared with each other in terms of Internet backbone as we moved into the new millennium (Figure 6.3). The top four metro areas are not the Sunbelt, multinucleate locations posited by Rondinelli and his colleagues. Rather, they are cities that were dominant in the old economy and that appear to be adapting well to the new: New York, Chicago, Washington, DC, and San Francisco.

Perhaps the watershed document in the debate about the impact of technology on cities, on whether location is tending to matter more or matter less, appeared in *The Economist* issue dated September 30, 1995. This was Frances

Rank	Metropolitan area	Bandwidth in Mbps
1	New York	234,258
2	Chicago	221,738
3	Washington, DC	208,159
4	San Francisco	201,722
5	Dallas	183,571
6	Atlanta	149,200
7	Los Angeles	140,649
8	Seattle	109,510
13	Boston	75,044
14	Philadelphia	74,167
19	Phoenix	45,868
21	Las Vegas	42,414
22	Miami	42,138
31	Minneapolis	29,734

Figure 6.3 Telecom bandwidth in selected MSAs

Source: Data excerpted from Malecki, 2002

Cairncross's "The Death of Distance: A Survey of Telecommunications." Cairncross expanded this into a book in 1997, which was so widely read that "a completely new edition" was published in 2001.[34]

Intriguingly, Cairncross enumerates several attributes pointing to a rebirth of cities. There is a global premium for skills in the knowledge-based economy, and companies will migrate to the best places for attracting skilled, productive workers. The Internet amplifies the value of the creative use of information, and workers meeting that criterion are a much scarcer resource than authors anticipating unremitting economic entropy imagine. Cairncross anticipates that most cities do face a hollowing out of their centers, now and in the future. But cities with downtowns combining strong residential, cultural, and entertainment choices, and a stimulating lifestyle, will attract and retain these high-value-added workers. In this, her conclusions are similar to Florida's "creative class" observations about spikiness. Even Joel Kotkin acknowledges that, despite the anti-urban forces afoot, the digital age favors dynamic and culturally rich urban environments.

Mitchell Moss and Hugh O'Neill, in a provocative but prescient paper, suggested that an information-intensive urban environment provides greater impetus for innovation.[35] Urban scholar Manuel Castells concurs with the observation that information networks, upon study, debunk the "futurologist myth" that the Internet Age portends the end of cities. He presents evidence showing that the Internet allows metropolitan concentration and global networking to proceed simultaneously.[36]

Despite the gathering weight of evidence that "place matters," that "the seduction of place"[37] remains a powerful lure for humans, in the popular mind there remains the illusion that technology must be bad for the viability of cities. In March 2015, I was privileged to deliver the "wrap-up talk" at the huge MIPIM trade show in Cannes, where 90 nations exhibited the latest and greatest in real estate development and architecture for nearly 20,000 registered attendees. I asked my audience to focus on the pressing need for intelligent property development, especially for cities, noting that the world's urban population is expected to increase by 80 percent in the next 35 years.[38] A young man challenged my premise, saying, "In a world where we all have everything we need on a portable device, why would you think that we'll be seeing more and bigger cities in the future?" My response was brief: "Because the world is richer in experience than your cell phone can accommodate."

Environmental pressures and the variable impact of/ on cities

In 1970, the US Congress passed the Clean Air Act. Two years later, the Clean Water Act became law. In large measure, the quality of life in America's cities, if not their very viability, owes a tremendous debt to these visionary pieces of legislation. Conversely, cities – especially dense cities – have proven to be true "Friends of the Earth."

Automobile travel has been a much discussed air-pollution topic and is a real bone of contention between those promoting greater urbanization and those defending the suburban house-and-three-car-garage version of the American Dream. Admittedly, there is a lot more to the environmental question than the internal combustion engine. But, with 253 million cars and trucks on American roads,[39] it is a really good place to start the discussion.

In *The Triumph of the City*,[40] Edward Glaeser cites research that he and Matthew Kahn have undertaken in recent years on the environmental performance of cities.[41] Some of their top-level results help set the stage for looking at cities and environmental sustainability. Density counts, a lot. In census tracts with population densities greater than 10,000 persons per square mile, households average gasoline usage of 687 gallons per year. In tracts with densities lower than one thousand per square mile, that figure leaps to 1,164 gallons – 69.5 percent more. As a gallon of gas produces approximately 22 pounds of carbon dioxide (CO_2), a dangerous greenhouse gas, the effect of density on air quality is immensely positive. If you need to get in your car for most of your daily tasks, and that's the result of the configuration of where you live and work, that configuration is part of our climate problem.

Of course, mass transit uses energy and emits CO_2 as well. But public transportation is far more efficient. The New York City transit system delivered 2.6 billion trips to its riders annually (2006 data), an average output of 0.9 pounds of CO_2 per trip, or less than 10 percent of the average automobile trip. And cities with strong Walkscores[42] – a measurement that calculates the density of amenities, which has been supplemented in recent years with "Transitscores" and "Bikescores" – allow people to do many of their daily journeys on foot, personally healthier and environmentally more beneficial.

Glaeser and Kahn calculated the annual cost of carbon dioxide emissions at the metro area level for 66 large urbanizations and then disaggregated the costs between the central cities and their suburbs. Figure 6.4 summarizes their findings for the ten largest metropolitan areas, as presented in their 2008 Kennedy School report. Unsurprisingly, the authors find strong correlations between automobile-oriented cities and high household emissions costs. Their analysis looks at a variety of other emissions sources, though, including home heating/cooling, differences in types of fuel used, and electricity consumption. Cities such as Boston and New York that rely on oil for home heating have higher emissions than cities such as Chicago that rely more heavily on natural gas. California's tough measures to encourage energy-efficient appliances help San Francisco's score, as does its moderate climate, a factor setting it off from Sunbelt cities, where air conditioning is used much more intensively.

As both the intra-metropolitan trend of suburbanization and the inter-metropolitan trend of Sunbelt growth proceeded, the effect was a pattern of progressively lower urban densities. Kotkin and others have noted that this may be interpreted as the market voting with its dollars and with its feet (although the feet are often attached to the gas pedal of the automobile). "Sprawl" has become the pejorative label associated with the dispersal trends.

Metro area	Annual cost per household at metro level ($)	Rank out of 66 areas	Suburb–city difference in annual cost per household ($)	Rank out of 48 areas
New York, NY	893	6	303	1
Los Angeles, CA	820	3	(36)	47
Chicago, IL	1,163	31	38	33
Boston, MA	1,058	19	214	4
Philadelphia, PA	1,248	42	185	5
Detroit, MI	1,292	53	(88)	48
Washington, DC	1,180	32	180	6
Houston, TX	1,334	62	158	7
San Francisco, CA	813	2	142	11
Atlanta, GA	1,313	57	220	3

Figure 6.4 Annual CO_2 emissions costs for largest metro areas

Note: Figures in brackets (in original research) reflect conditions whereby emissions per capita in the suburbs are lower than in the central city, owing to two factors: the distribution of population is heavily weighted toward suburbs for those metro areas, and their automobile dependency is especially high

In response, an anti-sprawl movement has been established among architects, designers, and urban planners. This is most closely associated with the Congress for the New Urbanism, established in 1993, with the founding leadership including Peter Calthorpe, Andres Duany, Elizabeth Module, Elizabeth Plater-Zyrbeck, Stefanos Polyzoides, and Daniel Solomon. The symptoms of sprawl, in design terms, are primarily identified as:

1 housing subdivisions as homogeneous entities;
2 shopping centers, as opposed to street-front retailing;
3 office parks as monotonic land uses, with total auto dependency;
4 civic institutions (town halls, churches, schools) without pedestrian access and so dispersed from housing they demand driving;
5 roadways designed to maximum size and in configurations of cul-de-sacs and collectors.[43]

Much blame is laid on single-use zoning. Duany et al. offer the tight-grained zoning of Coral Gables, Florida, as an alternative. Impacts of sprawl include dystopic outcomes such as income segregation, environmental degradation, inefficient resource allocation, and the hollowing out of urban centers. Moreover, under such conditions, traffic congestion increases, even with lower population density, and social consequences include ennui, anomie, obesity, and other physical malaise. Anthony Downs of the Brookings Institution cites the problem set similarly, albeit from a more economic and less sociological perspective: excessive travel, lack of affordable housing, unfair infrastructure cost allocation, difficulty of LULUs, inefficient subsidies, and the absorption of too much open space.[44]

Air quality is just one of the environmental issues affecting our cities, not just in the US, but worldwide. Water is another. Global demand for fresh water is projected to exceed supply by 40 percent by 2030, and water scarcity is a reality in much of the developing world – where 780 million people have no access to clean water, 2.5 billion have no access to modern sanitation, and more than 3 million die each year from water, sanitation, and hygiene-related causes. Some projections estimate 1.8 billion people living in regions with confirmed water scarcity by the year 2025.[45] The implications for real estate are enormous, as access to healthy water affects land value, community desirability, future viability, and investment. Consider also that China is home to 20 percent of the world's population, but only 7 percent of its fresh water. Water may become a global political issue, as well as a health issue, in a relatively short timeframe.

The US will likely experience serious water shortages as well. In fact, that is already a factor for many cities in the western and southern states. Aging water infrastructure, droughts, and reduced water deliveries to agriculture have the potential to cause severe economic problems. A number of places face severe water challenges: Las Vegas's Lake Mead, which supplies 100 percent of the city's water needs, is projected to have a 50 percent chance of drying out by 2025. A 2013 US government report showed that groundwater depletion in the US for the years 2000–2008 was nearly three times greater than the average rate of depletion for the preceding 108 years – from 1900 to 2008.[46]

Here is what the External Affairs Committee of the international organization, The Counselors of Real Estate,[47] has to say about impacts in the United States:

> The fastest growing states are ones where water scarcity is the most serious issue (California, Nevada, Arizona, New Mexico, and Texas).[48] Water access and cost may become a big enough factor in regional economies to influence geographic real estate investment allocations. US Environmental Protection Agency (EPA) recognizes these threats and has just released a study titled "The Importance of Water to the US Economy" which finds that "negative impacts to the quality and quantity of water . . . have significant ripple effects throughout the economy."[49]
>
> Water cost is primarily an issue for water-intensive businesses, whose concentrations in an economy can be measured. Water access and uncertainty considerations, brought on by weather related events, loss of water rights, growing resource scarcity, and lack of political/fiscal will to build, repair, and replace needed water infrastructure are more difficult to analyze.

> Measurement and monitoring of regulator and tenant concerns and the resulting actions regarding water issues may alter investment allocation decisions, as capital providers seek to manage downside risk.[50]

Just as in the cases of globalization and technology, the powerful forces of environmental change have left no locality unaffected. But all cities are manifestly not equally affected, nor have they equally been responsive to the challenges. This is another instance of the spikiness of urban economic geography, a way that cities have acted and continue to act to differentiate themselves.[51]

The Economist Intelligence Unit (EIU) has issued a report entitled the *US and Canada Green City Index: Assessing the Environmental Performance of 27 Major US and Canadian Cities.*[52] The Index is a multifactor rating of 27 cities, 22 in the United States and 5 in Canada. The variables studied included emissions, energy use, land use patterns, "green buildings," transportation patterns and infrastructure, water use, waste management, and air-quality policies (including implementation). Figure 6.5 displays the overall Green Index scores for each of the US cities and their rank among all 27 North American cities rated by the EIU.

Seven cities score above 70 index points, on a scale where 100 is the maximum possible score. Those cities are San Francisco, New York, Seattle, Denver, Boston, Los Angeles, and Washington, DC. Here are some observations about these top cities, based upon the EIU's analytical work:

- *San Francisco* has been a leader in the recycling movement and has one of the strongest performances in the built environment, transportation, and air-quality sectors. It is a small, compact city, and its temperate climate makes fewer demands on energy use than places subject to extreme hot and cold weather.

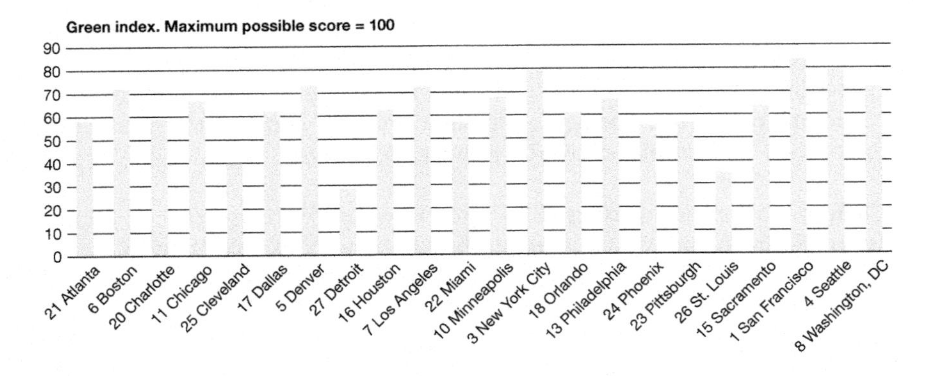

Figure 6.5 Environmental performance acts as an urban differentiator

Source: EIU

- *New York City* is far and away the largest and densest American urban center. It is the top-ranked city in transport, land use, and environmental governance. Despite its density, New York has a tremendous volume of parkland and scores high in both air and water quality – a vast improvement over the past half-century.
- *Seattle* has been in the forefront of adopting LEED (Leadership in Energy and Environmental Design) standards and ranked first among the 27 cities on this index variable. Like San Francisco, Seattle has been aggressive in promoting waste-management advances, and its city council has voted to adopt carbon neutrality as a key municipal priority.
- *Denver* is a mid-sized metro area in population, with 2.6 million in and around the city itself. Denver rates at the highest level for energy and environmental governance. It has been alert to all opportunities to enhance air and water quality and has invested in congestion-reduction tranporta-tion, although the billion-dollar T-REX project has failed to boost transit use much above one-half the national average (which is meager in its own right).
- *Boston* has roughly the same population as San Francisco, but its profile in New England's climate is significantly different. Boston scores excellent ratings for electricity consumption per dollar of economic output. It also does well in local policy initiatives in energy, and it has the second-lowest per capita water usage, after New York. Excellent land use planning is a plus, but the region's "T," Boston's mass transit system, is venerable (i.e., *old*) and has inadequate capacity for its commuter base.
- *Los Angeles*' ports and its still-large manufacturing sector pose carbon emission challenges, but the long-fought battle against smog is showing signs of encouraging progress. The local utilities are particularly helpful contributors in their energy-consumption and carbon-reduction programs. LA has also developed some innovative financing mechanisms by pledging future tax revenues to support current investment in mass transit.
- *Washington, DC*, as the nation's capital, has attempted to lead by example in promoting the green building initiative, especially through the policies of its local Green Building Act (mandating LEED compliance for both public and private buildings). Waste and water management are its weak points, owing to a mediocre recycling program and a water system that loses 14 percent of its flow daily owing to leakage.

Below the top tier of cities, EIU rates a dozen US cities between 50 and 70 on its Green Index scale, ranging from Minneapolis (67.7) and Chicago (66.9) at the high end, to Pittsburgh (56.6) and Phoenix (55.4) with weaker scores. The mid-tier rankings are a mélange of old cities and new, Frostbelt and Sunbelt, coastal and interior. Policy and practice are found to be areas where the mid-tier compares unfavorably with the leading cities. Climate and urban configuration (especially sprawl) are also factors that have impact. Management toward better environmental conditions needs to be supported by financial commitment, and, for many of the cities, a countervailing commitment to small government and low taxes means waste, water, and air-quality management get starved for funds.

The very bottom tier – those cities scoring below 50 – represent a steep fall-off from the mid-tier. There are three US cities that fit this description – Cleveland (39.7), St. Louis (35.1), and Detroit (28.4). All three are historical manufacturing centers that have seen their economies shrinking in the de-industrialization era and have experienced massive population loss. This hobbles their ability to act on good intentions, even if public officials have a policy preference for greater environmental sustainability. St. Louis has the added problem of a regional authority governing its water system, and Detroit's bankruptcy has put checks-and-balances power in the hands of the state of Michigan. So, sadly, as the more prosperous cities at the top of the Green Index can fund even further advances in environmental quality, those at the bottom of the list are likely to fall relatively further behind.

The prospect is for greater spikiness, not greater flatness, over the course of the foreseeable future.

Final observations on the big trends

Macroeconomic factors such as globalization, technology, and the environment affect us all, and no city escapes their influence. All too often, though, the so-called "view from 40,000 feet" gives rise to assumptions that the influence affects the universe of places to the same degree and in the same ways. Such a viewpoint gives rise to the expectation of a flat world, or increasing homogeneity across the economy. This is expressed in terms of local advantages being competed away, a reversion to the mean (cities tending toward the national average in their local measures), and sheer entropy by convergence in costs, or by the sprawling of cities outward in a search of lower density and scale.

The evidence presented in this chapter argues that the 40,000-foot view is wrong, and that distinguishing differences in the set of US cities should be apparent to all willing to look at the numbers.

I'd like to correct one possible misimpression, though. Some commentators would like to turn the "flat" versus "spiky" debate into an either/or proposition. Neither Thomas Friedman nor Richard Florida is so shallow a thinker as to believe that this is the case. Friedman is clear that there are many "unflat" manifestations to be found in the global economy, and even entitles one of his chapters, "The Unflat World." Florida acknowledges that the "world is flat" hypothesis is not completely misguided, and that some leveling of peaks and valleys in economic geography is not only possible, but in some senses desirable.

But it remains the dominant case that, when it comes to the set of large US cities, there are meaningful, persistent, and economically significant distinguishing differences – enough to suggest that there may be a taxonomy that says that, in the genus "American cities," there are co-existing several species. That is the subject to be explored in the remaining chapters of this book.

Forms of change: Disruption

The four forms of change – cycles, trends, maturation, and change of state – discussed in earlier chapters have had something in common. They have been forms of continuous change. They all have structure, patterns than can be studied, and, therefore, have a certain amount of predictability to them. That's a fortunate attribute, as the hard work needed to understand those four forms of change brings a reward. Paying close attention to the patterns of continuous change helps us to prepare for what's ahead. And such preparation helps us in managing risks.

But anyone who has been awake to the alterations in the world, and who has the slightest degree of appropriate humility, knows this: the one thing we know for certain about the future is that we know nothing for certain. Folk wisdom is encapsulated in a saying: "People plan, and God laughs." Scientists murmur, "We live in a stochastic universe." And I teach my students that, "There is no reason to do research unless you are willing to be surprised." Our knowledge is imperfect, incomplete, and fallible. Events in the world remind us of this, each time we become complacent.

One of my philosophy professors at the New School for Social Research in New York was the renowned political thinker Hannah Arendt. In her book *On Violence*, she wrote this: "Significant events, by definition, are occurrences that interrupt routine processes and routine procedures; only in a world in which nothing of importance ever happens could the futurologists' dream come true."[53] We acknowledge this even in our most sophisticated economic models, where the equations contain a lovely Greek letter – epsilon (ε) – which signifies "the error term" representing those events that do not conform to systematic performance and, therefore, can confound ordinary expectations.

Nassim Nicholas Taleb, who is now a Distinguished Professor of Risk Engineering at NYU Polytechnic, has famously written about "black swans," or low-probability/high-impact events and has written what amounts to a four-volume meditation on the subject of uncertainty.[54] Although many are put off by Taleb's polemical style, laced with sarcasm and almost hubristic in its sense of superiority over conventional approaches to risk, he has been vindicated again and again by those who lapse into what some call a definition of insanity attributed to Albert Einstein: doing the same thing over and over and expecting different results. Our thinking is imprinted by patterns, and those patterns shape our expectations. As Nobel Prize winner Daniel Kahneman describes our human tendencies, the experience of patterned behavior leads us to shortcuts in thinking, heuristic anticipations, which can often leave us unprepared for the surprises life throws our way.[55]

Crisis

Alan Greenspan's long tenure as Chairman of the Federal Reserve System was marked by a series of successes that earned him the nickname of "The Maestro."

But, in the aftermath of the great financial crisis of 2007–2010,[56] Greenspan's reputation was sorely diminished. The lasting image of the former Chairman is his testifying to confusion about the inability of markets to self-correct. The dominant economic theory predicted such self-regulation, but failed to deliver at the critical moment. The utter magnitude of the dislocation, its stunning speed, and the inability of conventional fiscal and monetary policy to intervene left the by-then-retired head of the nation's central bank agape. He was simply lost for words of explanation. As he later confessed, in an article whose candid title belied a still dogmatic belief in the ultimate rationality of markets, he never saw it coming.[57]

In a sense, isn't that what we mean by a "crisis"? We harken back to the Greeks, who spoke of the "critical moment," or *kairos*, which is above all a time when judgment and decision must come into play, when traditional patterns must be put aside, and the "eternal recurrence of the same" must be transcended by creative thinking.[58]

The financial crisis of 2007–2010 was disruptive; there is no debating that. But it can hardly be claimed that it was a "sudden onset" event. We might trace it back to the Depository Institutions Deregulation and Monetary Decontrol Act of 1978, the first in a series of measures that undid fundamental banking structures that had supported US finance since the New Deal era. One result was substantially increased volatility in the economy, together with greater frequency in external shocks, as globalization became a fact of life in the capital markets.

A look at the period from 1990 to 2015 shows the unraveling – and incidentally highlights Greenspan's adroit crisis management over much of this period. Figure 6.6 traces the movement of both short-term and long-term US Treasury rates over a quarter-century, as well as the inflation rate as measured by the CPI, with significant financial market events. During Greenspan's tenure, the Fed needed to address the capital insufficiency of the US banking system following the collapse of the thrift industry; the collapse of the Mexican peso, which pushed one of the nation's wealthiest jurisdictions (Orange County, California) toward bankruptcy; the epic default of Russia on its sovereign bonds and the virtually concurrent threat to the South Asian and eventually the Latin American economies, as currency issues weakened their banking system; and, back in the US, the speculative euphoria of the dot-com bubble and its eventual bursting after 2000 (Y2K).

Recalling the threats posed by each of these crises is a helpful reminder of Greenspan's great skill in policy management for each episode in turn. It is how he earned his reputation as a maestro. But, unfortunately, he never seemed to grasp that recurrent crises – on the order of one event every 3 years or so – was a signal that the system itself was betraying structural weakness. That weakness was brought on – it must be said – by the triumph of the "efficient market theorists" of the Chicago school of finance over those who understood that regulation served to preserve markets by protecting them from their own fallibility and potential for failure.[59] Bluntly stated, if the Fed's bedrock *raison*

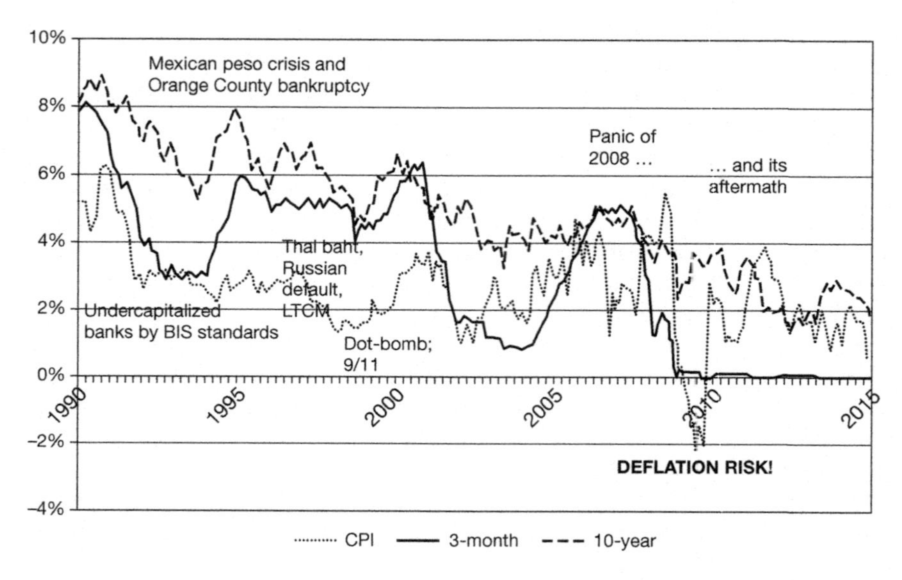

Figure 6.6 Recurrent crises betray structural weakness of the financial system: 1990–2015

Sources: Bureau of Labor Statistics (CPI); Federal Reserve H.15 reports (Treasury rates)

d'être is to guard the safety and soundness of the financial system, its Chairman should have asked the question of why crises were multiplying as markets became more unfettered.

It is still far from clear whether the disruption in the financial system will lead to a recasting of the relationship between markets and government as we move deeper into the 21st century. In the shocked aftermath of the post-Lehman Brothers meltdown, calls for systemic reform gained the upper hand for a time. But there is a powerful lobbying force being brought to bear that is essentially a call for a return to the *status quo ante*. This itself is a reminder that economic forces, even those with systemic impacts, are not impersonal forces of nature. It is a form of magical thinking to assert that so-called economic laws exist somehow independently of human will and action and are, therefore, beyond our control.

There are lessons for the governance of cities to be drawn from this, in times of crisis and also in intervals of normalcy.

The 9/11 lesson: "This changes everything"

Arguably, the al-Qaida attacks of September 11, 2001, were the defining psychological event for a generation of Americans – and for many beyond the borders of the United States as well. The images of jetliners slamming into the Twin Towers of New York's World Trade Center, the penetration of the walls of the Pentagon, and the heroic struggle of passengers bringing down a fourth jet into a Pennsylvania field, a jet apparently destined to destroy the US Capitol in Washington, DC, are seared into the collective awareness of Americans, probably indelibly.

The phrases, "Things will never be the same," and "This changes everything," were heard over and over again in the weeks following 9/11. And, in many ways, that has proven true. US citizens have become accustomed, if not inured, to limitations and restrictions that were hitherto virtually unimaginable. Inconveniences at the airport are the least of it. The Patriot Act suspended liberties that most had thought were matters of right and constitutionally protected. Intrusive surveillance by the National Security Agency and by local police has made any putative right to privacy more theoretical than real. In many office buildings around the country, the days of coming through the revolving doors and heading straight up in the elevator are over: photo IDs are checked, a bar code is issued to admit one to the building's interior, and signing out is expected as well. Repetition has made such experiences "the new normal." If this regimen is not embraced by all, it is perforce accepted. What else can you do?

In some ways, however, many anticipated changes post-9/11 did not materialize in the very cities that were the sites of the terrorist attacks. For instance, commentators predicted that companies and residents would flee New York and Washington in droves, propelled by the fear that the "target-rich environments" of these cities made them ripe for a reign of terror. Repetitive attacks would be difficult to predict or to prevent – in the streets, in the subways, in the shops, theatres, and sports arenas. Businesses would never again lease space on the top floors of skyscrapers. Firms would seek to avoid the existential risk entailed by having most executives and a large quantity of staff in a single location that could be subject to bombs, chemicals, or biological weapons. That all seemed logical. Yet those projections proved to be wrong.

My in-depth study of the city,[60] performed for New York's Civic Alliance in the year following the attacks, examined the location choices made by firms involuntarily displaced on 9/11, both tenants in the World Trade Center itself and those in surrounding buildings, directly damaged or rendered unusable by the toxic smoke that burned at Ground Zero for months afterward. Unexpectedly, the firms neither scattered willy-nilly around the region, seeking submarkets with high volumes of immediately available space, nor did they gravitate to the perimeter of the region, where the most inexpensive space could be found. Rather, they chose regional centrality, where they could access the entire metropolitan labor force, locations that had ample transportation choices, the most modern available buildings, and those areas providing the most significant agglomeration benefits.

Here are some of the major points made in the Executive Summary of that report:

- The terrorist attacks of September 11, 2001 traumatically disrupted Lower Manhattan. The destruction of the World Trade Center not only took a toll in human life and in grave emotional damage, it also physically eliminated more than 30 percent of the modern, Class A office stock in the Downtown market. Much of the remaining inventory in the market consists of buildings between 75 and 100 years old, not well suited to the operational needs of the competitive environment of the 21st century. This space will need to be re-supplied to the market if the key export industries in finance and advanced business services are to grow in Lower Manhattan.

- The former World Trade Center site, in terms of both its geographic location and its critical hub location for regional transportation (especially as presently envisioned in preliminary planning to link MTA and PATH systems), is best suited to provide the needed commercial space in an anticipated market recovery later in this decade.
- Space for a memorial to the events of September 11 can and should comprise part of the re-use of the site. However, it is too much to ask a mere sixteen acres to support all the expressed needs of Downtown: for residential, cultural, educational, and recreational land uses. Lower Manhattan, rather than the Trade Center site itself, provides the appropriate planning scope for the menu of desired land uses.
- A revitalized office economy constitutes Lower Manhattan's indispensable means to the end of more fully realizing its potential as a 24-hour community, with the possibility of joining Midtown Manhattan, Chicago's Near North Side, Washington's Georgetown area, and San Francisco's Nob Hill as thriving districts where strong commercial uses support vibrant residential neighborhoods, arts and entertainment, tourism, and an exciting street life.

Time has vindicated that perspective. Market forces, solid governmental leadership, a commitment to excellence on the part of the real estate community, and the manifest support of citizens behind the determination to make redevelopment a statement of a free people's tenacity and resiliency in the face of irrational hatred and violence are combining to take a horrible and destructive event and turn it from an incident of intended disruption into an opportunity to create something new, beautiful, and vibrant. I've been privileged to play a small part in that turnaround.[61]

Disruptions, even extremely ugly instances of destruction, need not only be instances of injury: they can also become catalysts of transformation.

New Orleans and New York

Charles Dickens famously opened *The Tale of Two Cities* with the line, "It was the best of times, it was the worst of times." He was speaking of the French Revolution, disruptive change if ever there was disruptive change. It was the beginning of the end, not only for the French monarchy, but for monarchy as Europe had known governance since feudal times. It was true revolution: a world turned upside down.[62]

Figure 6.7 also shows two cities – New Orleans and New York – over a period of many disruptions. The quarter-century since 1990 has seen waves of offshoring, financial deregulation, and technological change sweep across the American landscape. In addition, New Orleans and New York have seen the inundations of literal waves – flooding from massive storms – wash through their streets. Hurricane Katrina struck New Orleans with monstrous power in 2005. In 2012, another fearsome storm, Sandy (so powerful, it was named a "superstorm"), bore down on the New York metro area. Damage from Katrina was estimated at $148 billion across the range of its path and caused 1,833 deaths. Sandy wreaked havoc estimated at $71 billion in a region with more sophisticated infrastructure and a better-executed emergency plan, limiting deaths to 132 after its landfall just south of New York City, on the New Jersey shoreline.

However, the comparative tale shown in Figure 6.7 is most notable for the disproportionately steep economic contraction experienced by New Orleans, a sudden shrinkage of its employment base from which it has still not recovered. New Orleans' population was displaced, with entire neighborhoods (wards, in the local terminology) rendered uninhabitable and almost completely depopulated. Most of those relocated in New Orleans' emergency evacuation never returned.

New York City, by contrast, was undeniably subject to all the economic stresses of industry change, globalization, technology, and financial turmoil that the 1990–2014 period had to offer. The nation's most important banking center, New York was vulnerable to the dislocation of the loss of the thrift industry and the emergent competition of regional banking as the financial industry consolidated in the early 1990s. As the key global city in the country, the roiling circumstances of international finance – sovereign bond defaults, currency fluctuations, widening current account deficits in America's trade balance – strained New York throughout the 1990s and, not coincidently, led to the

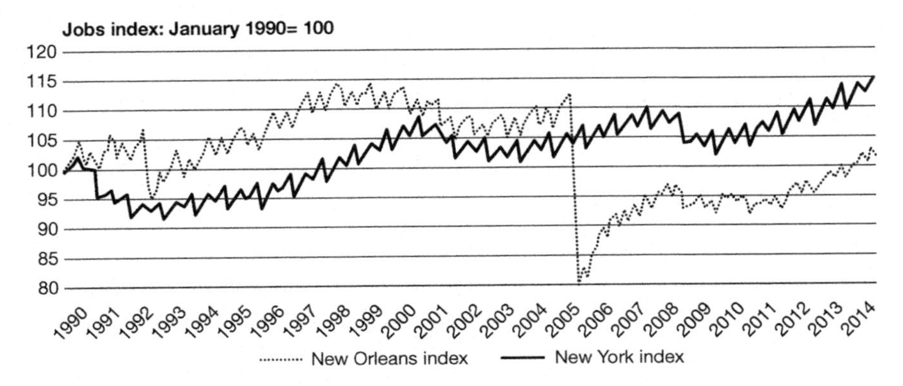

Figure 6.7 Exposure to discontinuous change: Economic depth not proof against recessions, but extremely important in event of disruption

Source: Data from US Bureau of Labor Statistics; indexation by author

spectacular collapse of Wall Street's largest hedge fund, Long-Term Capital Management.[63] And then, of course, there was the post-Lehman Brothers financial meltdown.

However, at no time during this period did New York ever experience the catastrophic loss of population and jobs triggered by Katrina in New Orleans. New York, at the end of 2014, had grown its jobs base a full 15 percent from 1990, whereas New Orleans had just about drawn even with its 1990 job total and was still substantially below the employment level enjoyed pre-Katrina, nearly a decade before.

Some might argue that New York had already seen its disruptive contraction in the 1970s. In absolute numbers, New York's experience in that decade was indeed wrenching. But it transpired over the course of a decade and, compared with its base, was less than the sudden loss of 27 percent of all employment in New Orleans over just 6 months. The 2012 US Census Bureau estimate for the City of New Orleans' population placed it 24 percent below the pre-Katrina tally of 484,674 in the year 2000. Between 2000 and 2004, New Orleans had already dipped 4.7 percent in population and had lost 6.2 percent from its 2000 jobs base. Katrina, in other words, struck an urban economy that was already in decline.[64] Sandy, on the other hand, hit New York at a time when it had already made great progress in reinventing itself – not only in the structure of its economic base, but in its infrastructure and in its governance.

What might account for New Orleans' greater fragility, or, conversely, New York's greater resilience? New Orleans' economy rested on a three-legged stool, in the words of the Bureau of Labor Statistics – tourism, port activities, and educational services. This seems to have been an inadequate base to withstand the impact of the storm and to rebound fully in its aftermath. New York, on the other hand, not only had reinvented itself, but had done so by diversifying its economy, advancing to a point where its creative-class base was already one-third of total employment, and had important experience in coping with a whole panoply of earlier challenges that provided both confidence and know-how in dealing with a major new emergency.

The lesson about "black swans" from this comparison is fairly simple. If we cannot truly predict highly improbable but impactful events, we *can* prepare for them.

Creative destruction

"Disruptive technology" has seemingly become the most widespread term of art for the second decade of the 21st century. A 2013 report from the McKinsey Global Institute identifies twelve "potentially economically disruptive technologies" that have an estimated range of annual economic impact of $14–33 trillion by the year 2025.[65] The key to the concept is that such technologies transform and ultimately supplant existing processes and products, become embedded in both physical and human capital, and, once adopted, act as a one-way valve that inhibits reversion to the pre-existing status quo.

McKinsey quite frankly acknowledges the inspiration of Joseph Schumpeter's work on the evolution of capitalism and his immortal phrase "creative destruction." Under the discipline of the profit motive, a capitalist economy relentlessly innovates in order to gain the rewards of higher productivity. But Schumpeter is clear that this comes at a cost to those older technologies, and the jobs that depend upon them, as they pass into obsolescence. As two economists at the Dallas Federal Reserve Bank once put it, "capitalism's pain and gain are inextricably linked. The process of creating new industries does not go forward without sweeping away the preexisting order."[66]

This much is now economic orthodoxy, and examples of the process are so numerous that they now sound trivial. Railroads replaced horses and canals; the airplane replaced railroads and ocean liners; the Internet is supplanting everything, from the US postal service to physical libraries to shopping malls. And so it goes.

For at least two generations, it appeared that cities themselves were subject to creative destruction – under the forces of deindustrialization, globalization, and technology discussed earlier in Chapter 6. Frank Lloyd Wright's quip that the future will be a race between the elevator and the automobile, and only a fool would bet on the elevator, sustained the weight of evidence through most of the late 20th century. The death of distance was announced in the 1990s, and it was observed that the freeway-networked, outwardly expanding, multinodal city, of which Los Angeles was the avatar, had definitively triumphed over the old-style, dense-core, mass transit-enabled urban form epitomized by New York, Chicago, Boston, and San Francisco.

But that seemingly irreversible pattern, the triumph of sprawl, apparently has hit a limit, and the entropic impacts of deindustrialization, globalization, and technology have found exceptions in some big cities. Those cities are prospering despite the "old bones" of subway systems, walk-to-work downtowns, and urban centers where shopping, recreation, civic activities, culture and religious structures share compact, multiuse districts.

We are reminded forcefully by the so-called "back to the city" movement that this is the urban form that has stood the test of time, not just at the turn of this millennium, but now for the 3-millennium span of urban history. Los Angeles, Dallas, Atlanta, Phoenix, and cities like them are the novel experiment. And, recently, Frank Lloyd Wright's bet seems poorly placed.

Sometimes, new is indeed "new and improved." But, often, it means "new and unproved."

Surprises are ever coming over the horizon. The most successful investor of our time, Warren Buffet, went out in 2009 and bought the Burlington Northern Santa Fe railroad – a 19th-century technology according to the Schumpeterian perspective. Over the following 5 years, the railroad's revenues rose 57 percent, earnings doubled, and the investment returned the entire $15.9 billion in cash and stock Buffet paid to acquire his stake.[67]

So it may be with cities. To borrow Mark Twain's wry remark, even with disruptive technology abounding, a proven form such as the compact, multiuse urban center – now dubbed the live–work–play city – may find reports of its death greatly exaggerated.

Novus ordo seclorum

If disruptive technology is the buzzword of today, it has an honored precursor in the phrase "paradigm shift." It is likely that many who rolled that phrase off the tongue until it became a cliché were unaware that it came from a slim volume that was probably the most influential contribution to the philosophy of science in the entire second half of the 20th century, Thomas Kuhn's *The Structure of Scientific Revolutions*.[68] Kuhn held that science normally proceeds incrementally, in small steps, until it reaches a point where the existing conceptual framework no longer provides accurate, parsimonious explanations for what is observed and, even more critically, fails to produce hypotheses that anticipate events that can be confirmed experimentally. Rapid progress and recurrent breakthroughs are part of the process of normal science. Scientific revolutions are much talked about, but the kind of shift that occurred when Copernicus' account of the universe was accepted, setting aside the previously dominant theory of Ptolemy, doesn't happen every decade, much less every year. The revolution in physics of the brilliant generation of Einstein, Planck, Heisenberg, and their colleagues challenged the mechanical physics of Newton – and Newton's greatest work occurred in the 1660s.

The burden of a claim for disruption, for truly discontinuous change, should have a high bar of proof. "This changes everything" is a very profound assertion, as is the claim that a set of technologies will "transform life, business, and the global economy" – matters that are at the very heart of what cities are and what cities do. I believe such claims should be met skeptically, critically, and with an appreciation that cities are very adaptable organisms, able to absorb and integrate technology, rather than merely react to it.

On the US $1 bill is an image of the Great Seal of the United States. The motto on that seal says, *Novus ordo seclorum* – "a new order of the ages." The Founding Fathers of America meant for the new nation to be discontinuous with the past, even as they looked to pre-monarchial arrangements in ancient Greece and Roman for inspiration in establishing a novel form of constitutional government. Like scientific revolutions, true political revolutions – the establishment of something radically new – are very rare. What passes for discontinuous change is most often merely the shift of the same conception of power from one party to another.[69]

For cities, it may very well be that fundamentally discontinuous change, especially over a sharply constricted period of time, is a sign of unhealthy weakness, rather than sudden progress. Cities may best be understood along with biological models, as adaptive organisms in the process of evolution.

Linnaeus passed on to us the Latin phrase, *Natura non facit saltus*, "Nature does not make leaps." Alfred Marshall used the phrase as an epigraph to his *Principles of Economics*. Cities function as the repository of culture and civilization, as much as they are sites of commerce. Technology and other forces that confront the city with potential disruption may find the successful city incorporating the innovative into its existing social framework, turning it into living history. That seems to be the dynamic at the heart of the US urban revival of the past quarter-century.

Notes

1 Martin Tolchin, "Obituary of Thomas P. O'Neill," *New York Times*, January 7, 1994.
2 Thomas L. Friedman. Published by Farrar, Straus & Giroux (New York, 2005). Friedman expanded upon his themes in his 2008 book, *Hot, Flat, and Crowded*, from the same publisher.
3 Joseph E. Stiglitz, *Globalization and its Discontents*, W.W. Norton (New York, 2002).
4 See Roger Lowenstein, *When Genius Failed: The Rise and Fall of Long-Term Capital Management*, Random House (New York, 2001).
5 The statement was made during a 1992 televised presidential debate. The moment is captured on YouTube, online at www.youtube.com/watch?v=jZYSL26zGTo
6 Beth Macy, *Factory Man: How One Furniture Maker Battled Offshoring, Stayed Local – and Helped Save an American Town*, Little, Brown (Boston, 2014).
7 Joseph E. Stiglitz, *Making Globalization Work*, W.W. Norton (New York, 2007), p. xii.
8 The hypothesis of absolute economic convergence among nations, which would be expected by economists and policy makers anticipating a leveling of outcomes as markets adjust disequilibria, has thus far not been supported by the preponderance of evidence. See Matthew Cole and Eric Neumayer, "The Pitfalls of Convergence Analysis: Is the Income Gap Really Widening?" *Applied Economics Letters 10: 6* (2003), pp. 355–357, and Panagiotis Artelaris, Paschalis Arvanitidis, and George Petrakos, "Convergence Patterns in the World Economy: Exploring the Non-Linearity Hypothesis," in No DYNREG32, Papers (2008), Economic and Social Research Institute (ESRI).
9 The source is the US Central Intelligence Agency website, online at www.cia.gov/library/publications/the-world-factbook/rankorder/2172rank
10 This is the reason that Kurt Vonnegut's incredibly cinematic passage in the novel *Slaughterhouse 5* is so powerful. Vonnegut imagines a literal reversal of the firebombing of Dresden at the end of World War II. In his narrative, the fires are gathered up into steel containers that then fly upward from the ground until they enter the bellies of airplanes, which then fly backwards to England. Not only is the law of gravity reversed, but also the law of entropy, insofar as it holds that nature proceeds from order to disorder.
11 See Peter Coveney and Roger Highfield, *The Arrow of Time: A Voyage through Science to Solve Time's Greatest Mystery*, Fawcett Columbine (New York, 1990), especially pp. 147–181.
12 Richard Florida, "The World is Spiky," *The Atlantic Monthly* (October 2005), pp. 48–51.
13 One excellent example of a growing literature on this subject is Michael Batty, *Cities and Complexity: Understanding Cities with Cellular Automata, Agent-Based Models, and Fractals*, MIT Press (Cambridge, MA, 2005).
14 Source is a Landauer Associates consulting study, led by the author in 1991, for a non-US client; it examined in detail 128 metropolitan areas in the US, Canada, and Mexico. The study predates the coinage of the term "24-hour city" by *Emerging Trends in Real Estate* but considered a set of variables wholly consonant with my subsequent research on this topic.

15 Beginning with von Thunen in the 19th century, and extended by Losch and Christaller in the 20th.

16 Two venerable studies by Brian Berry and William Garrison should be mentioned. In "The Functional Basis of the Central Place Hierarchy" (*Economic Geography 34:2* [1958]), they examined the distribution of economic activity in Snohomish County, WA, which is in the Seattle–Everett metropolitan area. Then, in the *Annals of the Association of Amercian Geographers* (*48:1* [1958]), these authors published "Alternative Explanations of Urban Rank-Size Relationships." The latter study, interestingly, explicitly considers the factor of entropy and finds it relevant to the topic of urban systems. The authors reject as implausible and theoretically unsatisfying explanations of urban economic differences based upon differences in local occupational structures, agglomeration, or industry distribution – in other words, aspects of heterogeneity observable among cities. They are in search of a formal solution that would subsume differences into a single "model." The premise of the current book is that, while acknowledging the real contributions of scholars such as Berry and Garrison, one size does not fit all in the realm of urban studies.

17 Saskia Sassen, *The Global City: New York, London, Tokyo*, Princeton University Press (Princeton, NJ, 1991).

18 David Clark, "Interdependent Urbanization in an Urban World: An Historical Overview," *The Geographical Journal 164:1* (March 1998), pp. 85–95.

19 Jane Jacobs, *Cities and the Wealth of Nations*, Random House (New York, 1984).

20 Kenichi Ohmae, *The End of the Nation State: The Rise of Regional Economies*, Free Press Paperback (New York, 1996).

21 Kenichi Ohmai, *The Borderless World: Power and Strategy in the Interlinked Economy*, HarperCollins (New York, 1991).

22 Kenichi Ohmae, *The Borderless World.* Harper Business (New York, 1999), p. 136.

23 The rush of foreign investment into emerging nations, for instance, has provided individuals and corporations with some significantly painful lessons over the past 25 years. Transparency and rule of law are now considered a basic go/no-go gate for investment decisions. In many places, investors found it very difficult to repatriate money or to enjoy even rudimentary liquidity when it came time to execute an exit strategy. Some have termed this the "Roach Motel" conundrum, after a US television commercial for a pest-control product where "roaches check in, but they don't check out."

24 See Robert Fishman, *Urban Utopias in the 20th Century: Ebenezer Howard, Frank Lloyd Wright, LeCorbusier*, Basic Books (New York, 1977), and *Bourgeois Utopias: The Rise and Fall of Suburbia*, Basic Books (New York, 1987).

25 Joel Kotkin, *The New Geography: How the Digital Revolution is Reshaping the American Landscape*, Random House (New York, 2001).

26 Robert D. Atkinson, "Technological Change and Cities," *Cityscape: A Journal of Policy Development and Research 3:3* (1998), pp. 129–170, US Department of Housing and Urban Development.

27 Dennis A. Rondinelli, James H. Johnson, Jr., and John D. Kasarda, "The Changing Forces of Urban Economic Development: Globalization and City Competitiveness in the 21st Century," *Cityscape: A Journal of Policy Development and Research 3:3* (1998), pp. 71–105, US Department of Housing and Urban Development.

28 Jane Pollard and Michael Storper, "A Tale of Twelve Cities: Metropolitan Employment Change in Dynamic Industries in the 1980s," *Economic Geography 72:1* (1996), pp. 1–22.

29 See Stephen S. Cohen and John Zysman, *Manufacturing Matters: the Myth of the Post-industrial Economy*, Basic Books (New York, 1987).

30 Peter Dreier, John Mollenkopf, and Todd Swanstrom, *Place Matters: Metropolitics for the 21st Century*, University of Kansas Press (Lawrence, KS, 2001).

31 Ronald L. Mitchelson and James O. Wheeler, "The Flow of Information in a Global Economy: The Role of the American Urban System in 1990," *Annals of the Association of American Geographers 84:1* (1994), pp. 87–107.
32 Kevin Kelly, one of the founders of *Wired* magazine and its former executive editor, argues that such a concentrating effect (which he calls negentropy or exotropy) is a persistent characteristic of successful technologies. See his *What Technology Wants*, Viking (New York, 2010), pp. 62–63, where he writes, "Everything we find interesting and good in the cosmos – living organisms, civilization, communities, intelligence, evolution itself – somehow maintains a persistent difference in the face of entropy's empty indifference."
33 Edward J. Malecki, "The Economic Geography of the Internet's Infrastructure," *Economic Geography 78:4* (2002), pp. 399–424.
34 Frances Cairncross, *The Death of Distance: How the Communications Revolution Will Change our Lives*. Orion (London, 1997).
35 Mitchell L. Moss and Hugh O'Neill, "Reinventing New York," Working Paper of the Taub Urban Research Center, New York University, 1991.
36 Manuel Castells, *The Internet Galaxy: Reflections on the Internet, Business and Society*, Oxford University Press (Oxford, UK, 2001).
37 The phrase is taken from Joseph Rykwert, *The Seduction of Place: Cities in the 21st Century*, Vintage (New York, 2002).
38 Here's the math: we now have a world population of just over 7 billion, of which 50 percent or so live in cities (so, 3.5 billion people). By 2050, the world population is projected to be 9 billion, with 70 percent urbanization (or 6.3 billion city dwellers).
39 Data are from the US Bureau of Transportation Statistics, accessed online at www.rita.dot.gov/bts/sites/rita.dot.gov.bts/files/publications/national_transportation_statistics/html/table_01_11.html
40 Edward Glaeser, *The Triumph of the City: How Our Greatest Invention Makes Us Richer, Smarter, Greener, Healthier, and Happier*, Penguin Press (New York, 2011). See especially pp. 206–210 for this discussion.
41 Summarized in a policy brief issued by the John F. Kennedy School of Government at Harvard University. Edward L. Glaeser and Matthew Kahn, "The Greenness of Cities," Harvard University (Cambridge, 2008). Data in Figure 6.4 are from this policy brief. A more formal presentation of the research was subsequently published in the *Journal of Urban Economics 67* (2010), pp. 404–418.
42 A valuable tool for measuring accessibility at the city, neighborhood, and site level; see online at www.walkscore.com/
43 Andres Duany, Elizabeth Plater-Zyberk, and Jeff Speck, *Suburban Nation: The Rise of Sprawl and the Decline of the American Dream*, North Point Press (New York, 2000). It might be noted how closely these symptoms correspond to Joel Garreau's *Edge City* descriptors: 5 million square feet or more office space, at least 600,000 square feet of retail leasable area, more jobs than bedrooms, perceived as a single location, brand new since 1961, in a suburban or exurban location.
44 Anthony Downs, "How American Cities Are Growing: The Big Picture," *Brookings Review* (Fall 1998), The Brookings Institution, Washington, DC.
45 "Charting our Water Future, Economic Frameworks to Inform Decision-making," Water Resources Institute (2009).
46 US Geologic Survey, "Groundwater Depletion in the United States 1900–2008."
47 I chaired the Counselors' organization in 2014.
48 M. Chan, "Water Resources and Population Growth, 2000–2020," US Deparment of Energy (July 2002).
49 Accessible online at http://water.epa.gov/action/importanceofwater/upload/IOW_Synthesis_ Highlights.pdf

50 Extracted from the Counselors of Real Estate External Affairs Committee Alert of January 2015, online at www.cre.org/external_affairs/#EAC_January_2015
51 Benjamin Barber, of the City University of New York, says in his book, *If Mayors Ruled the World: Dysfunctional Nations, Rising Cities* (Yale University Press, New Haven, CT, 2013): "Cities can do more than just lobby and advocate; they can directly affect carbon use within their domains through reforms in transportation, housing, parks, port facilities, and vehicles entirely under their control."
52 Published and sponsored by Siemans (Munich, 2011) and accessible online at www.usa.siemens.com/entry/en/greencityindex.htm
53 Hannah Arendt, *On Violence*, Harcourt Brace (New York, 1969).
54 *Fooled by Randomness: The Hidden Role of Chance in Life and Markets* (2001); *The Black Swan: The Impact of the Highly Improbable* (2007); *The Bed of Procrustes: Philosophical and Practical Aphorisms* (2010); *Antifragile: Things that Gain from Disorder* (2012). All published by Random House (New York).
55 Daniel Kahneman, *Thinking, Fast and Slow*, Farrar, Straus & Giroux (New York, 2011). See also John Brockman, *Thinking: The New Science of Decision-Making, Problem-Solving, and Prediction*, Harper Perennial (New York, 2013).
56 Rough dates that, like many attempts to mark a historical era, are just an approximation. I choose 2007 to begin, as that is the year in which Bear Stearns hedge funds began to unravel, as toxic home mortgages began to poison Wall Street and Main Street, leading the National Bureau of Economic Research to declare the onset of recession by the end of the year. I see 2010 as the first year of the comeback, as the economic stimulus package passed in 2009 began to take hold, and the economy began to eke out employment gains.
57 Alan Greenspan, "Never Saw It Coming," *Foreign Affairs* (November/December 2013). As I confessed myself in "The Morphology of the Credit Crisis," *Real Estate Issues 34:3* (2010), I was also surprised by the contagion that spread the embedded risk in subprime mortgages – a small sliver of the US housing market – across the financial system and indeed the global economy. But I draw starkly different lessons from the experience than does Greenspan.
58 See my essay "Judgment: Imagination, Creativity, and Delusion," in the philosophical journal *Existenz 3:1*.
59 See Justin Fox, *The Myth of the Rational Market*, Harper Business (New York, 2009), for a lucid and comprehensive history of the success of the efficient market hypothesis in becoming economic orthodoxy, with a critical examination of the limitations of this conceptual framework. And, for a thought-provoking discussion of cases where market-determined pricing must be questioned as a valuing mechanism, see Michael J. Sandel, *What Money Can't Buy*, Farrar, Straus & Giroux (New York, 2012).
60 Hugh F. Kelly, "The New York Regional and Downtown Office Market: History and Prospects after 9/11," Report prepared for the Civic Alliance to Rebuild Downtown New York, August 9, 2002.
61 See *Downtown 2020*, a research report by a multidisciplinary team assembled by the Steven L. Newman Real Estate Institute at Baruch College, the City University of New York (2008), and my follow-up report, "Going Long on New York" (2009). Both are accessible in PDF form online at www.baruch.cuny.edu/realestate/research-publishing/archives.html
62 Here is what Hannah Arendt comments on the subject:

> [The Declaration of the Rights of Man] rests upon man's natural rights . . . prepolitical rights that no government and no political power had the right to touch and to violate . . . the very content as well as the end of government and power. The *ancien régime* stood accused of having deprived its subjects of these rights – the rights of life and nature rather than the rights of freedom and citizenship.
>
> (Arendt, *On Revolution*, Viking Compass, New York, 1963)

63 See Roger Lowenstein's illuminating account, *When Genius Failed: The Rise and Fall of Long-Term Capital Management*, Random House (New York, 2000).
64 Michael L. Dolfman, Solidelle Fortier Wasser, and Bruce Bergman, "The Effects of Hurricane Katrina on the New Orleans Economy," US Bureau of Labor Statistics *Monthly Labor Review* (June 2007).
65 McKinsey & Co., *Disruptive Technologies: Advances that Will Transform Life, Business and the Global Economy*, McKinsey Global Institute (May 2013). The technologies are the mobile Internet, the automation of knowledge work, the Internet of things, cloud technology, advanced robotics, autonomous or near-autonomous vehicles, next-generation genomics, energy storage, 3D printing, advanced materials, advanced oil and gas exploration and recovery, and renewable energy.
66 "Creative Destruction," *The Concise Encyclopedia of Economics*, online at www.econlib.org/library/Enc/CreativeDestruction.html; article authors are W. Michael Cox and Richard Alm of the Dallas Fed economics staff.
67 Bloomberg News, accessed at www.bloomberg.com/news/articles/2014-11-10/buffetts-15-billion-from-bnsf-show-railroad-came-cheap
68 Published by the University of Chicago Press (Chicago, 1962).
69 See Arendt, *On Revolution*, p. 277.

7 What makes the 24-hour city different?

Supreme Court Justice Potter Stewart penned a remarkable concurrence opinion joining the majority decision in the High Court's obscenity case *Jacobellis* v. *Ohio (1964)*. He memorably wrote, "I can't define hard-core pornography, but I know it when I see it."[1] For quite a few years after the term "24-hour city" was coined in the 1995 edition of *Emerging Trends in Real Estate*, that level of descriptive definition was the standard for discussion of urban centers and their commercial property markets.

During those years, five cities were repeatedly alluded to as 24-hour places. New York was paramount among them, but Boston, Chicago, San Francisco, and Washington, DC, were frequently discussed in *Emerging Trends* under the 24-hour rubric. Other cities came into the conversation as their nighttime activities became increasingly prominent. Miami, for instance, saw the revitalization of its South Beach Art Deco district capture international attention, and its reputation as "the capital of Latin America" gained increasing prominence. Similarly, the neon glitz of Las Vegas that lit up the desert with the glow of the Strip, lined with casinos that as a matter of business practice kept clocks from the view of customers, epitomized a venue where "day for night" represented not only reality but a cultivated image for a worldwide audience.

Over time, especially as the 24-hour city discussion advanced for the real estate industry, five cities often were presented as contrasting 9-to-5 locations. These were Atlanta, Dallas, Los Angeles, Phoenix, and Minneapolis. Although these were "the usual suspects," other areas such as Philadelphia and Seattle were cited as metro areas where a similar dynamic was at work. Joel Garreau, in *Edge Cities*, presented Los Angeles as the paradigm of the spread-out, multinodal city where dispersal of population, economic activity, and physical development kept downtown as a place where workers commuted in to offices in the morning and crawled home in traffic jams in the evening rush hour. But Garreau chose King of Prussia, Pennsylvania – a suburb of Philadelphia – as his first vivid example of the kind of multifunction suburban node, giving rise to the title of his 1991 book.[2]

In this and the following chapters, I shall be using these two sets of cities as groups to discuss the typology of 24-hour and 9-to-5 places. Although the lists are drawn from the industry literature, I'd like to make a few general points before going into depth.

First of all, because all that was available when I began this research was a descriptive definition and comments from a broad but unscientific survey, the groupings could not be considered more than a clustering of cities with hypothesized common features. The point was to test the strength of those similarities.

Next, the attributes of 24-hour cities that *Emerging Trends* first set out in the mid-1990s as the "I know it when I see it" features had very little to do explicitly with 24-hour-ness, if that meant the measurement of diurnal activity. The attributes had more to do with the assumed conditions that would allow places to flourish deep into the evening hours than with the actual identification of such activity. That left a research challenge to tackle: to identify sources of information that did track what happened in urban places over the course of a day. Such time-specific information could advance us beyond a descriptive definition (what a place appeared to be, as a phenomenological matter) to an operational definition (what happens in a place, in some measurable way, as a function of temporal activity).

Third, both 24-hour city and 9-to-5 city needed to be understood as terms of convenience, rather than as hard-and-fast categories. On the one hand, in each grouping, it was expected from the outset that the cities would array along a spectrum: more or less intensely "24-hour" and more or less distinctively "9-to-5." One of the objectives of the research was to see if a bright line could be found between the two clusters of cities, labeled *ex hypothesi* as similar. They were to be seen as polarities in the way America's cities had evolved, but as living organisms as well, not specimens to be pinned to a conceptual corkboard. And, even if a bright line might be discovered in the measurements, it would not necessarily exclude the emergence of some intermediate, transitional state whereby a city could evolve from one category to another.

Lastly, as living organisms, all these cities (and others as well) need to be viewed as changing. That is to say, they are to be considered as subject to all the forms of change discussed earlier in this book: subject cycles, trends, maturation, state change, and disruption – as well as the complicated ways in which these five primary types of change combine over time.

The descriptive definition

From its coinage in the 1995 edition of *Emerging Trends in Real Estate*, a 24-hour city was considered to have several recognizable attributes: "attractive residential neighborhoods proximate to or integrated with the central commercial district; convenient shopping opportunities close to the workplace; a safe and secure environment; excellent mass transportation; and, recreational, cultural, and entertainment amenities."[3]

There are certainly advantages to such an easily understood description. It provides us with some signposts that can be generally recognized as we consider whether a city is a likely or unlikely candidate for inclusion as a 24-hour marketplace. Each of the elements, although qualitative, has the capacity to be tested with statistical measures that are (more or less) readily available. And, through repeated use, *Emerging Trends* contributed a common vocabulary to this theme, a very useful checklist that helped in the early adoption of the 24-hour/9-to-5 distinction[4] within the real estate community, among planners and architects, and urbanists who were grappling with esthetic, functional, and environmental issues in America's cities.

As always, though, a convenient checklist has its difficulties. Attributes can come to be considered a kind of shopping list for cities. "If you want to be a 24-hour city (which seems to be a good thing, at least to the *Emerging Trends'* experts, here is what you need to have." Perhaps it is helpful to keep this principle in mind: a list of ingredients is not the same as a recipe. Going to a market for gourmet foods does not guarantee a memorable banquet. In the hands of a master chef, those ingredients interact with each other in very special, sometimes spectacular ways. A great meal must have balance, contrast, complexity, and presentation to succeed. It often has the capacity to surprise and delight as well. Dimensions such as these have largely been absent from the discussion of 24-hour cities, and we will need to consider them as well as the conventional attributes as we proceed.

Housing

Separation of uses had become so canonical in the field of planning and zoning (and underscored by the definition of the CBD) that, even late in the 20th century, an assessment of downtown redevelopment strategies, in the *Journal of the American Planning Association*, did not list housing as one of the seven key approaches to be evaluated. The pros and cons of pedestrianization, indoor shopping centers, historical preservation, waterfront development, office development, special activity generators including convention centers and sports facilities, and transportation enhancement were considered, individually and in combination. But housing was relegated to afterthought status, among "other redevelopment strategies" that warranted only a single paragraph collectively.[5]

Such short shrift for housing is not just a matter of academic blindness or dogmatism among professional planners and government administrators. The emergence of the business improvement district (BID, also known as "public improvement district," "special improvement district," "economic improvement district," among other labels) has been a powerful contribution of the private sector to an effort to revitalize downtowns. There are over 400 BIDs in 43 states and the District of Columbia, and in most of the largest cities, including Chicago, New York, Los Angeles, Philadelphia, Seattle, and Washington, DC. According to a study published in the *Economic Development Quarterly*, BIDs are active in promoting capital improvements, marketing the downtown, securing economic development incentives, improving maintenance and cleanliness in the public areas, enhancing parking and transportation, providing greater security, and occasionally assisting in social issues. Housing is conspicuously absent from the BID agenda in most localities.[6]

Friedan and Sagalyn write extensively of CBD revitalization strategies in *Downtown, Inc.*, but include not a single chapter on housing or downtown residential needs or influences. Indeed, the terms "housing" and "residential" are not even included in the book's index.[7] Birch notes that planners have elected to concentrate on expanding office districts, retail areas, transportation access (especially highway connections and ample parking), and, eventually, special activity generators.

During the 1990s, downtown housing received greater attention. Notably, Birch's research included direct queries addressed to decision-makers in the public and private sectors, and this uncovered areas of discussion that had not yet penetrated the formal literature. As early as 2002, Birch was writing about the 24-hour city phenomenon: "Urban leaders were realizing that cities would remain economically and demographically unbalanced and have lifeless, ghost-town centers if nothing were done to substitute an 18- or 24-hour downtown for the current 8-hour workday." This was one of the early instances of the adoption of the real estate industry vocabulary into wider use in serious social and economic studies.

Cities ranging in size from the typically large locations studied in this book to much smaller urban areas such as Lexington, Kentucky, Chattanooga, Tennessee, and Des Moines, Iowa, were examined by Birch. She found that, though many acknowledged that downtown housing in support of a 24-hour urban core was seen as a critical effort, only 38 percent of the 45 cities she examined had higher downtown residential populations in 2000 compared with 1970. She termed housing a symbiotic influence in urban revitalization, not a "silver bullet," and confessed "no one really knows the proper composition of a balanced downtown."

In 2005, Eugenie Birch of the University of Pennsylvania proposed a "downtown typology" that distinguished five clusters of city centers: fully developed downtowns, emerging downtowns, downtowns on the edge of takeoff, slow-growing downtowns, and declining downtowns.[8] These are presented in Figure 7.1, labelled "Birch's downtown typologies," with the 24-hour cities hypothesized for this thesis marked with an asterisk (*) and the 9-to-5 cities indicated by a crosshatch (#). For New York, Birch's analysis considers Midtown and Lower Manhattan separately, and it separates Mesa from Phoenix, though both were identified within the same MSA until the most recent disaggregation by the Federal Office of Management and Budget (OMB).

The five fully developed downtowns are home to nearly half the nation's downtown households, and all five sustained positive household growth in each of the three decades from 1970 to 2000. The emerging downtowns account for another quarter of all downtown households and were more volatile in their residential growth, typically showing significant population increase only in the final decade of the 20th century. Downtowns on the edge of takeoff had drastic declines in population between 1970 and 1990, but robust growth in the 1990s

Fully developed downtowns	Emerging downtowns	Downtowns on the edge of takeoff	Slow-growing downtowns	Declining downtowns
Boston*	Atlanta#	Chattanooga	Albuquerque	Cincinnati
Chicago*	Baltimore	Dallas#	Austin	Columbus, GA
Lower Manhattan*	Charlotte	Miami*	Boise	Des Moines
Midtown Manhattan*	Cleveland	Milwaukee	Colorado Springs	Detroit
Philadelphia#	Denver	Washington, DC*	Columbus, OH	Jackson
	Los Angeles#		Indianapolis	Lexington
	Memphis		Lafayette	Mesa#
	New Orleans		Phoenix#	Minneapolis#
	Norfolk		Pittsburgh	Orlando
	Portland		Salt Lake City	San Antonio
	San Diego			Shreveport
	San Francisco*			St. Louis
	Seattle#			

Figure 7.1 Birch's downtown typologies

that outpaced the citywide growth in their respective urban areas. About one-half of Birch's sample is counted in the slow-growing or declining groupings, including Phoenix (and Mesa) and Minneapolis from the 9-to-5 study set.

Density appears to play a distinct role for the established downtowns. The fully developed downtowns average 23 households per acre, whereas none of the other groupings achieves a density higher than 5 households per acre. Of the emerging downtowns, only Washington, DC, has a residential density of 10 households or more per acre.

Urban revitalization data from the 1990s, at the level of the city (as defined by political jurisdiction), shows that, with the exception of New York and Chicago, there was no reversal of the suburbanization trend from prior decades. Sprawl continued to dominate metropolitan area growth, favoring places that are warm, lower-density, and automobile-dominant. Higher-density cities that did grow were distinguished by strong concentrations of human capital. Not only New York and Chicago, but also Boston, Philadelphia, Washington, DC, and San Francisco can be included in such a list of high-density cities. This research underscores the theme that urban revitalization is a phenomenon of a select set of cities, rather than a case of a rising tide lifting all.[9]

Coming at the question of housing from another angle, a later influential study argued that some cities and towns themselves have become scarce luxury goods, and the house prices are a good measure of this phenomenon. The most desirable, or "superstar," cities are characterized by inelastic land supply (because of either physical or political constraints), coupled with the growth of high-income populations. In such places, there is an exceptionally high home-price-to-rent ratio. The national increase in the number of high-income households creates a kind of crowding-out phenomenon in such cities. This is a model of competitive gentrification and may have international aspects as well. International high-income households become acquirers of housing units in the superstar cities, in this view, and anecdotal evidence in cities such as New York, Washington, Miami, and San Francisco appears to lend credence to the story.[10]

The age of a city's housing stock is not fixed, of course, and, when obsolescence or even abandonment removes older stock from the downtown inventory, to be replaced by renovated or new housing units, the balance between suburb and central city may shift. As gentrification is often observed taking place in older neighborhoods, older cities may have greater potential for attracting more affluent households seeking the architectural features, classic materials, and eclectic configurations in cities whose growth dates back into the 19th century.[11]

For the largest cities, downtowns have become the locus of fairly dramatic income ranges, as described by Birch, rather than the monotonic home of the wealthy posited by Gyouko and his colleagues. Studies of income inequality consistently depict 24-hour cities such as New York, San Francisco, Washington, Boston, and Miami as having particularly high Gini coefficients. Recently, Harvard's Edward Glaeser has presented the case that income disparity should not be viewed as a dystopic urban attribute, if lower-income households can be found to be upwardly mobile economically across generations.[12] Intriguingly, the joint Harvard–Berkeley Equality of Opportunity project points to significantly higher income mobility in the 24-hour cities mentioned above than in 9-to-5 cities such as Atlanta, Dallas, and Phoenix. Studies of poverty and race have found that diversity of income levels in urban areas facilitates upward income mobility in those born to poorer households.[13]

Housing, then, appears to be correctly identified as an important, but not exclusive, factor in the equation for strong downtowns. In terms of classical logic, it would be termed a "necessary but not sufficient condition" for a 24-hour city. The planning literature, particularly until very recently, underestimated its relevance. Commercially oriented downtown advocates such as the BIDs likewise stressed other strategies at the expense of housing. Urban economists, too, were long persuaded by their standard model that housing choice was a key factor in the hollowing out of US cities. The *Emerging Trends* respondents, unburdened by allegiance to such positions, were early in noting that market forces were calling theory and presuppositions into question and drew attention to the important change that was occurring in a few cities: the synergistic recovery of residential and commercial markets in a handful of the nation's largest metropolitan core districts.

Qualitative research has been a key component in the effort to establish metrics for 24-hour cities. A survey of 242 senior real estate executives, including those in asset and property management, investment, real estate transaction services and brokerage, development, and consulting, yielded results that add depth to documentary and statistical research findings. Survey respondents were asked to state the degree to which various characteristics of cities are contributions to commercial real estate investment success. One question addressed the topic of "attractive residential opportunities within or immediately adjacent to downtown," as a characteristic of 24-hour cities. The "multiple-choice" options provided the survey respondents ranged from "essential for success" through "not an important factor." Figure 7.2 displays the range of responses.

Once the top two options, "essential" and "very important," were considered, the top combined rank was "housing for middle managers," with 170 total votes (71 percent), comprised of 53 "essential" responses plus 117 "very important" votes. Next was housing for professional/technical staff (163 combined votes, or 69 percent), comprised of 58 "essential" and 105 "very important" responses. "Housing for top-level executives" closely followed in the combined tally (160 votes, or 67 percent), with 59 "essential" and 101 "very important" votes. "Housing for services and support staff" garnered 123 combined votes (41 percent), with 43 "essential" and 80 "very important" responses.

So the housing mix might be viewed as being tilted toward office workers, particularly relatively high-income managers and professional/technical staff, rather than affordable housing for lower-paid workers typical of the retail, restaurant, or hotel industries.

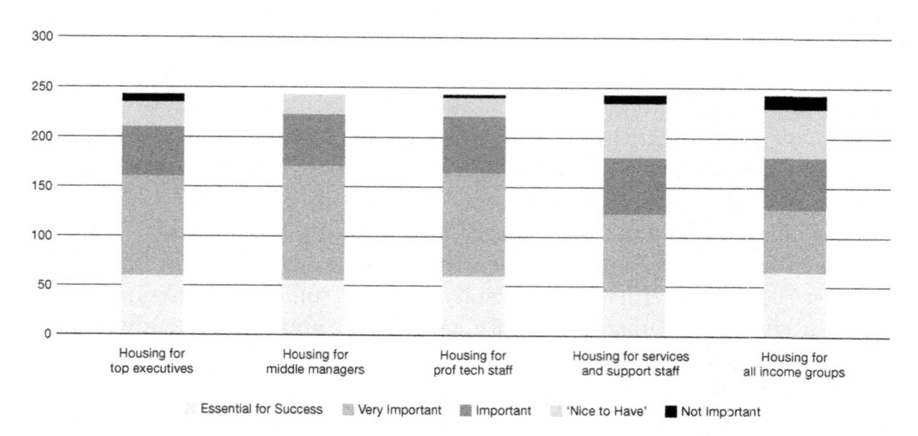

Figure 7.2 Relative importance for 24-hour cities: Residential opportunities in or near downtown

Source: Author's survey of top real estate executives

The tilt toward high middle- to upper-income housing can be related to the density required for 24-hour cities. Intensive land use implies strong land prices, and consequently high development costs, making residential building feasible only at comparatively high prices to the consumers of housing. Transportation, which will be considered in a later section of this chapter, is the obvious facility addressing the affordability gap for workers whose income is insufficient to support the highest-density residential land costs in or near downtown.

Interview respondents ratified the importance of a residential presence to vibrant downtowns, with one service sector executive saying, "The full time resident is most critical." The principal of a destination marketing company stressed that housing demand is linked to the utilization of a wide variety of urban amenities, such as restaurants, that are associated with high quality of life. The capsule description "live–work–play" begins with a nice place to live.

Housing choices for successful 24-hour cities

Interviewees often used the term "livability" as a positive attribute for cities and applied this term to both the 24-hour and 9-to-5 metros. But they sharply distinguished the livability of downtowns, with positive comments about 24-hour cities such as Chicago, Boston, and New York, and negative reactions to housing opportunities in Atlanta, Dallas, and Phoenix. Two executives specializing in bringing offshore capital to US real estate markets called good-quality downtown housing the "galvanizing force" behind the emergence of 24-hour cities.

Downtowns that have not been able to foster such housing have "faux downtowns," in the words of an institutional investor, who compared San Diego unfavorably with Chicago and Miami in terms of its center-city residential environment. Interestingly, Philadelphia is generally thought to have very good housing near its CBD, and there is something of a puzzle about why this has not elevated it into 24-hour-city status in the eyes of the executives I surveyed and interviewed, or indeed the *Emerging Trends* respondents over the years.

Retail stores and shopping

Urban historians have marked the establishment in 1923 of Country Club Plaza in Kansas City, located 4 miles outside the downtown district and primarily oriented toward automobile-dependent customers, as the beginning of the suburban challenge to the hegemony of downtown retailing. But it was only with the development of the enclosed, climate-controlled shopping mall in the 1950s that downtown's function as the commercial hub for consumers as well as for employers began to suffer noticeable deterioration.[14]

With the rise of the great industrial cities of the 19th century, the CBDs across the United States had boasted the multilevel department store as their retailing centerpiece. These emporia were more than sales centers for merchandise and functioned as civic jewels, whose tearooms were congregating places for social interaction and who often provided such amenities as musical entertainment. Holidays were not only occasions for intensified consumption, but for special events, as witnessed by the Thanksgiving Day parades sponsored by Hudson's in Detroit and Macy's in New York. Families brought their children to see Santa Claus and the Easter Bunny, and thus the department stores became centers of shared experience and generational continuity in the civic culture, a key feature in the life of the urban middle class.[15]

The demise of the downtown department store was, therefore, more than just an issue of competition among merchandising formats. It was symptomatic of the waning health of the core city itself. This was not an isolated phenomenon. A very partial list of notable store closings might begin with Boston's R.H. White in 1957 and Chicago's Mandel Brothers in 1959. Just a few more names illustrate the pervasiveness of the trend: Stewarts (Baltimore, 1979), Hudson's (Detroit, 1983), Foley's (Houston, 1984), Gimbels, and B. Altman (New York, 1986 and 1989, respectively) all shut their doors, as consumers took their business elsewhere.

A simple count of department stores or a résumé of the history of their closings only scratches the surface of the retailing picture relative to central cities. Still, the evidence even at this level indicates how much more robust the 24-hour city cluster is relative to the 9-to-5 cluster. The seven hypothesized 24-hour cities have a total of 42 remaining downtown department stores, whereas the seven 9-to-5 cities have just ten department stores in their core areas.[16]

The assertion of the *Emerging Trends* survey respondents and commentators, that a feature of 24-hour cities would be strong urban retailing, was a bold departure from the conventional wisdom and a multi-decade pattern of decline. For most US urban areas, strong downtown retailing was no more than an aspiration at the beginning of the 21st century. The US Department of HUD looked to a mixed-use strategy, including retailing, as the formula for center-city revitalization – often explicitly stated under the rubric of the "24/7 business zone" or "24-hour downtown."[17] Practically speaking, though, planning officials commonly think within the conventional framework of use-specific zoning areas, rather than unpacking existing land use schemes to allow truly mixed-use districts.

Orlando, San Diego, Phoenix, and Jacksonville have been among the cities seeking to develop 24-hour downtowns. Orlando's Church Street Station entertainment and retail district abuts the downtown office area and is also close to, but separated from, revitalized housing neighborhoods. San Diego's Gaslamp District and East Village are replete with restaurants and clubs, and the site of its major sports stadium, Petco Park, but isolated from the office district by wide streets and characterized by open-air parking lots that dilute the area's density considerably. Phoenix, too, has enormous sports venues – Chase Field and US Airways Center – but little in the way of synergistic commercial activity at street level (this is perhaps not a surprise in a city where the temperature exceeds 90° Fahrenheit 157 days per year). Jacksonville's

Downtown Development Authority pursues a nominal strategy seeking to combine retail, housing, and sports facilities as complements to its office core. But, Jacksonville resembles Phoenix in lacking street-level commercial synergy and is like San Diego in devoting acres of downtown land to surface parking in support of its stadiums and arenas.[18]

Although the integration of downtown retailing as a component of a 24-hour city core is a laudable objective, this is apparently easier to aspire to than to achieve. It is critical to have an existing foundation to build upon and a clear understanding of how agglomeration can work in the context of the consumption, as well as the productive, economy.

Although agglomeration is most frequently considered as a feature of the production economy – the clustering of industry and labor to maximize output – consumption also agglomerates.[19] This is familiar to those who take advantage of the variety of urban "districts" featuring largely homogeneous outlets for goods and services – for example, a theatre district, flower district, jewelry district, and so on (using identifiable areas of Manhattan), or even a "red-light district."

The homogeneity of the merchandise or services is not truly critical, though. Urban "high streets" can and do feature a variety of retailing establishments, all catering to a concentration of consumers gathered by residential or employment proximity. New York's Fifth Avenue, Boston's Newberry Street, Chicago's North Michigan Avenue, and San Francisco's Union Square readily come to mind as avatars of such high-street shopping in the US. Indeed, the suburban shopping mall that so decimated the retailing vitality of many downtowns can be considered a real estate product designed to create deliberately the consumption agglomeration features that urban high streets evolved through market mechanisms.

In the final decade of the 20th century, it could still be said that, "the economic distress of America's inner cities may be the most pressing issue facing the nation."[20] Attempting to deploy the analysis he had successfully utilized on an international scale, Michael Porter turned his attention to urban questions, especially in the case of the long-established pattern of decline in old industrial core cities. He noted that "clusters" (to use his preferred term for agglomerations) can be based upon customer relationships as well as on the synergies of firms competing for employees, employees seeking optimal job opportunities, or the economic benefits accruing from informal knowledge-sharing when firms colocate.

Porter notes that market size and population density are positive agglomerative features for establishments seeking to profit by penetrating retail market opportunities. Even in the poorer neighborhoods of cities such as Boston and Los Angeles, there is considerable spending power per acre. Furthermore, as many retailing jobs are appropriate for comparatively low-skilled workers, central cities can access a labor pool with cost advantages that may not be as readily available in more affluent suburban locations. A relatively affluent customer base of high-income office workers and upscale downtown residents may demand a variety of consumption choices where lower-income workers predominate. Restaurants, hotels, grocery stores, and convenience retailing fit such a profile.

Many economic studies have passed over such consumption-based agglomeration, as this particular cluster of business is viewed as a "local industry," that is, a segment of the economy that is less important than the "base" or "export industry" cluster. Traditionally, the latter has been the focus, because of its putative capacity to trade locally originated goods and services with the remainder of the nation or the world, bringing capital into the local market as a result.

Like Jane Jacobs[21] before him, Porter discusses how street-level retail in urban settings helps create a dense and sustainable community fabric, fostering local identity and improving safety and security by assuring that there are "eyes and ears" throughout the neighborhood. Citing examples from Chicago, Boston, and New York, Porter depicts a pattern that is easily recognized as creating positive economic externalities. Such externalities, if achieved, nicely match the desiderata of the "creative class" outlined by Florida.[22]

Glaeser, Kolko, and Saiz[23] note that, as real incomes rise, households increasingly seek attributes such as a variety of goods and services, pleasant physical surroundings, good public services, and convenience. These authors examine housing prices arrayed geographically from the urban center as an economic measure of a preference for urban amenities, including increased social interaction, shorter commutation distance, wider range of choice (large cities may afford a broader range of cuisine), and scale economies (cities must reach a given size to support certain functions, such as a comprehensive art museum, the opera, or a Major League Baseball franchise). Theoretically, such externalities could be capitalized into the price of housing and indicate the agglomerative benefit to consumers of such urban attributes. Like Richard Florida, Glaeser, Kolko, and Saiz interpret the data as meaning that cities in the future will need to attend to the desires of consumers, rather than employ economic development strategies primarily oriented toward employers.

One of the major causes of urban resurgence in the 1990s was a correlation between density and consumer amenities.[24] Whereas the decades of the 1960s and 1970s present strong evidence of a majority preference for lower densities,[25] it now appears that those preferences have reversed, at least for selected cities, in the minds of many people. Cities are not just about production, these authors stress, but about many qualitative variables, including "buzz"[26] and "pace."[27] Whereas a wage premium was required to motivate urban residence in the latter decades of the 20th century, compensating workers for increasing city crime and deteriorating quality of life, by 2000 there was a downward-sloping relationship between real wages and the large, dense cities. This suggests a need to compensate workers for the diseconomies of living in lower-density cities and in traditional suburbs.[28]

Glaeser and Gottlieb support their analysis by reviewing survey data from the consultancy DDB Needham showing that, around 1985, the preference for cities over small towns shifted from a negative to a positive response. For the purpose of the present discussion, this is an interesting point of inflection. It is about this time that commercial real estate risk-adjusted returns begin to move in favor of the 24-hour city cluster.

Retailing encompasses a vast array of shopping venues, everything from a corner newsstand to a full-service department store. But which shopping opportunities for residents and workers are most important in a 24-hour downtown? That question was directly posed in the research survey of real estate professionals.

The preferred characteristics indicated by survey respondents reflected the daily needs of a large resident population. The most frequently selected "essential for success" retail feature was indicated as "supermarkets and convenience stores" (102 responses, or 43 percent of total); this option only received 185 combined votes in the "essential" and "very important" categories, though. This figure was surpassed by the combined 194 votes for "a selection of moderately priced restaurants," which had 91 "essential" votes (38 percent) and 113 "very important" votes (47 percent). Interestingly, the low scores for retail features were "many national chain stores" (just 81 combined "essential" and "very important" votes in total, or 34 percent) and "one or more large department stores" (106 combined votes, or 45 percent).

The local character of supermarkets and mid-range restaurants, versus national retailing "brands," appears to validate research by Markusen and Schrock on the value of "regional distinctiveness" as a positive feature for cities.[29] It also runs contrary to the image of 24-hour downtowns as merely playgrounds for the elite, urban upscale ghettos catering to the highest income brackets only. Supermarkets and moderately priced restaurants are the basic consumption choices of the households described in the housing question on the survey (cited above), the middle managers and professional/technical workers employed in CBD office towers. This is also the kind of shopping needed by families choosing to remain in the city center during the child-rearing years, something that will emerge as important in the survey responses on schools (reviewed below in the discussion of amenities).

Although few respondents rose to the aspirational level articulated by a senior government official – "The vision of a great urban place with retail on wide tree-lined streets and mixed uses above is most important" – many implicitly concurred with the list proposed by industry-based executives: cafes, pharmacies, grocery stores, and bars. One partner in an investment intermediary called grocery stores "the litmus test" for 24-hour areas. A private equity investor indicated that restaurants in locations that can have early and late seatings for dinner are more productive and more profitable and bring activity to their surroundings later into the evening.

A hospitality industry executive offered the Palm restaurants as an indicator of the highest-functioning 24-hour cities. In his judgment, this brand does well in New York, Las Vegas, Los Angeles, Chicago, Washington, DC, and Boston. Dallas has "so-so success." And the Palm has had trouble with its locations in Northbrook (suburban Chicago), Detroit, Columbus, San Diego, Minneapolis,

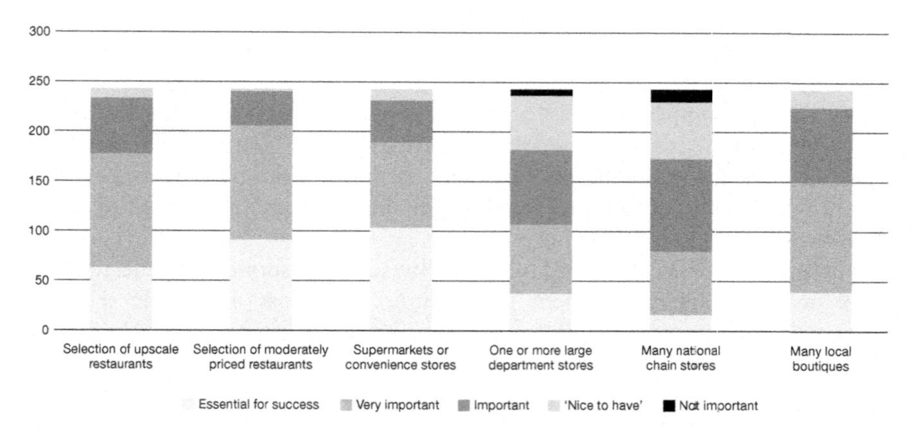

Figure 7.3 Preferred retail opportunities for 24-hour downtowns
Source: Author's survey of top real estate executives

Louisville, and Seattle. The thriving locations include virtually all the 24-hour cities, plus Los Angeles. The weaker locations mentioned include three of the remaining 9-to-5 markets, even though the proposed clusters being analyzed were not revealed to the interviewees.

A proprietary database called Walk Score (walkscore.com) has geo-coded more than 10,000 neighborhoods across the country, including the 14 cities being particularly examined in this book. Walk scores provide a measure for establishment density in categories of retail business, personal services, and various amenities. These are reasonable proxies for locally based consumption opportunities and afford the opportunity to compare the relative availability of such opportunities in CBDs, as well as in cities as a whole.

Figure 7.4 arrays the walk scores for the 14 cities, showing the score for the city as a whole and for the neighborhoods where commercial office space is most concentrated.[30] Establishment variables include merchandise retail outlets such as bookstores, clothing and music stores, hardware stores and drugstores; food services such as restaurants, coffee shops, and bars; personal service providers such as fitness centers; and entertainment, cultural, and recreational amenities, such as movie theatres, schools, parks, and libraries.

The resurgence of certain US cities in the late 20th century (and into the 21st) has been investigated under a number of rubrics. Markusen and Schrock note that the regeneration of the urban core involves consumption variables. Signs of diminished urban distinctiveness (increased homogeneity) include intensifying presence of chain retailers and consumption of inexpensive imported goods, the replication of shopping and entertainment formats common across the nation, and a preference for commoditized housing. Signs of enhanced distinctiveness include unique, locally provided culture and services, high levels of diversity, and a preference for historically or locally specific elements in the

City	Overall score	Neighborhood (score)	Neighborhood (score)
Walk Scores for 24-hour cities			
Boston	79	Beacon Hill/Back Bay (97)	Central (95)
Chicago	76	Loop (98)	Near North Side (97)
Las Vegas	55	Downtown (78)	Meadows Village (84)
Miami	72	Downtown (87)	Little Havana (82)
New York	83	Midtown Manhattan (100)	Financial District (100)
San Francisco	86	Financial District (99)	Downtown (98)
Washington, DC	70	Dupont Circle (99)	Downtown (97)
Walk Scores for 9-to-5 cities			
Atlanta	52	Midtown (87)	Downtown (81)
Dallas	51	West End Historic District (96)	Oak Lawn (80)
Los Angeles	67	Downtown (90)	Mid-Wilshire (88)
Minneapolis	70	Downtown West (95)	Loring Park (91)
Philadelphia	74	Center City East (98)	Center City West (98)
Phoenix	50	Encanto (62)	Central City (62)
Seattle	72	Downtown (97)	First Hill (98)

Figure 7.4 Walk Scores for cities and central neighborhoods

built environment. These authors note that relative concentrations of occupations, although commonly understood as a production variable, can reflect the level of demand from consumers in categories such as restaurant services, arts and entertainment, and discretionary personal care, such as spas and fitness centers.[31]

Markusen and Schrock calculate the regional distinctiveness index (RDI) for the 50 largest US metros, including all the areas in the hypothesized clusters that are the subject of this thesis. Figure 7.5 presents the ranking of the 14 cities, their 24-hour or 9-to-5 status as used in this study, and their RDI. It is notable that the 24-hour cluster, with the exception of Chicago, has uniformly high RDI scores, whereas, with the exception of Los Angeles, the 9-to-5 cities score lower on the distinctiveness scale.

In the area of urban retailing, as in the other attributes investigated in this chapter, the survey respondents and the commentators in *Emerging Trends* identified a shift that was not widely appreciated at the time that publication first raised the subject. Furthermore, it was generally only later that academic

Metro area	RDI rank (of 50)	RDI score	Hypothesized cluster
Washington, DC	3	5.99	24-hour
Las Vegas	7	4.01	24-hour
San Francisco	10	3.23	24-hour
Los Angeles	12	3.19	9-to-5
Boston	16	3.04	24-hour
New York	18	2.90	24-hour
Miami	19	2.77	24-hour
Seattle	26	2.30	9-to-5
Phoenix	34	1.58	9-to-5
Atlanta	42	1.27	9-to-5
Chicago	43	1.18	24-hour
Dallas	44	1.17	9-to-5
Philadelphia	45	1.14	9-to-5
Minneapolis	49	0.95	9-to-5

Figure 7.5 Regional distinctiveness ranks and scores for cities
Source: Markusen and Schrock (2006) for RDI scores

researchers would dig into the empirical and theoretical bases for the phenomenological change.

Despite the widely reported decline in urban retailing, in the early 21st century, the 24-hour cities had four times the number of remaining downtown department stores and much higher values for the geo-coded Walk Scores that measure the proximity of smaller-scale outlets for goods, services, and public amenities. Moreover, the 24-hour cities, as a rule, rate far higher on measures of locational distinctiveness. The question of whether 24-hour cities are clearly distinguishable from 9-to-5 cities, in measurable ways, again seems to prompt an affirmative reply.

Transportation

From the first time that *Emerging Trends* discussed 24-hour cities in 1995, it included transportation characteristics as a key element in its descriptive definition. Central business districts with 9-to-5 profiles were deemed to be "big losers" for investors as the 20th century turned into the 21st. Initially, *Emerging Trends* attempted to include some suburban areas within its working definition of 24-hour districts (though this argument received successively less attention over time), and so it included "commuter highway access" among the key 24-hour-city attributes. Mass transit by rail or bus, however, has been a consistent theme, and the transportation access question was linked to housing as early as 1996:

> Thriving residential communities rooted in and around business districts are the key to preserving 24-hour environments. The boss and his employees can live near the office. Not only can city dwellers walk or take public transportation to work; they can find a market or a place to eat around the corner. That's a luxury few suburbanites can attain.[32]

The US Bureau of the Census has collected journey-to-work data since 1960. Much of the scholarship relating transportation patterns to economic geography has sourced these journey-to-work statistics. The Census Bureau itself published summary findings from the 2000 decennial census, highlighting the modal split illuminated by the results of its 2000 survey, which covers a one-in-six sample of all US households. The pattern of commuting in 2000 showed little variation from the 1990 Census results: about three out of four workers commuted to work by private car, driving alone, and another 12 percent carpooled. Nationally, fewer than 1 in 20 workers used public transportation, and less than three percent walked to work.[33]

Data are available for the decennial census at many levels of geography, and the Bureau's brief provides some detail for the 50 states and for selected metropolitan areas. The incidence of driving alone was highest in the state of Michigan (83.2 percent), followed by Alabama (83.0 percent) and Ohio (82.8 percent). New York had the lowest incidence of solo automobile commutation (56.3 percent), with Hawaii (63.9) and Alaska (66.5 percent) also well below the US norm. At the metropolitan area level, several MSAs displayed public transportation utilization well in excess of the 4.7 percent US average. Such MSAs included New York (24.9 percent), Chicago (11.5 percent), San Francisco (9.5 percent), Washington, DC (9.4 percent), and Boston (9.0 percent). These top five MSAs in the use of public transit are all on the list of hypothesized 24-hour cities.

Like the decennial census, the 2009 American Community Survey (ACS) also reports findings at many levels of geography. A look at the 14 urban areas, and their principal or core cities, unveils a very broad range of data (Figure 7.6). The seven MSAs posited in the 9-to-5 set of cities have journey-to-work characteristics that are very similar to the US as a whole, suggesting that these seven metro areas are fairly typical of US urbanized areas. The 24-hour MSA set, however, is approximately 16 percentage points lower in automobile usage. Public transit commuting in the 24-hour MSAs is 12.9 percent, compared with 5.2 percent in the 9-to-5 MSAs, with the US as a whole having a 5.0 percent transit modal share. In addition, 4.0 percent of those in the 24-hour MSAs walk to work, compared with 2.9 percent nationally and just 2.4 percent in the 9-to-5 metro areas.

Name of place	Auto – driver only	Auto – carpool	Transit	Bicycle	Walk	Other or work at home
Metropolitan area data: 24-hour cities (figures as percent of total)						
Boston	68.5	8.0	12.2	1.0	5.0	5.3
Chicago	70.9	8.8	12.2	1.0	5.1	2.0
Las Vegas	78.9	11.6	3.4	0.4	1.9	3.8
Miami	77.7	10.4	3.5	0.6	1.8	6.0
New York	50.4	7.0	30.5	0.4	6.3	5.4
San Francisco	61.7	10.2	14.6	1.5	4.4	7.6
Washington, DC	66.1	10.6	14.2	0.6	3.2	5.3
Average for 24-hour metros	67.7	9.5	12.9	0.8	4.0	5.1
Metro data: 9-to-5 cities (figures as percent of total)						
Atlanta	77.2	10.6	3.7	0.2	1.4	6.9
Dallas	81.2	10.5	1.5	0.1	1.4	5.3
Los Angeles	73.6	10.8	6.2	0.9	2.6	5.9
Minneapolis	78.1	8.8	4.7	0.9	2.4	5.1
Philadelphia	73.7	7.9	9.3	0.7	3.8	4.6
Phoenix	76.2	12.0	2.4	0.9	1.8	6.7
Seattle	69.4	11.1	8.7	0.9	3.6	6.3
Average for 9-to-5 metros	75.6	10.2	5.2	0.7	2.4	5.8
US average	75.6	10.5	5.2	0.7	2.4	5.8
City-level data: 24-hour cities (figures as percent of total)						
Boston	37.0	7.7	34.5	2.1	14.1	4.6
Chicago	50.8	9.9	26.5	1.1	5.9	5.8
Las Vegas	77.9	10.9	3.4	0.3	2.6	4.9
Miami	69.4	10.4	11.6	0.4	3.4	4.8
New York	25.5	5.3	54.9	0.6	10.3	3.4
San Francisco	38.9	7.4	31.8	3.0	10.3	8.6
Washington, DC	36.5	6.7	37.1	2.2	11.1	6.4
Average for 24-hour cities	48.0	8.3	28.5	1.4	8.2	5.5
City-level data: 9-to-5 cities (figures as percent of total)						
Atlanta	67.5	8.3	12.3	0.7	4.2	7.0
Dallas	78.5	10.7	3.9	0.1	1.9	4.9
Los Angeles	67.1	10.5	11.3	1.0	3.4	6.7
Minneapolis	61.8	9.4	13.5	3.5	6.7	5.1
Philadelphia	51.3	8.5	24.9	2.2	8.7	4.4
Phoenix	74.5	13.2	3.2	0.9	2.0	6.2
Seattle	52.9	9.6	19.5	3.0	7.7	7.3
Average for 9-to-5 cities	64.8	10.0	12.7	1.6	4.9	5.9
US Average	75.9	10.5	5.0	0.5	2.9	5.2

Figure 7.6 Journey-to-work data (ACS)

Focusing on the central cities within the metro areas, the divergences from the US norm are magnified. In the 9-to-5 sample set, automobile usage is a smaller 64.8 percent of the total. For such cities, public transit use represents a 12.7 percent share, and walking to work grows to 4.9 percent. For the 24-hour sample set, though, automobile commutation drops to just 56.3 percent, public transit use grows to 28.5 percent, and walk-to-work commutation is 8.2 percent.

All figures are simple averages from the sample. Individual cities are even more distinctively different. Fewer than one in three New York City residents commute to work by auto, with 54.9 percent using public transit and an additional 10.3 percent walking. All five of *Emerging Trends*' core 24-hour cities score above 25 percent in mass transit use. Among the 9-to-5 cities, only Philadelphia approaches the 25 percent mark in transit commutation. At the other end of the spectrum, all seven 9-to-5 cities exceed 60 percent in their automobile commuting, and five of the seven surpass that threshold in "driver alone" trips. Miami and Las Vegas, from the 24-hour cluster, also have more than 60 percent solo driver cars in their journey-to-work data.

The academic literature has, in the main, found support for the standard entropic models in the journey-to-work statistics. The A–M–M model proposes that density/distance gradients descend outward from the center, and that, because the time value of upper-income households outweighs out-of-pocket costs for such households, they are early adopters of "faster commuting options" and choose perimeter locations where they can optimize their housing choices on lower-cost land. In the simplest form of this model, lower-income households are left at the center, ironically using higher-cost land, but left with higher-density multifamily housing units and greater dependency on older and slower commutation options. This is commonly how US cities have developed, even such historically important areas as Boston and Philadelphia.[34]

More recently, though, scholars have recognized that the trend toward "re-gentrification" in several central cities may pose a challenge to the A–M–M model.[35] The density/distance gradients critical to the A–M–M model flatten considerably as new forms of transportation become less costly and are adopted more widely across the income spectrum. As costs flatten, upper-income households gain comparative advantage close to the center, especially if the housing-cost differential between suburban and center city has itself narrowed. Lower transportation costs induce poorer households to shift to lower-cost land in suburban areas. Upper-income households, meanwhile, see the time-value component of travel as having greater weight than the out-of-pocket commuting costs and, hence, may find that proximity to work again justifies the higher land values in the urban center. The result would then be "regentrification" pressure.

Here, a theoretical foundation that measures the interaction between housing and transportation variables can be seen for the 24-hour city hypothesis. Modeling of urban systems is by nature a simplifying procedure, and the transportation issue is complex. Scholars have explored many dimensions beyond the basic variables. These include innovations and changes in the urban transit system,[36] the impact of the built environment,[37] the more frequently seen instance of polynucleated cities,[38] the heterogeneity of urban economies,[39] and the relation of the economic health of the CBD to the value of accessibility.[40]

One of the key elements separating 24-hour cities from others is density: in 2000, at least one-quarter of all commuters lived within 2 kilometers of rail transit in Boston, Chicago, New York, San Francisco, and Washington, DC. Of all other US MSAs, only Philadelphia and San Diego display this attribute outside *Emerging Trends*' classic group of five 24-hour urban centers.[41]

When household-income patterns are examined, rail transit appears to increase migration of poorer households to suburban areas. However, Voith notes that the new train ridership in Washington, for example, is unlikely to be the same as the existing bus ridership of two or three decades ago. And, with the time-value component of transportation counting more for the upper-income households, the greater success of transit investment in higher-density areas seems well matched to Brueckner and Rosenthal's findings.[42] Baum-Snow and Kahn find weak returns on transit investment in metro areas suffering long-term decline, including Buffalo, Baltimore, and St. Louis, or in low-density metros such as Dallas, Atlanta, and Sacramento. Based on their analysis, these authors suggest that transit investment undertaken after 2000 in Minneapolis, Phoenix, and Seattle (among our 9-to-5 sample set) is unlikely to show desirable returns. New York, which they do not mention but which has directed much capital spending into its subway system since 2000, would be well positioned for positive impacts from rail transit expansion.

Using New York's region as a study area, Chen, Gong, and Paaswell have sought to study the role of the built environment in transportation choice. Employment density at place of work was found to be the most significant factor influencing the mode of transportation chosen, outweighing other factors, including the distance to transit access from home and from work. Interestingly, socioeconomic variables and travel time-and-cost factors are less significant in explaining the choice between transit and automobile options. This study is useful, not only because New York is the largest MSA and the locality with the most abundant transportation options, but also because it has a large, fully supported set of home-to-work travel options (all modes can accommodate intra-city, suburb-to-city, city-to-suburb, and suburb-to-suburb commutation), and the density range is exceptionally broad, from 268 persons per square mile in Sussex County, New Jersey, to 45,499 persons per square mile in Manhattan.

This research offers a counterexample to the presumption that, as metropolitan areas develop multiple subcenters, the core must suffer. From the earliest literature on polynucleated cities, this had appeared to be the case. Greene noted this phenomenon in a study examining changes in employment concentration in five metropolitan areas during the decade of the 1960s. Those metro areas were Atlanta, Baltimore, Buffalo, Denver, and Fort Worth. As the study period was one of intense suburbanization, it is perhaps unsurprising that both residential- and employment-density increases were found in various places throughout the metro areas.

This would later be highlighted as the "edge city" phenomenon by Garreau. In none of Greene's subject cities was there evidence of a thriving core. Chen et al. highlight a recursive interplay within the built environment, specifically between the density and complexity of land uses in the core. Where the city center has sufficient mass, it would appear that the competition between core and suburban nodes need not be modeled as a zero-sum game.

Greene's study, in retrospect, hints at phenomena that would become increasingly important over time, for both a few cities that have sustained strong cores (the 24-hour cluster as the key examples) and for many polynucleated metropolitan areas (including many in the 9-to-5 market cluster). There are clustering effects at locations with transportation advantages such as highway intersections and airport access, effects that interrupt the "declining density gradients" assumed in the conventional model of urban configuration. Government facilities also appeared to play a catalytic role. Regional shopping malls played that role as well in the late 20th century, although the siting of malls may itself simply have reflected the pre-existing advantage of the transportation conditions.

Whatever the causality, it seems apparent that transportation characteristics can alter density gradients from their monotonic slope of decline, creating anentropic patterns within urban areas that can represent the kind of "emergent self-organization" predicted in complexity theory.[43]

Agglomeration benefits arise for firms as the new nodes are established. Those advantages do not seem to grow from inter-firm linkages, though, but from multipurpose trip benefits available to commuters, benefits not available when businesses are isolated or randomly distributed. Nodes, once established, have a recursive and reinforcing relationship with transportation choice. As nodes grow, they concentrate transportation demand and develop local population density gradients within the metropolitan area. This compromises the classical pattern of land use anticipated by land and transportation planners in the most common approaches to urban growth in the post-World War II era.

The development of subcenters associated with good transportation access to the CBD may not be detrimental to downtowns, if agglomeration economies are sufficiently strong to allow growth in the central core, albeit at rates slower than suburban growth. Voith studied the impact of improvement in fixed-rail access to Center City, Philadelphia. He found that the improved access dramatically increased suburban housing values and was highly correlated with CBD employment growth, but not correlated with changes in suburban employment.

Increased economic activity in the core need not be inconsistent with strengthening suburban residential values, as long as sufficiently convenient transit access is available. Nor do the studies suggest that this confluence of factors implies, necessarily, the residential hollowing out of downtown. Birch's inclusion of Philadelphia on her list of "fully developed" downtowns underscores this point. (The evidence of the housing and transportation studies does raise a question as to why Philadelphia's attributes fail to make it one of the 24-hour cities identified by *Emerging Trends*, though.)

The transportation literature, then, appears to support the intuitions underlying the 24-hour-city hypothesis. Journey-to-work statistics offer suggestive empirical data that differentiate the proposed set of 24-hour cities from both the 9-to-5 cluster and the overall US averages. The canonical model for urban form, the A–M–M model, remains a powerful explanatory instrument for the entropic patterns found in most metropolitan areas. More recent research has increasingly found exceptions when multiple commuting options are available and when economies of scale and scope evolve within metropolitan areas.

Most particularly, examples of emergent self-organization occur around transportation nodes, altering the A–M–M's monotonically declining density gradient. This indeed finds form in the polynucleated metropolitan areas increasingly common in the US. But it also finds form in the regentrification of downtowns, where the time cost of commuting, rather than the dollar cost of travel, encourages upper-income households to locate close to the concentration of high-income employment opportunities. The impact of economic agglomeration remains powerful, especially in a comparatively small subset of US cities, most frequently those included in the 24-hour-city cluster.

Like so many other factors, the impact of transportation technologies has had many-faceted consequences for cities. Once cities outgrew their "walk-to-work" size in the 19th century, transport enabled business districts to grow beyond the limits set by the population living within walking distance of the center. Beginning with streetcars and commuter rail, and intensifying in the automobile era, transport enabled populations to reside outside the city entirely. This, as much as zoning regulations, encouraged the separation of uses that helped hollow out scores, if not hundreds, of US cities over the past half-century.

Although approximately 85 percent of all Americans commute to work by car (and the vast majority of trips are a lone driver without passengers, as seen in Figure 7.6), our survey and interviews of upper-level real estate executives produced an unequivocal endorsement of public transit as a *sine qua non* condition for 24-hour downtowns. Sixty-two percent "strongly" agreed that "a 24-hour downtown must have high-quality public transit service," and an additional 30 percent selected the "somewhat agree" option, leaving only 8 percent in any level of disagreement or with "no opinion."

To help specify more exactly the features of transportation infrastructure and service most important for 24-hour cities, the executives were asked to evaluate the importance of highway access, parking, rail and bus service, pedestrian capacity, and truck-related congestion on city streets. Figure 7.7 displays the array of responses.

Consistent with the overwhelmingly affirmative position on public transit, the survey panel accorded the highest rank to "high capacity rail mass transit" as the key transportation infrastructure and service feature for 24-hour cities.

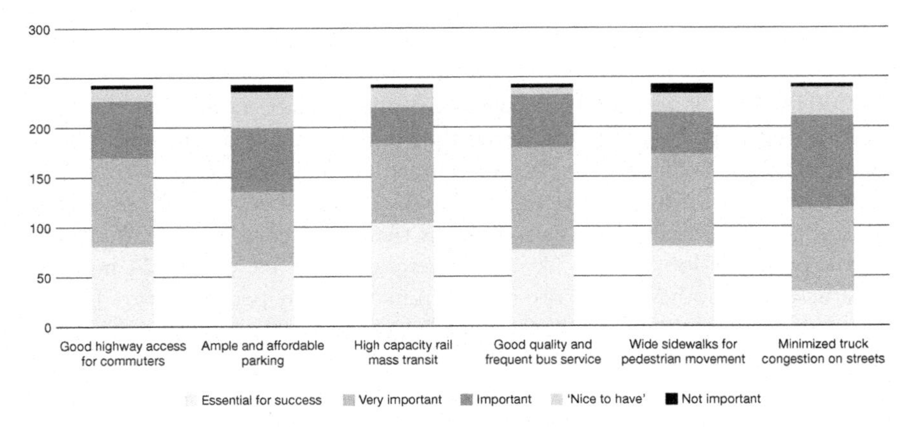

Figure 7.7 Importance of transportation modalities for 24-hour downtowns

Source: Author's survey of top real estate executives

Rail mass transit captured 104 (43 percent) "essential" and an additional 77 (32 percent) "very important" responses. "Good quality, frequent bus service" ranked second, with a combined 74 percent of top answers (77 "essential" votes, or 32 percent, and 101 "very important" votes, or 42 percent). Highway commutation was not ignored, with a combined 71 percent total for "essential" (34 percent) or "very important" (37 percent). But even this was not as high a score as "wide sidewalks for pedestrian movement," which collected 80 "essential" votes (34 percent) and 90 "very important" responses (38 percent).

A portfolio manager with a large institutional investor suggested that, "It is often the lack of good access via highway or high-speed transportation that drives the development of a 24-hour city, as people are more likely to live in the city the more difficult it is to commute." Such a perspective, though, omits a potential alternative: the edge city option of multiple employment centers scattered throughout the urban area. Nevertheless, the survey respondents and interviewees were very sensitive to the impact that excessive congestion had on the viability of downtowns, some of the negative effects of large highways on cities, and the fairly consistent presence of large fixed-rail networks, both intra-urban mass transit and commuter rail, in most of the large cities in the hypothesized 24-hour-city cluster.

The sentiment favoring mass transit was not universal. A developer, based in Texas but working for a national organization, demurred, "The presence of public transportation is overblown. The Woodlands, Texas (an "edge city" just outside of Houston) qualifies as a 24-hour city with very little public transportation." The Woodlands is a celebrated master-planned community that won the Urban Land Institute's Award for Excellence (1994) and is lauded by Glaeser in *The Triumph of the City* as a "leafy suburb" appealing to the middle class desiring relatively low-cost single-family housing, in a beautiful environment with many amenities.

The assertion that The Woodlands qualifies as a 24-hour city, however, may just be local self-promotion. At about 2,300 persons per square mile, it is far below the typical threshold of urban density. Indeed, with a population of 55,000 in 2000, it was listed as a census-designated place, rather than as a city. And, perhaps most significantly, given the cosmopolitan character of 24-hour cities, with the collateral diversity of population, it is especially odd to find the demographic distribution of The Woodlands' residents to be 92.4 percent white, less than 3 percent Asian, and less than 2 percent African–American. The Woodlands looks more like an enclave and less like a city, 24-hour or otherwise. As another survey respondent noted, "Many people believe [their] cities are 24-hour cities, and they are not." A property manager with a national real estate brokerage and management firm said of the major city of which The Woodlands is a satellite, "Despite the fact that Houston is the 4th largest city in the US, with a sizable and well-developed CBD, I would not classify it as a 24-hour city."[44]

A consensus view from interviewees was not hard to pinpoint: "Transit options need to be abundant. They cannot be limited only to "great bus service." [Cities] need streetcars, convenient connections to mass transit, bike lanes, and a safe pedestrian environment." An international developer repeatedly stressed the importance of in-place transportation infrastructure:

> It is not easy to "create" a 24-hour city. There is a need for "concentration" (i.e., density), mass transportation. The costs of major infrastructure today are just too high, especially when the payoffs take many years but political time-horizons are just four years long. China is the only exception . . . with the massive high-speed rail and other growth initiatives taking place there.

Interviewees frequently cited New York's subways, Boston's T, Chicago's Ls, San Francisco's BART system, and Washington, DC's Metro as supporting the 24-hour downtown. Investments in these systems have paid long-term dividends to the cities far-sighted enough to devote capital to infrastructure whose span of service is generational in extent.

Crime

From the moment of its introduction as a theme in *Emerging Trends*, the discussion of 24-hour cities considered the worrisome issue of crime. The 9-to-5 markets were seen as subject to increasing crime and other social ills, contributing to urban blight. By contrast, the *Emerging Trends* commentators observed greater levels of safety and security in the 24-hour markets, even in the early years of their discussion, and remarked that this appeared to be a persistent feature as they revisited the theme in subsequent issues in the first decade of the 21st century. They noted in 1998 that the US as a whole was experiencing a measurable drop in crime and identified the major 24-hour cities as particular beneficiaries from this trend. At the turn of the millennium, the respondents saw suburban markets as troubled by the very criminal behaviors that many of their residents had left the cities to escape.

Nowhere, it seems, is there a greater dichotomy between the cold, hard facts and the inflammatory reports of the popular media than in the realm of crime. Violent, even lurid, crime is a daily staple in print, in broadcast media, and over the Internet. Even as overall crime rates have dropped by about half (from 1991 to 2013, violent crime was down 51.5 percent, and property crime was down 46.9 percent; see Figure 3.3 in Chapter 3), the relentless 24-hour news cycle has, if anything, devoted more attention to the high-profile "index crimes" fraying the American social tapestry. As far as the arguably more significant criminal behavior in the white-collar sector, we hear relatively little, unless an ambitious prosecutor goes after fraud, market manipulation, bribery and corruption, or tax chicanery and orchestrates a perp walk from the executive suite for the benefit of the TV cameras.

In 1930, the US Federal Bureau of Investigation (FBI) began to collect, publish, and archive standardized data on crimes known to the law enforcement community. Recognizing local differences in crime categorization, the FBI issues a handbook to the now nearly 17,000 law enforcement agencies operating in the 50 states and the District of Columbia. Agencies are directed to report crime data according to the guidelines specified in that handbook, rather than simply passing on data as classified by state and local statutes. The result is the series of Uniform Crime Reports (UCR) that are the standard measure of crime rates for the nation, and its subdivisions of states, metropolitan areas, cities, and counties. The UCR sums two primary categories of crime: violent crime (murder, forcible rape,[45] robbery, and aggravated assault), and property crime (burglary, larceny–theft, motor vehicle theft, and arson). To normalize the data for size of place, crime rates are expressed as crimes per 100,000 of population.

The UCR data are not without problems. The FBI itself cautions against using the data to rank law enforcement agencies or their effectiveness. The Bureau acknowledges the complexity of the variables that mold crime in particular jurisdictions. In its annual UCR report, *Crime in the United States*, the FBI cites no fewer than 13 variables that interact with the crime data, including issues such as:

- population density and degree of urbanization;
- age-based demography;
- economic characteristics of the population;
- transportation patterns;
- cultural, recreational, and educational characteristics of places.

The Bureau's caveats are prompted by critiques in the sociological and criminological literature that go back many decades, including the failure of the reporting system to capture all crime (the UCR is only "crimes known to the police"); crudeness of categorization; exclusion of major classes of offense (drug crimes and white-collar crime) from the summary statistics reported; and the use of a simple counting device that masks information about the nature of the offenses.

Nevertheless, scholars in the criminal justice field consider the UCR to present sufficiently robust data on crime that it should be considered the best generally available measure of trends in crime and the relative incidence of criminality across US geography. Academic debates notwithstanding, the UCR remains the data standard for researchers into US crime, as it is the longitudinal series with greatest extent and the cross-sectional database with greatest penetration. As such, it is the source relied upon by researchers consulted in the preparation of this book.

Urban crime rates exceed US averages, and large cities are generally the drivers for the higher urban crime statistics. The FBI categorizes cities by size (cities, in the FBI parlance, denote incorporated municipalities of any size; this usage differs from other urban terminology that follows political jurisdictional labels such as "towns" and "villages") and disaggregates index-crime data according to size groupings. The level of index crime (per 100,000 population) tends to rise with size of city, as shown in Figure 7.8. The exception is for cities

Population group	**Violent crimes**	**Property crimes**	**Total crime rate**
All agencies	366	2,727	3,093
All cities	432	3,170	3,602
Group I (250,000 and over)	702	3,655	4,357
1 million and more (subset)	627	2,954	3,581
500,000 to 999,999 (subset)	800	4,370	5,170
250,000 to 499,999 (subset)	723	4,063	4,786
Group II (100,000–249,999)	454	3,531	3,985
Group III (50,000–99,999)	338	2,899	3,237
Group IV (25,000–49,999)	277	2,781	3,058
Group V (10,000–24,999)	256	2,699	2,955
Group VI (less than 10,000)	279	2,927	3,206
Metropolitan counties	247	1,924	2,171
Non-metropolitan counties	164	1,405	1,569
Suburban areas	238	2,186	2,424

Figure 7.8 Crime rates per 100,000 by population group: 2013

Source: FBI Uniform Crime Reports, *Crime in the United States 2013*

of more than 1 million in population, where the crime rate is lower than for cities in all size categories from 100,000 to 999,000 residents. The big-city data are heavily influenced by the exceptionally low crime rates of New York and Los Angeles.

All of the cities in the hypothesized 24-hour and 9-to-5 clusters analyzed in this thesis are included in Group I, cities of 250,000 or more in population. The 14 cities analyzed for this research are distributed in the following subgroups (see Figure 7.8): cities with more than 1 million in population (New York, Los Angeles, Chicago, Philadelphia, Phoenix, Dallas); cities between 0.5 million and 1 million (San Francisco, Seattle, Washington, DC, Boston, Las Vegas); and cities between 0.25 million and 0.5 million (Atlanta, Miami, Minneapolis).

Group I has the highest total crime rate (normalized by 100,000 of population) and the highest index for each of the major subcategories, violent crime and property crime. The greater incidence of crime in large cities has often been cited as an impetus for suburban population flight and the decreasing competitiveness of the urban core as a place to do business.[46]

From this perspective, the phenomenon of a large 24-hour city as a locus for growth and a target for investment based upon superior risk-adjusted returns must be considered surprising. Do the 24-hour cities differ from their Group I peers in the crime data? And do they differ from the hypothesized set of 9-to-5 cities? Figure 7.9 suggests that the cluster of 24-hour cities is below the norm for all Group I cities' total index crime rate of about 4,357 per 100,000 population, influenced by New York City's significantly lower crime rate. But,

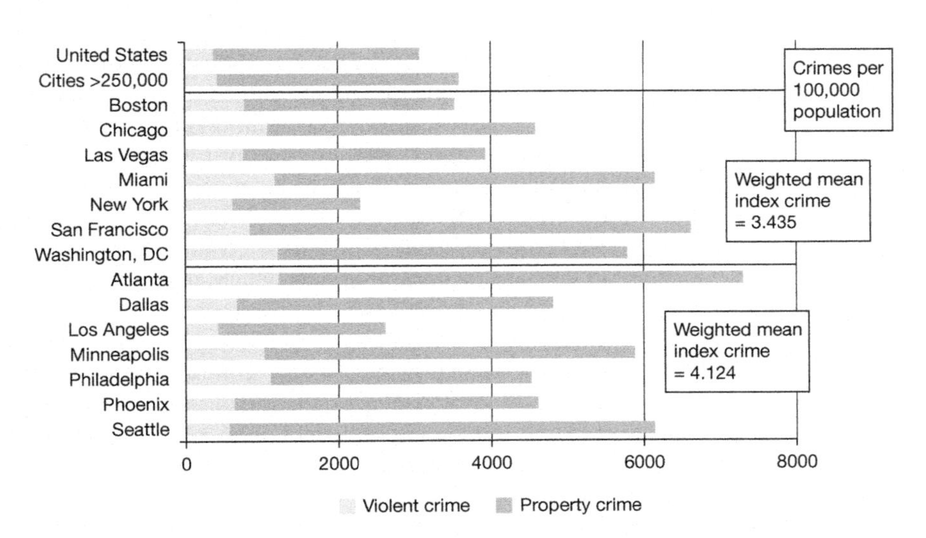

Figure 7.9 2013 index-crime rates: Selected cities vs. US and all large cities

Source: FBI Uniform Crime Statistics

if the 9-to-5-city cluster is considered as the competitive set of places for business and household location, the rate of crime in those cities, at 4,124 crimes per 100,000 population – 20 percent higher than the 24-hour group – emerges as a potential impediment to economic performance. In the 9-to-5 cluster, Los Angeles stands out, and perhaps it is not coincidental that William Bratton headed that city's police department in between stints as New York City's police commissioner.[47]

Prima facie, the implication that crime would bear some motivating force in the larger picture of interregional migration in the United States would seem to have a very weak basis. Demographers have long documented the patterns I have termed "entropic," including the spread of the US population from the densely populated Northeast region to the more sprawling cities of the South and West. But the UCR data published over decades by the FBI identifies the Northeast as the region with the lowest levels of index crimes, and the South and West as relatively high crime regions, once the data are normalized for population size. Other popularly supposed impacts relating crime and urban sociology also have a difficult time in standing up to the scrutiny of rigorous research.

Using cross-sectional and longitudinal data for a 2009 study of 278 cities and their surrounding suburbs, two NYU researchers, Ingrid Gould Ellen and Katherine O'Regan, found crime rates dropping more swiftly in central cities, both in the aggregate and relative to the suburbs associated with each core city.[48] The cities with the largest declines in crime tended to be in the Northeast. These researchers relied on an HUD compilation of the UCR data, and, based on their work, there may be a promising opportunity to test core city–suburb crime relationships specifically for the 9-to-5- and 24-hour-city sample in future research.

These authors also examined crime data for the foreign and native-born and appear to confirm the cross-city evidence on immigration and crime. Controlling for other demographic characteristics of cities, recent immigrants have no effect on crime rates in the first generation, and foreign-born youth are less, rather than more, likely than native-born Americans to be engaged in criminal activity. The relative improvement of Northeast cities and the halving of the crime rate differential between core cities and their associated suburbs seem to have enhanced the ability of cities to retain households which, in the previous era of rising crime, would have likely relocated to the suburbs.

As has been the case in the review of the housing, transportation, and retailing literature, then, the literature on crime provides another stone for the foundation of an argument that at least some cities may be identified as reversing the forces of entropy, establishing a self-reinforcing pattern of greater concentration. If so, a finding of superior risk-adjusted returns to at least a limited class of city would have improved plausibility.

Certainly, it can be said that top real estate executives believe this.

It would seem beyond question that lower levels of crime and higher levels of safety and security would be attributes sought by all population components, but conditions that could be economically afforded principally by the affluent. And, as crowding and poverty are commonly associated with the propagation of criminal activity, questions of density and income disparity naturally arise as elements germane to the 24-hour-city discussion.

Consequently, our survey of real estate executives sought levels of agreement and disagreement with the assertion, "Cities with 24-hour downtowns are safer than cities with 9-to-5 downtowns." Some 41 percent endorsed the statement "strongly," and 39 percent agreed "somewhat." Only 2 percent "strongly disagreed," and 11 percent disagreed "somewhat." Seven percent registered "no opinion." As the UCR statistics show, the empirical data support the respondents' perceptions unambiguously.

Apart from the considerations covered in the sociological literature, it seems useful to inquire what urban security measures are seen as most effective. The survey therefore asked for an evaluation of the level of importance of a variety of safe/security variables. The results are displayed in Figure 7.10.

A remarkable 80 percent of the survey respondents concurred with the prescriptions of Jane Jacobs and William A. Whyte[49] that a high level of outdoor activity providing "eyes on the street" was either "essential" (41 percent) or "very important" (39 percent). Next highest in the combined total was "bright lighting of streets," with 70 "essential" votes and 100 "very important" votes, for a total of 71 percent. This edged out the "visible police presence on the streets" with a total of 68 percent "essential" or "very important" votes. Despite the popularity of BIDs, just 26 percent of the respondents related "uniformed BID workers" as either "essential" (5 percent) or "very important" (21 percent) in terms of safety and security. Interestingly, too, investment in conspicuous security such as CCTV devices was regarded as "essential" by just 12 percent of the respondents and "very important" by 35 percent.

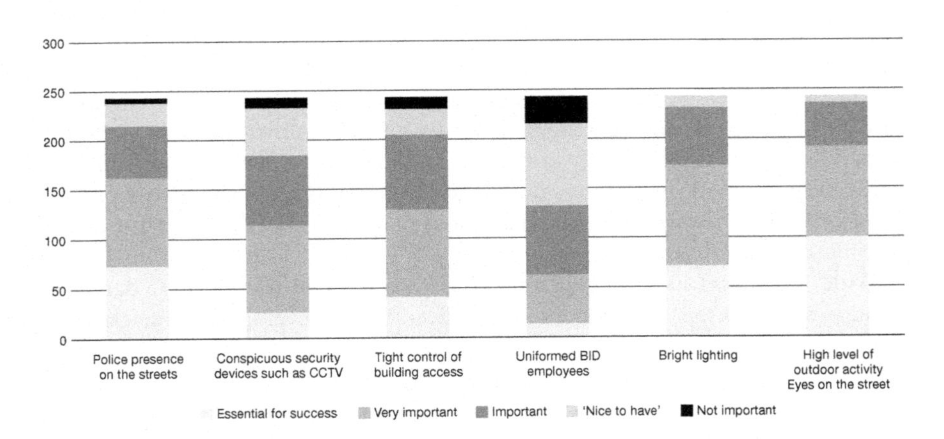

Figure 7.10 The most important safety features for 24-hour downtowns

Source: Author's survey of top real estate executives

So, not only the provision of pedestrian-friendly cityscapes, but also supportive land uses such as sidewalk cafes, brightly lit retail streets, and even zoning-influenced bright signage such as that mandated at New York's Times Square are perceived to contribute more to public safety than a police presence itself or the incremental benefits of technological devices or private security enhancements.[50]

Comments from the panel of interviewees generally concurred with the survey responses. The influence of Jane Jacobs is evident in several comments about the importance of "eyes on the street." A pension fund intermediary noted that hotels help by increasing urban pedestrian traffic after 6 p.m. A mortgage executive stated that, in the physical inspection of properties that is part of the underwriting process, his company explicitly looks to evaluate personal safety by observing the number of people on the street. He cited Cleveland as a place where "there's just not enough people around."

Policing itself was not discounted by the interviewees, though. Some cited the improvement in New York City under Mayor Giuliani and Police Commissioner William Bratton as "leading the way" in the 1990s. An intermediary for offshore investors singled out the gentrification of Manhattan's Soho neighborhood in the 1990s as the first to demonstrate the improved safety of close-in residential areas and believed that Washington, DC, has followed New York's lead. A developer credited Chicago's Mayor Richard M. Daley (the son of Mayor Richard J. Daley, who has been discussed previously) for having made neighborhoods such as Streeterville change for the better, a sentiment echoed by a fund manager who said, "You hardly ever hear of a crime in downtown Chicago." One investor stressed that, "you can make cities both safe and fun," and another noted that improved safety was critical to raising demand levels to the point where inner-city residential rehabilitation becomes economically feasible.

The decline in crime, although a trend that has been seen in virtually all US cities, still left the interviewees wary of some individual places. Miami was mentioned by more than one for the perceived danger of crime. Downtown Atlanta also received cautionary comments, as did Philadelphia. The Tenderloin district of San Francisco was singled out by an institutional investor as being impeded by crime, especially nuisance crimes against property and illegal panhandling. The Fort Worth side of the Dallas Metroplex has seen a private security force underwritten by the Bass family, a reflection of the particular desire for improved safety there.[51]

Wide social and income gaps may lead to a level of friction that damages quality of life. One chief operations officer for a development/investment company said bluntly, "Panhandling and the smell of urine in the street is a turnoff to investors." But both the literature and the responses to the survey and interviews confirm that population heterogeneity and urban density are not synonyms for crime, nor are they directly correlated with crime. Diversity can be consonant with safety and turns out to be an attractive force for 24-hour urban locations.

Amenities and "quality of life"

The fifth characteristic of 24-hour locations specified consistently in the *Emerging Trends* discussion is the availability of recreational and cultural amenities – and the contrasting paucity of such amenities in 9-to-5 downtown districts. The interaction of amenities with the core land uses of commercial and residential development creates the multifaceted environment that draws high-skilled workers to 24-hour urban centers. Such an environment gives the 24-hour downtown a competitive advantage over edge cities that do not have the critical mass to support an adequate amenity mix. It also blunts the putative competition of telecommuting, which limits the amount of social interaction when compared with more traditional workplaces.

Although places outside the cluster of markets hypothesized as the 24-hour cities for this study may attain some level of amenity mix – *Emerging Trends* has identified Atlanta's Buckhead, Seattle's Bellevue, Washington, DC's Bethesda and Reston suburbs, and Walnut Creek outside San Francisco as such places – on balance, these are pale imitators of the major 24-hour downtowns. The "new urbanism" efforts to recreate "traditional neighborhoods" with such features also pay tribute to the synergies that have evolved in cities such as New York, Boston, and Chicago. But the new urbanism, to be frank, has more to do with small towns than it does with actual cities.

The key is the gestalt of the 24-hour downtown. Replicating this is not simply a matter of adding pieces: an aquarium here, a convention center there, an arts district on the side. High-human-capital workers of the creative class gravitate to distinctive neighborhoods such as Washington, DC's Georgetown, New York's Greenwich Village, and Boston's Back Bay.[52]

Interestingly, such a perspective runs contrary to the conventional approach to economic development undertaken by many cities seeking to revitalize their downtowns precisely by the strategy of subsidizing convention centers and sports stadiums. It also omits discussion of the quality-of-life feature most often viewed as significant in the academic literature reviewed below: weather/climate. In the many years of *Emerging Trends* commentary on 24-hour and 9-to-5 cities, "weather" has been cited precisely once, in its discussion of San Diego in 2005, and "climate" is never suggested as a variable impacting investment performance. Under the pressure of both short-run (extreme weather patterns of heat, cold, and storms) and long-run (rising sea levels, greenhouse gas effects, resource scarcity) effects of climate change, that omission is likely to be corrected, and corrected dramatically.

Rating and ranking cities has become a national publishing pastime in the United States. *Business Week* publishes an annual "best places to live" list. So do *Kiplinger Magazine*, *Forbes*, and *Relocate America*, to go with the annual surveys conducted by the accountancy Grant Thornton and the Mercer Consultancy. Beginning in 1981, a book-length treatment appeared with the title *Places Rated Almanac*,[53] which attempted to quantify the ranking process in considerable detail. This popular book quickly became an easy reference that was widely used in corporate relocation studies, real estate evaluations, and other analytical exercises where it was felt desirable to augment the standard economic base and market analysis with data beyond the usual discussion of demographics, employment and income, rents, and vacancy rates.

Recreation and culture are specifically mentioned by the *Emerging Trends* respondents and commentators as relevant variables for real estate investment performance, and it is helpful to recognize the multitude of potential data sets that are subsumed in these categories. Attention to such items is not new. During a major site search for a large heavy manufacturing facility in the 1980s, the corporate executives brainstormed attributes of a desirable location. In addition to economic and engineering considerations, they realized that they needed a place that served the needs of both executives and production workers. Their shorthand for such a mix was "a place with both country clubs and bowling alleys."

A number of researchers have confirmed the aptness of quality-of-life measures as tools to address cities' competition for capital, first in relation to the production of goods and services and, next, in the competition for the spatial array of consumption activities discussed in the retail section earlier in this chapter. A considerable literature indicates that quality-of-life issues enter into business location decisions and are even discounted into the wage differentials separating city-by-city income levels. If, let us suppose, upper-income households have greater mobility and locational choice, quality of life would help determine the pool of disposable personal income concentrated place by place, as wealthier households seek the set of attributes defining a more desirable quality of life. Such an argument underscores a connection developed by many authors: that a self-reinforcing relationship may exist between human or social capital and the concentration of physical and financial capital in high-quality-of-life cities.

Glaeser and Saiz explore such connections in their treatment of "the skilled city." Addressing the question of why educational levels predict agglomeration, they propose that human capital can be conceived of as a kind of amenity in itself: that skilled people prefer to interact with other skilled people. Moreover, if the salient feature of dynamic cities is information exchange, highly skilled workers who specialize in ideas benefit from a dense population of their peers. Third, if a Darwinian approach to urban affairs is taken, and cities only survive by their adaptability to change, high human capital predicts growth because it enables more facile adaptation to change.

The topic of educational levels reminds us that access to school systems perceived to be superior (and, in fact, more generously funded) was a major impetus to the suburbanization era of 1950–1975. Educational attainment is a key (but hardly the only) measure of human capital. The amenity mix should reflect the preferences of populations that differ by educational level. Moreover, the very presence of colleges and universities may contribute to the 24-hourness of a city. College and university students are particularly untethered to a 9-to-5 diurnal rhythm. Also, as a city attracts a generation of college graduates, it will need a strong K–12 school system to retain that population as it passes through the family-formation and child-rearing years.

Glaeser and Saiz found that the number of colleges per 1,000 of population in 1940 is a strong predictor of late-20th-century urban population growth. This seems to indicate that persons will gravitate to places that have a solid basis for producing skills. One effect is that this higher-skill concentration drives higher housing prices. The higher housing prices, in turn, appear to be sustained by an observable effect: workers in "skilled cities" are not only receiving higher nominal wages, but are being paid more relative to equally skilled workers in "less-skilled cities." This means that – contrary to the assumptions of "least cost" approaches to economic development – an important cadre of workers moves toward higher-cost cities.[54]

These authors make a particularly important contribution in comparing human capital effects at the city and metropolitan-area levels. Cities have much denser concentrations of amenities than the more diffuse metropolitan areas that encompass them. Metropolitan areas are, by definition, integrated labor market geographies, but the central city may, in fact, be the driving force in quality of life, if this means an amenity mix of cultural and recreational facilities. If it is density that supports this amenity mix, this may imply that high-skill cities will have more intensive land use, less buildable land remaining, and, consequently, more regulated zoning. Such barriers to entry would restrict new competition from development, exerting upward pressure on real estate prices.[55] This economic configuration would be consonant with the 24-hour cities' hypothesized superiority as investment locations.[56]

In the section discussing retailing earlier in this chapter, I noted that supermarkets, grocery stores, and moderately priced restaurants were considered to be the most important shopping opportunities for downtowns aspiring to 24-hour-city status. Consonant with this (perhaps surprising) finding is the array of educational features and the most highly preferred recreational/cultural features, as indicated by the real estate executives surveyed and interviewed.

The respondents appear to envision a downtown population that will be stable, remaining in the central city through the child-rearing years (Figure 7.11). This is a stark departure from the suburbanization tendency of the returning World War II veterans, as well as their children in the Boomer generation.

"Good K–12 schools" was the most frequent selection among the educational emphases of the respondents, with 66 (or 28 percent) indicating this as "essential for success," and an additional 86 (or 36 percent) rating it "very important." There was almost a dead heat in the combined voting between K–12 and "well-regarded private and public colleges," which garnered 155 votes in those categories (52 "essential" plus 103 "very important"). Such "core" educational strengths outdistanced "major research universities" and "business schools with executive MBA programs." "Strong vocational/technical education" was in last place with just 69 (or 29 percent) combined "essential" and "very important" responses.

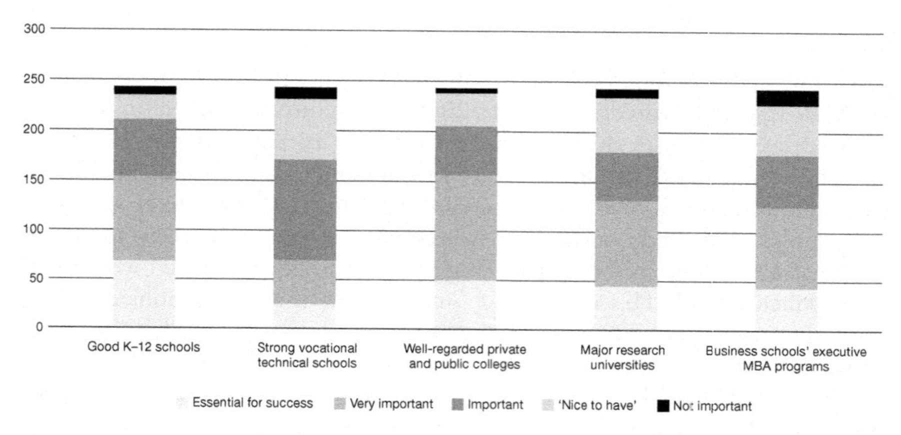

Figure 7.11 Key education components for 24-hour downtowns
Source: Author's survey of top real estate executives

This profile of a diverse downtown population led by family-with-children households is solidly in line with research done by Birch in "Who Lives Downtown?"[57]

The vitality college and university students generally contribute to urban areas was mentioned before. There is a statistically significant correlation of higher-education enrollment with producer services employment, a measure of human capital in the workforce.[58] What the responses to the survey question about education show goes beyond those important points. Attention to primary and secondary education is perceived to help cities retain that energy and that human capital through the child-rearing years, rather than seeing an exodus to suburbia, with attendant strains on metropolitan land use and transportation.

Non-shopping amenities (Figure 7.12) favored by the survey respondents were fairly egalitarian, perhaps a surprise to those who view 24-hour cities as playgrounds for the elite. The profile of a mixed-age downtown population also contributes, apparently, to the choice of "parks and open space" as the number one choice of the respondents, both as "essential" (79 votes, or 34 percent) and in combination with "very important" (summing to 181 votes, or 77 percent). "Theatre and performing arts venues" ranked second, with 65 "essential" votes (27 percent) and 176 votes when combined with 111 "very important" (combining to 74 percent).

High-visibility projects such as "sports stadiums and arenas" (19 "essential" and 69 "very important" votes combining for 37 percent) and "casinos/gaming establishments" (0 "essential" and 5 "very important" responses) were far lower in appeal for the survey respondents. "Zoos and aquariums" did not fare much better than casinos, with 2 "essential" and 16 "very important" votes, combining to 8 percent. Distinctive urban architecture, however, had a combined

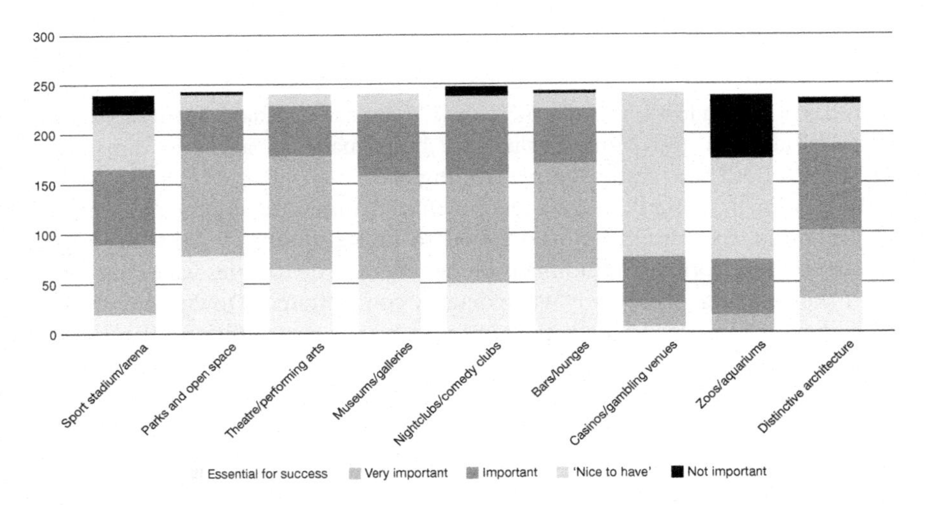

Figure 7.12 Preferred recreational/cultural amenities for 24-hour downtowns
Source: Author's survey of top real estate executives

101 votes for "essential" or "very important," slightly higher than the total for sports facilities.

If density is a positive characteristic of the 24-hour downtown, unrelieved density can be oppressively negative. Open space was valued highly by the survey respondents and interviewees. This is an excellent example of an "economic externality," and one that public policy directly influences by actions ranging from zoning maps to infrastructure investment. An investor/developer who specializes in suburban properties noted ruefully that he seeks to build amenities such as those provided by 24-hour cities in his projects, but that these are not externalities for him, but added costs.

One developer specifically cited Chicago's Millennium Park in explaining the revival of Chicago's historic Loop district, south of the Chicago River. He said, "Every good neighborhood has a park. People go to the opera, to the museum, to the ballpark – but these don't create 24-hour places in the same way." The Boston Common was favorably mentioned by others, as was Manhattan's Central Park and the more recent Hudson River lineal park along the West Side waterfront. Washington, DC's abundant open space, especially the National Mall, was also lauded. A corporate real estate specialist called open space "a difference-maker" for cities. The egalitarian nature of public parks was viewed positively, with one executive noting the similarity to mass transit: "There are no first-class seats on the New York City subway, and no reserved seating in Central Park."

Several interviewees considered art and culture one of the defining characteristics of 24-hour downtowns, with one institutional investor indicating that, "downtown revivals are amenities-based and arts-driven." A developer indicated that, "arts and culture make cities attractive" for residents and tourism and alluded to Boston's "world-class museums" and to Miami's "Art Basel" event. This has not been lost on the 9-to-5 cities. Atlanta and Dallas have arts districts, and Philadelphia and Minneapolis have both made advances in enhancing venues for both the plastic and performing arts.

Recreational, cultural, and entertainment amenities are among the "quality of life" indicators that have been found to be significant in attracting high-human-capital workers. Agglomeration benefits accrue to cities with clusters of high-skill workers, particularly when densely concentrated. Downtowns, rather than their encompassing metropolitan areas, have a more efficient distribution of these amenities, if the high-human-capital workers are considered as the consumer base. Once established, such a configuration poses a high barrier to future competition, because of the difficulty and cost of replicating both the facilities and the customer base. This appears to favor the 24-hour cities over the 9-to-5 cities.

Notes

1 The case record can be found online at http://laws.findlaw.com/us/378/184.html and is discussed in Bob Woodward and Scott Armstrong, *The Brethren*, Simon & Schuster (New York, 1979), pp. 193–200.
2 Op. cit., pp. 3 and 7.
3 *Emerging Trends in Real Estate 1995* identifies and discusses these attributes at pp. 1, 3, 21, 39, and 40.
4 Originally, *Emerging Trends in Real Estate* used the term "8-hour city" interchangeably with "9-to-5," but dropped that usage in later years.
5 Kent A. Robertson, "Downtown Redevelopment Strategies in the United States: An End-of-the-Century Assessment," *Journal of the American Planning Association 61:4*, pp. 429–438.
6 Jennifer Moulton, "Ten Steps to a Living Downtown," A Brookings Institution Center for Urban and Metropolitan Policy discussion paper, October 1999. Jerry Mitchell, "Business Improvement Districts and the 'New' Revitalization of Downtown," *Economic Development Quarterly 15:2* (May 2001), pp. 115–123. Lynn M. Ross, "Downtown Housing? Not without a Plan," *Downtown Idea Exchange*, October 15, 2006.
7 Bernard J. Friedan and Lynn B. Sagalyn, *Downtown, Inc.: How America Rebuilds Cities*, MIT Press (Cambridge, MA, 1989).
8 Eugenie L. Birch, "Who Lives Downtown Today (and Are They Any Different from Downtowners of Thirty Years Ago?", The Lincoln Institute of Land Policy (Cambridge, MA, 2005).
9 Edward L. Glaeser and Jesse M. Shapiro, "City Growth and the 2000 Census: Which Places Grew and Why," The Brookings Institution Center for Urban and Metropolitan Policy (Washington, DC, 2001).
10 Joseph Gyourko, Christopher Mayer, and Todd Sinai, "Superstar Cities," *American Economic Journal: Economic Policy* (American Economic Association) *5:4* (November 2013), pp. 167–199.
11 Jan K. Brueckner and Stuart S. Rosenthal, "Gentrification and Neighborhood Housing Cycles: Will America's Future Downtowns Be Rich?", CESifo Working Paper 1579, October 2005.

12 Edward Glaeser, *The Triumph of the City*, Penguin Press (New York, 2011).
13 Raj Chetty, Nathaniel Hendren, Patrick Kline, and Emmanuel Saez, "Where is the Land of Opportunity? The Geography of Intergenerational Mobility in the United States," *Quarterly Journal of Economics 129:4* (2014), pp. 1553–1623.
14 Garreau, op. cit. Friedan and Sagalyn, op. cit. Peter Hall, *Cities of Tomorrow: An Intellectual History of Planning and Design in the Twentieth Century*, Blackwell Publishing (Oxford, UK, 2002).
15 Jan Whitaker, *Service and Style: How the American Department Store Fashioned the Middle Class*, St. Martin's Press (New York, 2006).
16 Google Maps search for "downtown department stores" by city, accessed May 15, 2015.
17 US Department of Housing and Urban Development, *Strategies for Success: Reinventing Cities for the 21st Century*, US Government Printing Office (Washington, DC, 2001).
18 Robyne S. Turner, "The Politics of Design and Development in the Postmodern Downtown," *Journal of Urban Affairs* (Urban Affairs Association, Wilmington, DE) *24:5* (2002), pp. 533–548.
19 Takatoshi Tabuchi and Atsushi Yoshida, "Separating Urban Agglomeration Economies in Consumption and Production," *Journal of Urban Economics* (Academic Press, New York 2000) *48*, pp. 70–84. Edward L. Glaeser, Jed Kolko, and Albert Saiz, "Consumer City," *Journal of Economic Geography* (Oxford UK, 2001) *1*, pp. 27–50.
20 Michael E. Porter, "The Competitive Advantage of the Inner City," *Harvard Business Review* (Cambridge, MA; May–June 1995).
21 Jane Jacobs, *The Death and Life of Great American Cities*, Vintage Books (New York, 1963), and *The Economy of Cities*, Random House (New York, 1969).
22 Richard Florida, *The Rise of the Creative Class*, Basic Books (New York, 2002), and *Cities and the Creative Class*, Routledge (New York and London, 2005).
23 Art. cit.
24 Edward L. Glaeser and Joshua D. Gottlieb, "Urban Resurgence and the Consumer City," *Urban Studies 43:8* (July 2006), pp. 1275–1299.
25 Joel Kotkin, *The New Geography: How the Digital Revolution Is Reshaping the American Landscape*, Random House (New York, 2001).
26 Daniel Silver and Terry Nichols Clark, "Buzz as an Urban Resource," *Canadian Journal of Sociology* (Winter 2013).
27 Luis Bettencourt, Jose Lobo, and Geoffrey West, "Why Are Big Cities Faster? Universal Scaling and Self-Similarity in Urban Organization and Dynamics," *European Physical Journal B 63* (2008), pp. 285–293.
28 Glaeser and Gottlieb, art. cit.
29 Ann Markusen and Greg Schrock, "The Distinctive City: Divergent Patterns in American Urban Growth, Hierarchy, and Specialization," *Urban Studies 43*, pp. 1301–1323.
30 See Gary Pivo and Jeffrey D. Fisher, "Investment Returns from Responsible Property Investments: Energy Efficient, Transit-Oriented and Urban Regeneration Properties in the US from 1998–2008," Working Paper of the Responsible Investing Center, Boston College and University of Arizona, and the Benecki Center for Real Estate Studies at Indiana University (2009), for the application of Walk Scores as a metric in evaluating US real estate portfolios.
31 See "New York City: The Great Reset," a special issue of NYU/Schack Institute's *Premises* Magazine (Summer 2015), which examines detailed occupational data for such industries. The report, a special applied research study of NYU/Schack, is authored by Richard Florida, Hugh F. Kelly, Steven Pedigo, and Rosemary Scanlon.

32 *Emerging Trends in Real Estate 1996*, p. 22.
33 Clara Reschovsky, "Journey to Work 2000: A Census Brief," US Bureau of the Census (Washington, DC, March 2004).
34 David Harrison and John Kain, "Cumulative Urban Growth and Urban Density Functions," *Journal of Urban Economics 1:1* (January 1974), pp. 61–98.
35 Stephen F. LeRoy and Jon Sonstelie, "Paradise Lost and Regained: Transportation Innovation, Income, and Residential Location," *Journal of Urban Economics 13* (1983), pp. 67–89. Robert C. Steen, "Nonubiquitous Transportation and Urban Population Density Gradients," *Journal of Urban Economics 20* (1986), pp. 97–106. Alan Gin and Jon Sonstelie, "The Streetcar and Residential Location in 19th Century Philadelphia," *Journal of Urban Economics 32* (1992), pp. 92–107.
36 Nathaniel Baum-Snow and Matthew E. Kahn, "Effects of Urban Rail Transit Expansions: Evidence from Sixteen Cities, 1970–2000," *Brookings-Wharton Papers on Urban Affairs* (2005).
37 Cynthia Chen, Hongmian Gong, and Robert Paaswell, "Role of the Built Environment on Mode Choice Decisions: Additional Evidence on the Impact of Density," *Transportation 35* (2008), pp. 285–299.
38 David L. Greene, "Recent Trends in Urban Spatial Structure," *Growth and Change* (January 1980), pp. 29–40.
39 Jan K. Brueckner and Harris Selod, "The Political Economy of Urban Transport-System Choice," *Journal of Public Economics 90* (2006), pp. 983–1005.
40 Richard Voith, "Changing Capitalization of CBD-oriented Transportation Systems: Evidence from Philadelphia 1970–1988," *Journal of Urban Economics 33* (1993), pp. 361–376.
41 Baum-Snow and Kahn, art. cit.
42 Brueckner and Rosenthal, art. cit.
43 For an accessible introduction to this difficult literature, see Murray Gell-Mann's *The Quark and the Jaguar: Adventures in the Simple and the Complex*, W.H. Freeman (New York, 1994), John H. Holland, *Emergence: From Chaos to Order*, Addison-Wesley (Reading, MA, 1995), and Stuart Kauffman, *At Home in the Universe: The Search for the Laws of Self-Organization and Selection in Evolution*, Oxford University Press (New York, 1995).
44 The present author, however, must observe that Houston has made enormous strides in developing as a city of remarkable diversity over the past 30 years. In particular, Houston has transformed its downtown area from a zone that had many acres of vacant parcels and undesirable land uses in the 1980s to one that has achieved both the density and mix of uses of the best urban cores. But that is its downtown, not The Woodlands.
45 In 2013, the FBI amended its crime index reporting standards to seek all instances of rape. Rape is generally acknowledged to be significantly underreported in the crime data, as victims are sometimes reluctant to come forward to the police. By contrast, for example, vehicle theft is well reported, as a police report is required for auto insurance purposes.
46 James Q. Wilson and George Kelling, "Broken Windows: The Police and Neighborhood Safety," *The Atlantic Monthly* (March 1982); James Q. Wilson, "What, If Anything, Can the Federal Government Do About Crime?", US Department of Justice (1996); Peter Hall, *Cities in Civilization*, Pantheon Publishing (New York, 1998), pp. 977–978, 986; Julie Berry Cullen and Steven D. Levitt, "Crime, Urban Flight, and the Consequences for Cities," *The Review of Economics and Statistics 81:2* (May 1999), pp. 159–169.
47 W. Chan Kim and Renee Mouborgne, "Tipping Point Leadership," *Harvard Business Review* (April 2003), pp. 34–48.

48 Ingrid Gould Ellen and Katherine O'Regan, "Crime and US Cities: Recent Patterns and Implications," Working Paper of the NYU Furman Center for Real Estate and Urban Policy (New York, 2009).
49 Jane Jacobs, *The Death and Life of Great American Cities, passim*, but especially at Chapter 2. William H. Whyte, *City: Rediscovering the Center*, Doubleday (New York 1988), pp. 158–163.
50 In this, the respondents were well aligned with the observations on the "evening economy" of John Montgomery in *The New Wealth of Cities: City Dynamics and the Fifth Wave*, Ashgate (Farnham, UK, 2007).
51 Joe Patoski, "Wowtown!", *Texas Monthly* (April 1998).
52 Richard Florida, *Who's Your City?*, pp. 205, 246, 259, and 264, among many other places in the book.
53 Richard Boyer and David Savageau, *Places Rated Almanac*, Rand McNally (New York, 1981).
54 Edward L. Glaeser and Albert Saiz, "The Rise of the Skilled City," Brookings-Wharton Papers on Urban Affairs (Washington, DC, 2004).
55 Steven Malpezzi, "Housing Prices, Externalities, and Regulation in US Metropolitan Areas," *Journal of Housing Research 7:2* (1996), pp. 209–241, and "The Regulation of Urban Development: Lessons from the International Experience," Working paper of the University of Wisconsin (Madison, WI, 1999).
56 David Lynn, Bohdy Hedgcock, and Jeff Organisciak, "Supply Constrained Markets," *Real Estate Issues 35:2* (2010), pp. 20–27.
57 Art. cit.
58 The bivariate correlation is 0.89, significant at the 99 percent level.

8 Fourteen cities

In the preceding chapter, I presented a case that, based upon the descriptive definition of the 24-hour city, we could find the weight of evidence supporting two key claims. The first is the assertion that, based on housing, urban retailing, transportation, crime, and public amenities, the seven cities proposed as 24-hour places could be distinguished in measurable ways from the seven cities proposed as 9-to-5 locations. The second is that, once the measurements were taken, the 24-hour places were not only different, but statistically superior.

This is not said to offend any city, or to contend that any individual must prefer one class of city to the other. Rome, after all, bears the title "The Eternal City" and has given us some of the most durable of principles. One of those is *de gustibus nihil disputandum est*, or, "there is no debating matters of taste." Nevertheless, the whole point raised by the discussion of 24-hour cities in the real estate literature goes beyond taste. The contention for 24-hour cities is that they would prove better places for commercial real estate investors.

That is the challenging issue to be examined: that the qualitative aspects of the 24-hour cities would be valued by investors as a superior configuration, producing better outcomes for income-producing properties and, consequently, would become magnets for investment capital. Chapter 9 will examine that claim.

In this chapter, I hope to take the argument beyond the data presented in Chapter 7 and look at information bearing directly on diurnal patterns, measures of 24-hour-ness. This continues the amassing of evidence that we can use to distinguish the two sets of seven cities from each other. Further, I want to provide a discussion of the 14 cities, as I have evaluated them quantitatively, and as I have experienced them in my on-the-ground professional work in real estate. To be sure that this is not just "one man's opinion," I would also like to present the comments of top-level commercial property professionals gathered especially for this study.

In this discussion, it will become clear that I have found a spectrum of round-the-clock intensity. Although the weight of evidence distinguishes the 24-hour cities as a group from the 9-to-5 cities with differences sufficient to support their classification, it is also evident that a process of organic change is a feature of the urban landscape; I will indicate that, if I use a "bright line" distinction for the purposes of the present analysis – a distinction that will be supported in the real estate data presented in Chapter 9 – I will also identify those cities that, in recent years, have evolved beyond the constraints of the 9-to-5 stereotype and that might be termed "18-hour cities" in recognition of that evolution.

Diurnal measures

What possible way could we take the pulse of a city as it moves through the daily ebb and flow of activity?[1] If there is a clue about such energy, perhaps energy itself might provide a clue. Every modern city sustains some level of round-the-clock activity. Those inclined to see a distinction between cities based upon the degree of activity sometimes still demur at the 24-hour description, recognizing that even cities that have an abundance of nighttime bustle have periods of relative quiet in the pre-dawn hours. Nevertheless, the degree of the ebb in the wee small hours does differ.

Electrical consumption patterns give us a clue. Let's take two places, Minneapolis, from the set of 9-to-5 cities, and Washington, DC, from the 24-hour group. The peak period for energy usage in Minneapolis is the 6–7 p.m. hour and is sustained though 9 p.m. at 98 percent of peak, before falling to 91 percent of peak by 11 p.m. This indicates a surge in usage as local workers return to their residences. In Washington, DC, however, the peak is at 2–3 p.m., and usage fluctuates between 93 percent and the peak for each hour from 8 a.m. to 11 p.m., with the exception of the 7–9 p.m. period, when usage is roughly 91 percent of peak. Minneapolis' usage is more intense in the early evening, from 6 p.m. to 9 p.m.; Washington is closer to peak from 9 p.m. to 1 a.m., and also from 6 a.m. to 6 p.m.

The power usage data glimpsed in this comparison offer a suggestion of how tracking diurnal activity could produce a basis for establishing a measurable

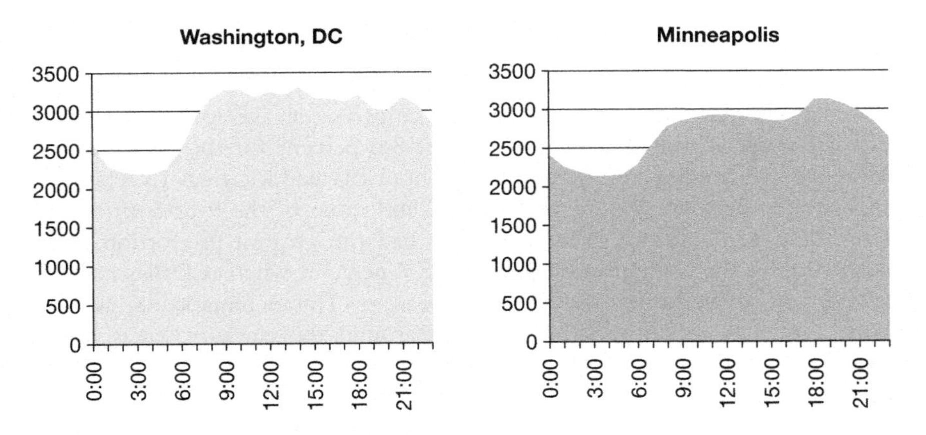

Figure 8.1 Pattern of electricity consumption in two cities

Sources: Potomac Electric Power (Washington, DC); North State Power Company (Minneapolis)

distinction between the hypothesized 24-hour metros and the 9-to-5 metros. Unfortunately, these data are not collected in a central repository of energy providers, and the electrical utilities in most of the metros declined to provide this information when I approached them individually.

Automobile traffic

Traffic data are available for all the cities, however, and provide some indication of the degree of mobility sustained over 24-hour periods. These data are collected by city and state departments of transportation and are based upon electronic monitoring of vehicular flow at various locations in each jurisdiction. For each of the 14 cities, I collected data for major thoroughfares (both large streets or avenues and highways) in or immediately proximate to the business district. Observations were restricted to the middle of the business week (Tuesday, Wednesday, or Thursday), to avoid the potential impact of traffic related to weekend trips and longer-distance travel. Holiday periods were avoided for the same reason.[2]

Traffic, of course, is a measure of automobile movement. Cars and car-related urban form are considered a feature more closely related to 9-to-5 than to 24-hour cities, although all cities have been shaped to some degree by the automobile for the past 100 years. Movement of people via mass transit or as pedestrians, by contrast, is characteristically associated with 24-hour cities, but automobile traffic is very much a part of the 24-hour city, as anyone who has suffered through cross-town traffic in Manhattan, the Kennedy Expressway in Chicago, or Dupont Circle in Washington, DC, can attest. Understanding the diurnal dynamics of traffic can help in exploring the overall volume of movement in the respective cities, as well as the relative efficiency of capacity utilization.

The first observation to be made is that the 24-hour cities see a greater percentage of overall automobile traffic occurring during the nighttime hours, between 9 p.m. and 5 a.m. (Figure 8.2). On average, the seven 24-hour cities tabulate 18.3 percent of their daily traffic in that 8-hour period, whereas the 9-to-5 cities have just 11.6 percent. Looking more closely at the 1–5 a.m. period, the relationship is the same, as the 24-hour cities have an average of 5.8 percent of their traffic in this 4-hour span, versus 3.6 percent for the 9-to-5 cities. Moreover, the ranges are discrete. No 24-hour city had less than 13.5 percent of its traffic between 9 p.m. and 5 a.m., and none of the 9-to-5 cities had more than 13.0 percent. New York City had the greatest proportion of its traffic during the 8 nighttime hours, at 25.3 percent, whereas Dallas had the lowest ratio of nighttime traffic, at 9.5 percent. This is remarkable, as New York City runs its entire subway system around the clock, whereas Dallas suspends its small light rail system overnight. In both cases, it appears that the mass transit decisions reflect the overall demand for mobility over the course of the day and night.

The traffic data compiled counted slightly more than 4.5 million vehicles. The objective in securing the data was to sample activity in and around the

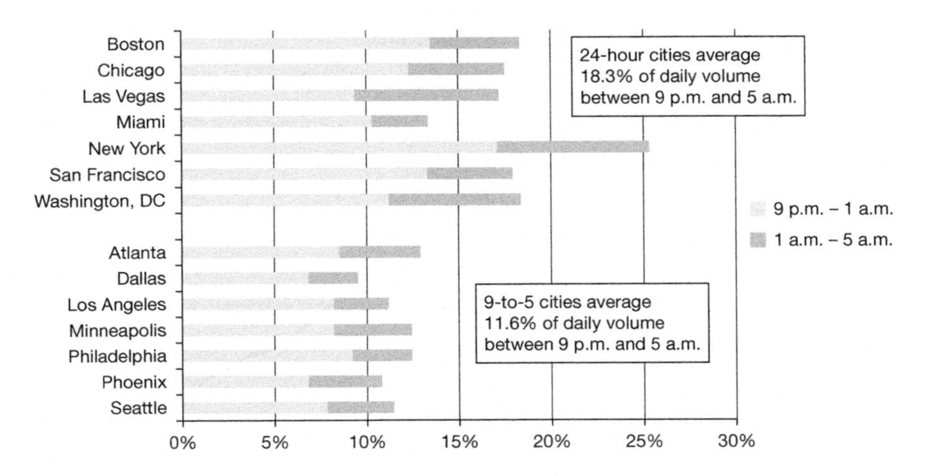

Figure 8.2 Traffic counts detect nocturnal activity: 24-hour cities see 13–25% of volume after 9 p.m.; 9-to-5 cities vary between 9 and 13%

Sources: State and local departments of transportation; data for major thoroughfares in and near the CBD on business weekday, in non-holiday season

urban core in cities where the size and configuration of the downtowns differed. In the 24-hour cities, the overall count was 1.66 million vehicles. For the 9-to-5 cities, the count was 2.85 million vehicles. This indicates greater traffic intensity in the 9-to-5 cities, as would be expected from the journey-to-work data presented in Chapter 7.

Figure 8.3 depicts substantially higher peaks of absolute traffic volume for the 9-to-5 cities in the morning and evening rush hours. The cluster of 9-to-5 cities tallies 6.92 percent of its daily volume between 8 a.m. and 9 a.m., and 7.85 percent of its daily volume between 5 p.m. and 6 p.m. Although the 24-hour cities also have their peak vehicular commutation at the same hours, those peaks are measurably lower than for the 9-to-5 cities. In addition, the span of trough to peak is more extreme for the 9-to-5 cities, with the peak for this cluster at 21 times the volume at the trough (7,466 at 5 p.m. versus 353 at 3 a.m.), compared with an eightfold peak-to-trough ratio (3,262 at 5 p.m. versus 401 at 3 a.m.) for the 24-hour cities, when controlling for the number of data-collection points accessed for each city.

The absolute traffic volumes in the 9-to-5 cities are 89 percent higher than for the 24-hour cities. At the morning peak, the 24-hour cities show 3,117 vehicles counted per collection point, whereas the 9-to-5 cities have 6,575 vehicles. At the evening rush hour, the contrast is even higher, with the 24-hour cities registering 3,262 vehicles and the 9-to-5 cities having 7,466.

There are two related effects observable in the data. The first is that the 9-to-5 cities see a much higher concentration of their automobile traffic during

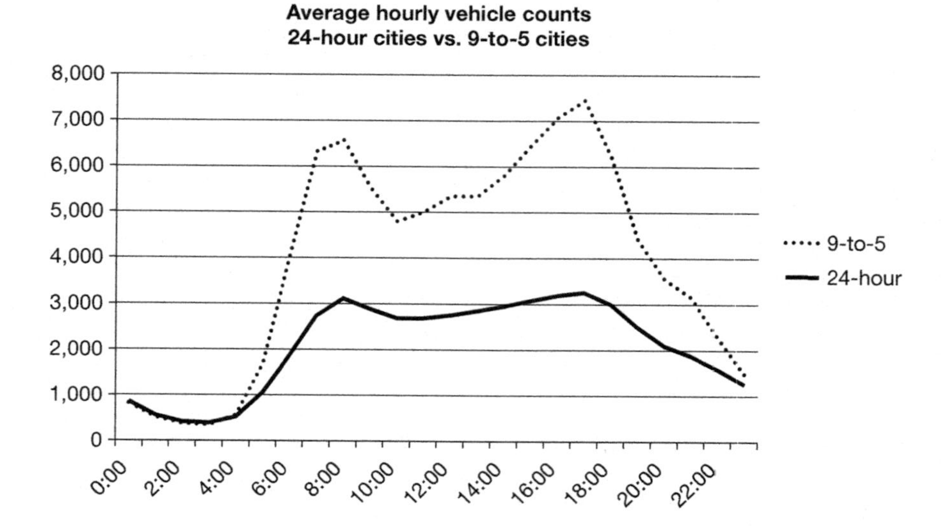

Figure 8.3 Daily traffic per data-collection point by hour of the day: 24-hour cities vs. 9-to-5 cities

Sources: Municipal and state departments of transportation

the daylight hours, leading to potentially severe congestion. At 6,575 vehicles per hour, assuming five lanes of traffic capacity and signalling efficiency of 0.8 (roughly a "green light" time of 4 minutes for every 1 minute of "red light"), this implies a car passing the data-collection point every 3.5 seconds – bumper-to-bumper conditions. At 7,466 vehicles per hour, the flow of traffic over the collection point is an even lower 3.0 seconds per vehicle. The implication, in practice, is very slow speeds for all cars in the peak hours and a very high risk of accidents, as virtually no spacing between cars in the queue is possible. At the evening peak for the 24-hour cities, given the same assumptions, traffic would pass the data-collection point at one vehicle very 7 seconds. But, as the lower volume would require fewer lanes to accommodate the flow, substituting four lanes of capacity in the 24-hour cities still would yield a more manageable flow rate of one car every 5.6 seconds.

The second effect is inefficiency in utilization in the 9-to-5 cities. Streets and highways must be built to accommodate the volume of peak traffic, but, if the peak is 21 times the trough, as is the case in 9-to-5 cities, for many hours the capital-intensive roadway infrastructure is hardly used. In the case of the 9-to-5 cities, for 4 hours between 1 a.m. and 5 a.m., the traffic volume is less than 10 percent of peak. Across the span of the entire day and night, the average utilization of the road capacity is just 53 percent. With lower average vehicle counts, the 24-hour cities require, on average, less of an investment in roadways. And the lower capacity is used much more efficiently. At no time during the night does capacity utilization drop below 12.3 percent, and average utilization for the entire day and night is 64.1 percent.

So, not only is a difference in diurnal pattern evident from the traffic data, their consequences imply not only difference, but advantage to the 24-hour urban centers.

Open at night

In the Introduction to this book, I highlighted the 24-hour neighborhood greengrocer near my home in Kensington, Brooklyn. Many US cities have a variety of businesses open on a round-the-clock basis: newsstands, health clubs, diners, fast-food outlets, gas stations, supermarkets, and other retail outlets (even bowling alleys!). Of course, there are many public services that must be available day and night: police and fire safety, hospitals, homeless shelters, and even public libraries in some fortunate places. We take it for granted, now, that media outlets are constantly beaming news and entertainment: the late-night "test pattern" on our TV sets is a memory for my generation, and a historical enigma to the Millennials. Public utilities providing electricity, gas, and water services never shut down. Neither do multinational businesses in finance, manufacturing, or global logistics. In truth, it would be hard to find a major city that did not sustain some level of 24-hour business activity.

It is a question of degree: how much 24-hour work is being done city by city? And it is also a question of local demand: which businesses remain in operation round the clock because the energy of the urban place makes it profitable to keep the doors open?

Research projects are often frustrated by the inaccessibility of information – even in an age that celebrates the Internet's ubiquity. But there is at least one happy exception in retailing, where good data on round-the-clock activity are easily available.

Drugstores

Formerly, pharmacies were highly specialized as apothecaries; they now serve a retail niche akin to what were once called "junior department stores." Contemporary drugstores sell a wide variety of goods, from cosmetics to toys, from office supplies to basic apparel, and from hardware to groceries. This makes them a one-stop-shop for many daily necessities, and round-the-clock operation can provide a monopoly niche in the local marketplace at hours of the night when other retail outlets are shuttered.

As drugstore chains have expanded into national retail outlets, data collection for their operations has become more standardized and centralized. The major chains, such as Walgreens, Rite-Aid, and CVS, have expanded both by new store openings and by the acquisition of smaller local competitors. For competitive reasons, they have built sophisticated websites that are searchable by geography

24-hour cities	Number of stores	9-to-5 cities	Number of stores
Boston	14	Atlanta	13
Chicago	23	Dallas	13
Las Vegas	28	Los Angeles	18
Miami	37	Minneapolis	16
New York	209	Philadelphia	31
San Francisco	13	Phoenix	14
Washington, DC	25	Seattle	10
Mean of sample	49.9*	Mean of sample	16.4
Median of sample	25	Median of sample	14

Figure 8.4 Number of drugstores open 24 hours (within 10 miles of city center)

Note: *Mean of 24-hour set would be 3.3 if New York City were excluded
Sources: Corporate websites; Google Maps

and by hours of operation. These chains are dominant in 13 of the 14 markets that are the focus of this study, Seattle being the exception. For the markets where the national chains have operations, a search for stores within 10 miles of the city center with round-the-clock operations was performed, and the results tallied. For Seattle, the largest local drugstore chain, Bartell Drugs – whose merchandise configuration is indistinguishable from the two national chains – was researched under the same criteria. And, for all cities, a search was undertaken on Google Maps to identify other independent drugstores open on a round-the-clock basis. Figure 8.4 presents the results.

The cluster of 24-hour cities has a substantially greater number of round-the-clock drugstores, 349, than does the 9-to-5 cluster, with 115 locations. New York dramatically skews the sample, but, even holding that city aside, the 24-hour grouping still has more round-the-clock drugstores in the remaining six cities (140) than the set of seven 9-to-5 cities tallies in total. Five of the 24-hour markets (Chicago, Las Vegas, Miami, New York, and Washington, DC) have 20 or more such establishments apiece, where the only 9-to-5 city to cross the 20-store threshold is Philadelphia. The median number of stores in the 24-hour cluster is 25, whereas the 9-to-5 set has a median of only a little over half that number.

Food and restaurants

The December 2010 issue of *Travel + Leisure* magazine reported the results of its annual reader survey of favorite US destination cities. The magazine reports ranks for 35 cities, relative to a variety of attributes. All 14 of the cities in the two study clusters for this thesis are included in the survey's questionnaire. This survey is a popular poll, rather than a scientific assessment, but it provides a measure of opinion from the magazine's upscale readership, which is sophisticated enough to subscribe to a periodical devoted to cosmopolitan topics of interest oriented toward those with significant discretionary income.

Figure 8.5 presents the rankings accorded by residents and visitors for the most well-known restaurants and for restaurants specializing in ethnic cuisine. Higher rankings (1 is best; 35 is worst) reflect superior standing. A mix of upscale and moderately priced restaurants is considered a vital component of a 24-hour downtown, as mentioned in Chapter 7. The ethnic restaurants are particularly interesting to consider, as population diversity appears to be a hallmark of 24-hour cities.[3]

As a group, the 24-hour cities achieve better scores in both restaurant classifications, as measured by the mean score and the median. Five of the 24-hour cities (New York [1], Las Vegas [2], Chicago [3], San Francisco [6], and Miami [10]) earned top-ten rankings for the upscale restaurants, whereas Los Angeles tied San Francisco for sixth position, and Dallas was rated ninth. The differential is wider in the category of ethnic restaurants, where four of the 24-hour markets (New York, San Francisco, Chicago, and Washington, DC) rate in the top ten, whereas only Los Angeles joins this group from the 9-to-5 cluster.

As greater choice among good restaurants is a motivation for workers to remain in the city after working hours, the stronger scores of the 24-hour cluster suggest why those cities may sustain economic activity further into the evening compared with the lower-scoring 9-to-5 cluster.

24-hour cities	**Top restaurants**	**Ethnic restaurants**	**9-to-5 cities**	**Top restaurants**	**Ethnic restaurants**
Boston	16	20	Atlanta	15	25
Chicago	3	6	Dallas	9	31
Las Vegas	2	27	Los Angeles	6	10
Miami	10	14	Minneapolis	25	21
New York	1	1	Philadelphia	20	19
San Francisco	6	3	Phoenix	11	24
Washington, DC	13	9	Seattle	25	14
Mean of Sample	7.27	11.43	Mean of Sample	11.14	20.57
Median of Sample	6	9	Median of Sample	15	21

Figure 8.5 Rankings (among 35 cities) for food and restaurants

Source: *Travel + Leisure*, December 2010

Culture and the arts

"Urbanity" is almost by definition related to the fine and performing arts.[4] Montgomery explicitly indicates that the evening economy revolves around entertainment and culture.[5] Hence, cities with superiority in the opportunity to enjoy concerts, other performing arts such as the theatre, and museums should have advantage in supporting the evening activities that extend vitality beyond the conventional 9-to-5 business day.

Travel + Leisure's survey polled its readership on their evaluation of cities' offerings in classical music, museums, and the performing arts. Figure 8.6 displays the findings. Within the 24-hour cluster, there is a striking dichotomy: New York, Chicago, San Francisco, Washington, DC, and Boston all earn very high ratings; Las Vegas and Miami are near the bottom of the list. But, among the 9-to-5 cities, scores are almost universally low: best for Minneapolis, Philadelphia, and Seattle in this cluster; mediocre to poor for Los Angeles, Atlanta, Dallas, and Phoenix. As in other measures, these data appear to suggest that the 24-hour/9-to-5 distinction is not a binary separation, but admits of stratification within each grouping and/or the potential for evolution between groups.

So, as in the instances of traffic pattern, data on structural elements of the urban economy oriented toward evening activity (drugstores, restaurants, cultural institutions) appear to show not only measurable differences between the two groups of cities, but differences that indicate that the advantage belongs to the set of 24-hour locations. It remains to review the more direct evaluation of the cities as centers of "nightlife," a trait that likely predominates in the mind of the general public when the term "24-hour city" is used.

24-hour cities	Class-ical music	Mus-eums	Per-forming arts	9-to-5 cities	Class-ical music	Mus-eums	Per-forming arts
Boston	5	9	13	Atlanta	20	25	25
Chicago	4	3	3	Dallas	23	26	27
Las Vegas	35	35	20	Los Angeles	22	27	23
Miami	32	33	32	Minneapolis	8	20	4
New York	1	2	1	Philadelphia	10	8	17
San Francisco	6	7	7	Phoenix	27	26	26
Washington, DC	7	1	9	Seattle	14	17	12
Mean	12.86	12.86	12.15	Mean	17.71	21.29	19.14
Median	6	3	7	Median	20	25	23

Figure 8.6 Rankings (among 35 cities) for the arts and culture

Source: *Travel + Leisure*, December 2010

Nightlife

Research is not always done at a computer, spending hours crunching numbers or downloading studies to be examined. Field research has long been an honored form of investigation. And it can be fun, as well as informative.[6] But not everyone agrees with this perspective.

When I was deep into the research I have been reporting in this book, I happened to be riding the subway in Manhattan and was going through the *Zagat's Survey of New York City Nightlife*.[7] A young lady, 30-something in age, sat down next to me and saw what I was reading. "What are you looking for?" she asked. When I answered that I was doing some research, she confided, "I can help. I know all the best places!" But when I explained what the research was about, she looked at me like I was the most boring person in New York City.

Restaurants, museums, and theatres may be useful in encouraging evening activity for urban areas, but "nightlife" is more closely identified with a recipe that consists of roughly equal parts alcohol,[8] live music, and young singles, such as my all-too-transitory subway companion.

As part of its 2010 readers' poll, *Travel + Leisure* magazine asked several questions relating to nightlife issues. These included the quality of the more sophisticated "cocktail hour" opportunities and the more egalitarian, singles-oriented "bar scene," as well as an evaluation of live-music venues for bands. Figure 8.7 summarizes the results. Interestingly, several cities not included in the two study clusters dominate the top spots in this segment of the *Travel + Leisure* survey. Those cities were Austin, New Orleans, and Nashville, where live music is a major feature of the economic base and the city's tourism marketing, with resorts such as San Juan, Puerto Rico, and college towns such as Providence, Rhode Island, also ranking quite highly.

The results for the clusters are consistent with those found in previous figures in this chapter, showing superior mean and median values for the 24-hour cluster when compared with the 9-to-5 cluster. New York, Chicago, and Las Vegas received top-ten ratings in all three categories, Miami ranked seventh for its bar scene, and San Francisco ranked ninth for cocktail hour. No 9-to-5 cities earned a top-ten ranking for any of the three categories.

The website city-data.com conducted a user-generated poll between March and December 2009 asking, "What are the top ten cities for nightlife in North America?" Readers' lists were solicited, and the responses were not required to be rank-ordered. Thus, an absolute count of the frequency of mention in the 200 responses to the query reflects fairly the respondents' opinions. The survey is one of the most consistently active polls on the website, and I have compiled the responses to the "top 10 for nightlife" discussion for the 3 years 2011–2013. City-data.com also took a more restrictive survey in 2013, asking specifically about US cities and limiting the responses to the "top five." I present the frequency of response for the 24-hour and 9-to-5 cities in Figure 8.8. (Ties are reflected in the rankings.)

24-hour cities	Cock-tail hour	Live music	Bar scene	9-to-5 cities	Cock-tail hour	Live music	Bar scene
Boston	29	24	23	Atlanta	24	24	19
Chicago	8	6	6	Dallas	28	34	30
Las Vegas	4	9	4	Los Angeles	20	15	17
Miami	13	26	7	Minneapolis	20	14	25
New York	3	5	4	Philadelphia	28	24	17
San Francisco	9	13	13	Phoenix	18	28	17
Washington, DC	18	28	22	Seattle	24	14	29
Mean	11.79	15.43	10.93	Mean	22.93	21.43	23.36
Median	8	13	7	Median	24	24	19

Figure 8.7 Rankings (among 35 cities) for nightlife
Source: *Travel + Leisure*, December 2010

The 24-hour-cities cluster have higher mean and median ranks in all three polls, with all but Washington, DC, ranking in the top 10 of the 34 cities receiving votes from the 2007–2009 survey respondents. Among the 9-to-5 cities, only Los Angeles (ranked third) and Philadelphia (ranked tenth) placed in the top ten cities for nightlife in that survey. Minneapolis and Phoenix ranked lowest of the 14 cities in the two study clusters, in a tie for 23rd place. The results are very consistent in the 2011–2013 sampling. All seven 24-hour markets are rated "top ten," and Atlanta joins Los Angeles and Philadelphia on the top-ten list. Phoenix and Minneapolis are again the laggards. As before, 34 cities received at least one vote in the city-data poll.

In the more restrictive 2013 "Top Five in the US" query, the medians were again 5 for the 24-hour city set and 11 for the 9-to-5 grouping. Boston and Washington, DC, fell out of the top ten in frequency of mention, whereas Los Angeles, Atlanta, and Philadelphia maintained top-ten positions. However, when limited to a top-five criterion, four of the 9-to-5 cities fall below the 15th rank: Minneapolis (16), Seattle and Dallas (tied at 17), and Phoenix (26). Thirty-six cities secured at least one vote in the top-five poll.

Although such online polling does not pretend to have scientific or academic rigor, it represents a sampling of viewpoints from users of the 80th most heavily trafficked website in the United States, according to Quantcast.[9] The user profile for this website is close to normal for the US population in income characteristics, but is 25 percent higher in adults aged 25–34 years and 12 percent higher for the 35–44 years age group. The user profile is 14 percent higher than the US average for the college-educated, but 26 percent higher than the US benchmark for graduate education. Thus, the views of site

24-hour cities	Poll: 2007–2009	Poll: 2011–2013	"Top five in US" poll (2013)	9-to-5 cities	Poll: 2007–2009	Poll: 2011–2013	"Top five in US" poll (2013)
Boston	7	10	14	Atlanta	13	8	7
Chicago	4	5	5	Dallas	17	14	17
Las Vegas	4	2	3	Los Angeles	3	4	4
Miami	2	1	2	Minneapolis	23	19	16
New York	1	2	1	Philadelphia	10	9	9
San Francisco	4	6	8	Phoenix	23	20	26
Washington, DC	17	7	11	Seattle	14	11	17
Mean	5.57	4.71	5.29	Mean	14.71	12.14	13.71
Median	4	5	5	Median	14	11	16

Figure 8.8 Rank order: Identification as a top North American city for nightlife

Source: www.city-data.com; ranks by frequency of inclusion on list of top places

respondents would tend to reflect a population of relatively well-educated persons in their working years, with an active interest in urban affairs.[10]

Summary of evening-economy indicators

Granting the deficiencies I've mentioned that preclude any claim of statistical significance, the preponderance of evidence reviewed in this chapter supports the continued use of the terms "24-hour city" and "9-to-5 city" as a meaningful classification protocol. Taken as a group, the variables reviewed as influences on diurnal patterns of activity or on the evening economy tell a story that is quite consistent, and Figure 8.9 summarizes and comments upon the data review.

For mean and median statistics, differences were classified as follows: if the difference of the small value divided into the large value was greater than 50 percent, the difference was considered "large"; if the difference was less than 25 percent, it was considered "small," and results between 25 percent and 50 percent were labelled "moderate." For example, in Figure 8.3, the average traffic count per collection point was 95,048 for 9-to-5 cities and 50,204 for 24-hour cities, an 89 percent differential that equates to a "large" mean difference.

To evaluate the overlapping of cluster ranges, the median of each range (i.e., the range for 24-hour cities and the range for the 9-to-5 cities) was observed. Then, the values in ranges were individually examined. Any observations from the lower range that exceeded the median in the higher range were noted, as were the converse cases. Those observations were then summed. If no such observations were found, the variable sets were considered discrete. One or two

Figure	Topic	Mean difference	Median difference	Range overlaps	Remarks
8.2	Traffic count (day/night)	Large	Moderate	None	Variable sets appear discrete
8.3	Traffic per collection point through 24 hours	Large	Large	Slight	Values are close for midnight–4 a.m. period, then diverge widely
8.4	24-hour drugstores	Large	Large	Moderate	San Francisco below 9–5 median
8.5	*Travel + Leisure* food and restaurants	Large	Large	Slight	Boston below 9–5 median and Los Angeles at 24 median in top restaurants; Las Vegas below 9–5 median in ethnic restaurants
8.6	*Travel + Leisure* arts and culture	Large for music and performing arts; moderate for museums	Large for all	Slight	Las Vegas and Miami are outliers, with very low scores in all categories; Minneapolis scores above the 24 median in performing arts
8.7	*Travel + Leisure* nightlife	Large for cocktail hour and bar scene; moderate for live music	Large for all	Slight for cocktail hour and bar scene; moderate for live music	Boston below 9–5 median for cocktail hour; Boston at 9–5 median, Miami and Washington below 9–5 median for live music
8.8	City-data. com nightlife	Large for all polls	Large for all polls	Slight in 2007–2009; moderate in later polls	Washington falls below the 9–5 median in two of three polls; Los Angeles is above the 24-hour median in all three polls

Figure 8.9 Synoptic table of diurnal variables

such observations indicated "slight" overlap; three or four were termed "moderate" overlap; five or six represented "large" overlap; seven would indicate that the two variable sets were in fact homogeneous. Thus, the data displayed in Figure 8.2 do not overlap (the highest nocturnal share of daily traffic in the 9-to-5 cities, for Atlanta, is lower than the lowest nocturnal share for 24-hour cities, for Miami), whereas, in Figure 8.4, 24-hour drugstores, San Francisco falls below the 9-to-5 median – indicating a "slight" overlap. In the Remarks column of Figure 8.9, I note all individual data points falling in the overlap categories.

Most of the variable sets have large mean and median differences when measured against the scale previously described. Two of the subsets in the *Travel + Leisure* readership survey had moderate mean differences, although the difference in medians was large for all variables examined from this readership survey. The moderate mean differences were found in the museum component of arts and culture and in the live music component of the nightlife category. In arts and culture, Las Vegas and Miami were distinct outliers, with very low scores in all categories evaluated. Minneapolis, on the other hand, scored very strongly in the performing arts component of art and culture.

Although 8 of the 14 cities are noted in the Remarks column of Figure 8.9, more often than not each city is found in its "expected" place when range overlaps were considered. The appearance of so many cities, but only occasionally, on the "wrong" side of the comparison of medians suggests that the data sets are reviewing variables that are truly different, and that the risk of a hidden variable would be relatively low (again, recognizing the nonscientific methods of data collection and not attempting to overstate the robustness of the data sets).

Nothing in the data examination in this chapter indicates that the proposed clusters of 24-hour and 9-to-5 cities are untenable on the basis of characteristics of the evening economy. On the contrary, given the limitations of the data, the results of the analysis indicate that it is appropriate to use these clusters as the basis for the more rigorous analysis to follow in Chapter 9.

Peregrinations: The 24-hour and 9-to-5 cities studied

Cities have a "feel," a signature ambience that might be termed a sense of place. This is, perhaps, part of what Markusen and Schrock were seeking to capture empirically in their measurement of "regional distinctiveness."[11] It's also (negatively, and datedly) what Gertrude Stein was speaking of when she made her classic put-down of Oakland: "There's no there there." For those who travel a lot, there's more than mere recognition upon returning to a place. Places are not just recognized, they are experienced – sights, sounds, smells, and even the enveloping temperature and humidity characteristic of seasons.

Yes, things change. According to Thomas Wolfe's memorable line, "You can't go home again." But, equally, you can. And, in my case, I must. I think that one of the reasons that I've never relocated from Brooklyn is that I do travel so much. So, when I come home, I want to feel at home. That sense of rootedness, of place, is part of my identity. And, as much as Brooklyn has changed since my growing-up years, I still love to get on my bicycle and revisit the now demographically altered but still familiar neighborhoods where I played stickball, walked to school, learned the subways and buses, and held my first jobs. Even now, when my flight path toward LaGuardia Airport banks northeast from the Verrazano Bridge, I can look down to find Prospect Park and Green-Wood Cemetery and see the landscape that has been so influential in my personal history. That's a key touchstone.

You'll find hometown pride wherever you go, and I want to be clear that I have no wish to denigrate any city in this book. I've been a Yankee fan in Fenway Park (a jewel, by the way) and admire rather than hate the fierce pride of Red Sox Nation. I have friends who feel that same loyalty and love for Miami, for Atlanta, for Dallas. Yet there *are* differences, and those differences matter. In fact, what those friends love about their cities are precisely the elements that distinguish them from New York. The world would be insufferably boring if every place were the same.

Thus far, my discussion in this book has been mostly about facts and figures, the milestones passed as a number of US cities have evolved into the early 21st century. I want to flesh the discussion out with some anecdotal details. These are based upon personal observation and upon comments from dozens of individuals interviewed for this book, in addition to some of the numbers I've analyzed. The brief capsule summaries cannot pretend to be exhaustive. Indeed, features highlighted for one city may very well be true of several others. My intention is simply to present some concrete elements that contribute to the lived experience placing some cities in the 24-hour grouping and others in the set of 9-to-5 places.

Atlanta

I have traveled to Atlanta consistently since the mid-1980s and have watched as this city has grown as the "Capital of the New South." Its civic spirit was one of the primary reasons it won the bid to be the host city for the Summer Olympic Games in 1996, when the world was invited to "come celebrate our dream," a poignant reminder of Dr. Martin Luther King's 1963 speech at the Lincoln Memorial in Washington, DC. Atlanta is a city of ambition, most certainly, and one with widely recognized potential.

But it is a long way from being a 24-hour city, and I must say that this appears to be by choice.

Atlanta has done just about everything possible to assure its downtown stays a 9-to-5 place. Interstates slice through its center, creating unnatural barriers north–south of I-20 and east–west of the I-75/85 "Downtown Connector," which is one of the nation's ten most-congested roadways, despite having between 10 and 16 lanes. The Downtown Connector carries upwards of a quarter-million cars per day.

Besides the physical barriers of the highways, Atlanta's downtown competes for business activity, as well as residential uses, culture and entertainment, and public amenities, with several other districts within the city: Midtown and Buckhead most prominently. This diffuses the city's energy and limits the agglomeration benefits that accrue to the nation's most vibrant CBDs.

City/suburban politics has kept the MARTA system skeletal, a lineal railroad without a complex capillary network that would allow efficient passenger access throughout the city – and keep so many of those quarter-million cars from congesting highways at peak periods.

Downtown, Midtown, Buckhead, and the Perimeter districts act as separate nodes. Midtown and Buckhead have the most complexity of use. Midtown has both the Arts District, with the High Museum and the Woodruff Performing Arts Center, and the energizing effect of the 21,000 students attending Georgia Tech. However, the university is west of the Downtown Connector, and the Arts District is east of the highway. Georgia Tech earns a Walk Score of 70, and the Arts District's is 77. Both are much better than Atlanta's overall "auto-dependent" score of 43, but far from the "walkers' paradise" scores more typical of such neighborhoods elsewhere in the country. I vividly recall returning to an Arts District hotel a few years ago in the early evening, walking from the nearby MARTA station – almost the only pedestrian on the street. Why? Virtually all those attending concert or museum events drive, using the 7,000-plus parking spaces available in the immediate vicinity.

Again, why? The auto-orientation reflects the relatively low density of Atlanta, 3,188 persons per square mile. Even within its city limits, Atlanta can have a suburban feel. This is true in places such as Buckhead and the Perimeter area – classic edge cities, built around the anchors of large shopping malls, with office buildings and residential towers filling out the mix of land use. And Atlanta continues to sprawl beyond its perimeter highway, the circumferential I-285. I once made a verbal misstep in addressing an Atlanta audience. I meant to say that Atlanta's manifest destiny seemed to be to grow north until it reached Chattanooga, Tennessee. But I said, "until it reaches Chicago," instead. No one in the audience seemed to believe this idea was totally ridiculous.

Atlanta, in fact, seems to be encouraging such sprawl, rather than attempting to rein it in. The latest evidence of this is the relocation of the local baseball stadium, to be known as Sun Trust Field, from downtown to the Cumberland area, just where I-75 crosses the I-285 circumferential. Whereas cities such as Baltimore, Denver, San Francisco, and Miami are bringing their sports facilities closer to downtown, Atlanta is going the other way. Moreover, local economic development officials have induced Mercedes Benz to move its headquarters from New Jersey to a location just an exit or two from Sun Trust Field, in a suburb called Sandy Springs.

Atlanta has great assets, including tremendous colleges and universities (ranging from Georgia Tech and Emory University to several elite historically black colleges such as Spelman and Morehouse). It is increasingly cosmopolitan and is rated exceptionally tolerant for the LGBT community – an attribute proudly pointed out to me by a cab driver, who showed where Elton John keeps his local condo. It has what residents describe as an enviable quality of life.

But my own observation – and again this is not meant to demean Atlanta – is that the numbers in this and the previous chapter aren't at all misleading. Atlanta is far more toward the 9-to-5 pole on the urban spectrum than it is to the 24-hour city.

Boston

The sports term "fan" is short for "fanatic," and perhaps no cadre of team supporters deserves the label more than the loyal followers of the Boston Red Sox, who suffered from a World Series championship drought from 1918 to 2004 but who retained the ardor of their faithful.[12] The city of Boston itself has been a comeback story, again and again, rebounding from the declines of industries ranging from textiles and shoes to transistors and mini-computers. Although there are those that dispute its claim to 24-hour standing, many cite its scale and livability, history, and culture as enduring attributes.

It is history that makes Boston a tourist destination. Of all 14 cities that are the focus of this book, Boston has the cityscape most evocative of a European settlement. The Freedom Trail allows a visitor to walk the sites of the Boston Massacre, Paul Revere's house, Old North Church, Bunker Hill, and the berth of Old Ironsides. That the Trail is marked on Boston's sidewalks is not incidental: this is one of the most walkable of cities. Walkscore.com rates downtown Boston 97 on a scale of 100: a "walkers' paradise."

Boston is both compact and focused. The city itself is less than 90 square miles in area (though it sits at the center of an urbanized area about 20 times that size), and the core city has a population density of 13,370 persons.[13] The focal point is easy to specify: it is the Boston Common and Public Gardens that sit at the juncture of the Financial District and Back Bay, just at the foot of Beacon Hill. Together, these urban jewels place 74 acres of green space at the disposal of the citizens of Boston and the more than 19 million visitors[14] coming annually. They are part of the "Emerald Necklace," a 1,100-acre system of parks and parkways laid out by the great 19th-century landscape architect Frederick Law Olmstead. The "Big Dig" has enhanced this with the Rose Kennedy Greenway, which sits atop the automobile tunnel that has replaced the aging elevated highway that once cut off the Boston CBD from its waterfront.

Eugenie Birch classifies Boston as a "fully developed" residential downtown.[15] Beacon Hill and the Back Bay are clearly walk-to-work neighborhoods, and the North End, with its tourist attractions and restaurants, is an easy walk to the Financial District and Government Center. Retailing along Newbury Street and Downtown Crossing (which recently attracted the Irish merchant Primark to the former Filene's department store site) has epitomized downtown shopping, but just as impressive is the cluster in Back Bay that includes Lord & Taylor, Neiman Marcus, Macy's, and Saks Fifth Avenue.

Boston's universities are legendary urban assets. Harvard and MIT, just across the Charles River in Cambridge, garner most of the worldwide attention. But Boston University, Boston College, Northeastern University, and Tufts University are joined by the New England Conservatory of Music and the School of the Museum of Fine Arts in drawing the creative class to Boston. Students enliven the 24-hour ambience of cities, and the Boston area has 152,000 undergraduate and graduate students attending its institutions.

An executive with a pension fund investment advisor (West Coast-based, by the way), noted, "Boston's education/business link is powerful, and it is a city with great design and is a great environment for investment."

Chicago

On the way to O'Hare Airport from Chicago's Magnificent Mile one morning, my cab driver, a Ghanian immigrant, pointed out the bumper-to-bumper traffic on the Kennedy Expressway headed toward downtown. "Those drivers think they have a traffic problem," he said. "What they really have is a community problem." He had my attention, and I asked, "What do you mean?" He went on:

> Most of those people work within a mile of each other, and most of them also live in the same neighborhoods. But they don't know each other, so each one gets in a car, all by himself, and then starts his day complaining that there are too many cars on the road. Maybe they should stop complaining and talk to the people they work with and the people they live near and make some connections. They'd be happier, and maybe everyone could get to work with less stress.

Utopian? Remember the data from Figure 7.6: only 10.5 percent of US commuting involves carpooling, even to the extent of one passenger plus the driver. And Chicago has even less carpooling than that, at 8.8 percent. That's ameliorated somewhat by Chicago's 12.2 percent transit commuting share (versus 5.0 percent nationally), but that cab driver had his finger on the problem and knew intuitively that Chicago had room for improvement – even if just toward the US average carpool share.

It is not that 24-hour cities are all about transit. The live–work–play paradigm works best if you can walk to work. And, for Chicago, 5.1 percent of journeys to work are on foot, second only to New York City, and substantially above the national average of 2.9 percent. The pedestrian character of 24-hour cities – coupled with mass transit – is a distinguishing difference from 9-to-5 places.

That relatively lesser dependence upon the automobile (compared with, say, Atlanta, Dallas, and Phoenix) unlocks other opportunities for cities, especially in nightlife. Not only Rush Street, the restaurant and club strip just west of North Michigan Avenue, but neighborhoods such as Streeterville, Wicker Park, and Logan Square (the latter two right along the Chicago Transit Authority Blue Line to O'Hare) are local hubs for socializing and for the creative class. In the city-data.com surveys mentioned earlier in this chapter, there was a rousing debate about the impact of transit on nightlife. A Chicagoan assertively argued that keeping intoxicated drivers off the road was a crucial element in developing and sustaining vibrant late-night areas.

Jane Jacobs long ago knew that crowded streets were safer streets, and much the same could be said of subways. Crowded roadways, not so much. My interviewees credited Chicago's relative safety as one of the key reasons for its attractiveness. One local developer mentioned "the threshold change Mayor Daley (the second one) made in cleaning up the city." Yet most of the kudos was specifically about the highly developed downtown and the gentrified communities. Communities such as Lawndale, South Austin, and Englewood are largely off the radar screen, and yet have been much in the news, as Chicago's urban distress has been capturing headlines.[16]

Even cities enjoying high vibrancy and solid scores on most of the variables reviewed in this book can't take their prosperity for granted. As the urban unrest that has risen to the surface in 2014 and 2015 in places as diverse as Cleveland, Brooklyn, Baltimore, and Ferguson, Missouri, has shown, inequality of income and of opportunity still fester and threaten progress in 24-hour cities and 9-to-5 cities alike.

Dallas

"Downtown Dallas 360" is subtitled "a pathway to the future." The plan outlined in this 2011 document is aspirational – speaking of Dallas's objective as one that would capitalize on the potential to become a 24-hour city. That alone is a recognition that Dallas is not there yet. The 360 plan identifies downtown Dallas as being within a loop of freeways, and this "identification by highway" is another clue to the challenge the city faces in seeking to move from 9-to-5 to 24-hour status.[17]

When most people think of Dallas, they are not, in fact, thinking of the city of Dallas.[18] Dallas means "the Metroplex" to those who consider urban economies. The Metroplex is the Dallas–Fort Worth metropolitan area, 13 counties covering 9,286 square miles, with a population of about 6.5 million, converting to a population density of just under 700 per square mile. The "center" of the Metroplex is the vast DFW airport, and the City of Dallas is the eastern "pole" of the region. Other than Dallas and Fort Worth, there are seven other "principal cities"[19] in the Metroplex: Arlington, Irving, Richardson, Plano, Denton, McKinney, and Carrollton.

Arlington is where the Texas Rangers of Major League Baseball play their home games. The Dallas Cowboys' stadium is in Coppell, immediately north of the DFW Airport, to the east of the city of Grapevine and northwest of the city of Irving. McKinney is 35 miles north of the city of Dallas, and Denton is

40 miles northwest of Dallas's downtown. A network of freeways and tollways constitute the ligaments holding the Metroplex together, with the inner loop around downtown Dallas just the first ring. I-635 (the LBJ) connects Richardson to Irving, and, further out, the George H.W. Bush Turnpike links Plano and Carrollton. The Sam Rayburn Tollway slices southwest from McKinney to Coppell and Grapevine.

It should be no surprise, given this configuration, that, at both the metro level and the city level, Dallas has the highest auto-dependency of the 14 cities in the journey-to-work data, and also has nearly the lowest Walk Score (Phoenix is slightly lower). In many ways, Dallas defines what urban entropy means – the dispersal of energy rather than its concentration. The major corporations calling the Dallas area home illustrate the diaspora:

- In Dallas itself, we find AT&T (telecommunications), Holly Frontier (oil refining), Energy Transfer (pipelines), Southwest Airlines (transportation), Dean Foods (prepared foods), Texas Instruments (electronics), Tenet Healthcare (hospitals), Celanese (chemicals), and Energy Future Holdings (electrical generation and distribution).
- Irving has Exxon Mobil (oil and gas), Fluor (industrial engineering), Kimberley-Clark (consumer products), and Commercial Metals (steel).
- Fort Worth is home to AMR, the parent company of American Airlines.
- J.C. Penney (retail department stores) is based in Plano, as is the Dr. Pepper Snapple Group (beverages).
- GameStop (electronics retailing) is in Grapevine.
- MetroPCS Communications (recently merged with T-Mobile USA) is based in Richardson.[20]

Dallas, then, has a lot going on in its economy, but it is literally all over the place. Without concentration, there is little opportunity to achieve critical mass. As one commercial property financier said, "This is a bright and optimistic 9-to-5 city, suffering from massive lack of discipline in its patterns of urban growth." That said, the ambitions of Downtown Dallas 360 should not be dismissed as impracticable. Just look at downtown Houston, which was largely empty land from the Allen Center to the East-Tex Freeway in the mid-1980s, but is now a fully functioning urban center. It can be done.

Las Vegas

On the list of 24-hour cities, Las Vegas is an anomaly. Yet, as I began to discuss what cities to consider as the right set of candidates, several people urged me to include this unusual place. One investment advisor said, "It is probably the ultimate 24-hour city," with neon lighting the sky until dawn. However, many of those I interviewed endorsed the inclusion of Las Vegas as a 24-hour city with a firm "Yes . . . but . . ." response.

I'll be candid. I am not a fan of Las Vegas. I have a personal rule: I will go to Las Vegas, but you have to pay me to do that. Nevertheless, despite my personal bias and the many obvious ways that Las Vegas is unlike the other 24-hour cities, it is appropriate on a purely diurnal basis to recognize its similarities as well.

Las Vegas's auto traffic patterns over 24 hours look more like Boston and San Francisco than like Dallas or Phoenix. Las Vegas represents the median spot among 24-hour cities when round-the-clock drugstores within 10 miles of the city center are counted. Although ranking poorly in arts and culture, Las Vegas ranks extremely high in the food and restaurant category and is also near the top in rankings for nightlife specifically.

How is it different? Las Vegas has a much lower population density than is typical of 24-hour cities. The characteristic mix of downtown residential and office space uses is also lacking, as is the diverse stew of industries and occupations most often seen in other 24-hour settings.

In a word, Las Vegas is much more an economic monoculture. That was partly the reason why this city posted one of the highest scores on Markusen and Schrock's RDI.[21] The casino industry is far and away the most important driver of the Las Vegas economy. This has had an interesting result, as gaming has become more prevalent across the United States. Far from reducing Las Vegas's gambling activity, it has enhanced it. Those enjoying a taste of the casino on riverboats or reservations look on a trip to Las Vegas as a chance to "play in the big league."

Not everyone buys into the stereotype of Las Vegas, or sees the monoculture as inevitable. Tony Hsieh, the CEO of Zappos, is spearheading an effort to reinvent downtown Las Vegas as a tech-venture haven for start-ups. He has been attracting purposefully "weird" retail and restaurant facilities for the entrepreneurial workforce that is the core of this version of the creative class.[22] His Project Downtown specifically plans space for community activities, and he has proposed measuring "ROC" (return on community) as a basic business output. If anything, Las Vegas's wide-open ethos makes Hsieh's over-the-edge experiment in urban reinvention seem, well, almost normal. Whatever its ultimate success or failure, it is bold and it is happening already – and seems oddly apt for a 24-hour environment.[23]

Los Angeles

In the early-to-mid-1980s, Los Angeles' downtown was little changed from its 1950s incarnation. This was brought home to me as I drove through the area and caught a glimpse, purely out of peripheral vision, of Los Angeles' City Hall from a few blocks away. I recognized it immediately, but then had to pull over to the curb when the reason for that flash of recognition hit me. I knew what the Los Angeles City Hall looked like, though it was my first visit to the city, because it was featured in the opening credits of the vintage

TV program *Dragnet*, a program so old that I'd viewed it on a black and white television set.

Over the next 10 or 15 years, I would be back in Los Angeles recurrently on business. I'd make a practice of going out for a morning run of 5–7 miles, which lets you see a lot of a city in just an hour or so. When I'd describe the path of my runs to my local colleagues, they would often shake their heads, believing I was taking foolish risks running through areas in and around downtown by myself around sunrise. I found the streets to be incredibly broad boulevards by New York standards, and I was certainly unimpeded by pedestrians as I ran. It felt safe enough to me.

Little by little, the vintage Los Angeles skyline started to take on a more modern flavor, but this was still a predominantly 9-to-5 venue when I began the research that has led up to this book. Based upon the commentaries in *Emerging Trends in Real Estate* through 2010, I did not hesitate much in including Los Angeles in my grouping of 9-to-5 downtowns, in contrast to the New York and San Francisco models. Most of my interviewees held similar opinions. A Texas-based institutional investor told me, "It is not a 24-hour city, even if it has all the bells and whistles." A specialist in international investment said, "LA is an exciting city with lots to do, but its downtown is dead." A pension fund intermediary from Chicago saw something happening, though, and felt, "while it is not a 24-hour city, it has a shot."

Based upon my most recent business in Los Angeles, I am convinced that this 9-to-5 downtown is indeed evolving, and is certainly worthy to be considered an 18-hour city in the making. From the Walt Disney Concert Center atop LA's Bunker Hill, down to the Macy's shopping center at 7th Street between Hope and Flower Streets, I found evening activity on the sidewalks that was of an order of magnitude higher than on my earliest visits. This change is a testimony to the desire of civic officials to make it happen, a handful of developers who saw an opportunity at hand, and visionary planning that refused to take it for granted that the automobile would never cede market share to mass transit in this most car-crazy of cities.

As a result, Los Angeles has about 77 miles of subway and light rail right-of-way serving nearly 350,000 customers daily. The siting of the Disney Center and the Museum of Contemporary Art near the office towers and hotels of downtown encouraged the subsequent development of rental apartments and condominiums. The live–work–play model was embraced and effectively marketed. It is a sign of change when some Angelinos complain about the "Manhattanization" of the LA skyline.[24]

As in Houston, the transformation of Los Angeles has taken decades, and the evolution toward 24-hour-ness is probably not completed. But the naysayers notwithstanding, the progress is a good thing. It is certainly amazing to find the center of Los Angeles with a Walk Score of 92, and a Transit Score of 100! Whodathunkit?

Miami

Miami is sometimes touted as the capital of Latin America, its international feel and flavor only growing over time. But being a hub of trade and finance oriented to the southern hemisphere is only one of the ways in which Miami epitomizes some of the key "trends behind the trends" discussed earlier in Chapter 6.

Globalization for Miami means far more than serving as a repository of flight capital from unstable economies in Central and South America. Exiles from Cuba and Haiti have flocked to Miami for both political and economic refuge for more than 50 years. Investment from Europe has been growing, and Miami typically ranks in the top ten cities in the annual poll conducted by the Association of Foreign Investors in Real Estate. Nor is the story only about real estate. International banking has been flourishing, with Espiritu Santo Bank of Portugal, BNP Paribas of France, Banco Santander and Sabadell Bank of Spain, and Scotia Bank of Canada among those with a presence.[25]

Miami is flanked by the Atlantic Ocean to the east and the Everglades to the west, so environmental issues are never far from public awareness. Florida is the state most vulnerable to sea-level rise – even more than Louisiana, because of Florida's vastly longer coastline. Miami has the largest amount of exposed assets (by value) and the fourth largest population at risk of rising sea levels in the world (after Mumbai in India, and Guangzhou and Shanghai in China), according to a recent OECD study.[26] Consequently, Miami-Dade County and other South Florida governments are vigorously studying the infrastructure needs posed by this threat. Miami truly understands that "climate change denial" is not a viable option for its public policy.

The housing bubble and its subsequent collapse late last decade were particularly spectacular in Miami, which saw an influx of "flipper" investors left holding unsold and often uncompleted condominium projects. What is most interesting locally is how this was turned to advantage in downtown Miami. Developers and condo owners, startled by the immense overhang of inventory and seeing no quick return of potential buyers, adapted by turning to the rental market. At the same time, fortuitously, several universities opened or expanded facilities in or near downtown. The schools included the University of Miami Medical School, Florida International University, Miami-Dade College, and the Miami International University of Art and Design. For many students, renting the unsold condos was the solution to their housing needs. Free mass transit via the "people-mover" monorail system was certainly a plus as well. The presence of thousands of college students unquestionably spurred a more 24-hour ambience locally.

Miami has invested heavily in arts and culture within its downtown and has an emergent creative class neighborhood in Wynwood, just north of downtown. Although the annual Art Basel Miami Beach, spurred by the revitalized Art Deco district across Biscayne Bay from downtown, kickstarted the arts scene in South Florida, this has now taken root in Miami proper. In many ways, the high nightlife scores for Miami – and its justified place on the 24-hour-city list – are a combination of the influences of international visitors, the subtropical climate, education, and the arts.

Minneapolis

An executive of an Asian sovereign wealth fund interviewed in my research said of Minneapolis, "This is a nice, livable city – but it is too cold." Amen, said several others. I first flew into Minneapolis one January evening in 1990, and, upon landing, the pilot announced, "It is not too bad tonight for the Twin Cities – minus 5° Fahrenheit." Brrr.

If global warming is a sea-level threat to Miami, an increase in temperature might be welcomed in Minnesota. But, like good cities everywhere, Minneapolis has been adaptive. In particular, it has developed its skyway system – over 11 miles of interconnected pedestrian bridges, climate-controlled to be sure, that link buildings across 69 square blocks downtown. At the second- and third-floor levels, office buildings, residential towers, banks, hotels, and government facilities access the system. The skybridges also connect to the Nicolette retail mall, the Target Field sports stadium, and the E-Block Entertainment District. Minneapolis has found a way to shield itself from a harsh climate and still promote the live–work–play urban vision.[27]

But the consensus is that Minneapolis is still far from 24-hour-city status. A representative of a U.K. investment firm called it "nice, but provincial." A hotelier described it as "a step-child to Chicago." An executive with a private real estate investment trust (REIT) based in Texas said, however, "It is one of my markets. I don't know why people live there, but people stay despite high taxes, high cost of living, and awful weather." A top real estate researcher, also based in Texas, agreed that there was "great vitality, especially downtown, even if it isn't a 24-hour location."

Part of the "niceness factor" for Minneapolis is its Midwestern ethos. It has strong educational attainment scores and has achieved acclaim for its waterfront revitalization along the upper Mississippi River. There is a sophistication here, too, with good polling scores for the performing arts and classical music. But, on nightlife-specific variables, including food and restaurants, bars, and live music, Minneapolis is mid-pack or lower among the city-data.com and *Travel + Leisure* surveys.

Although it is a compact city of only 58 square miles, its population density is just 7,414 per square miles, well below the densities of most 24-hour cities. The metro area encompasses more than 8,000 square miles, and its 3.5 million population equates to a density of 515 residents per square mile. The concentration factor militates against 24-hour-ness at such levels of sprawl. Birch counts Minneapolis among the "declining downtowns" in her 2005 study.[28] More recently, there have been signs of residential conversions and new apartment development here. The evolution is in the early stages. It certainly bears watching in the coming years.[29]

New York

New York City is an outlier. Many hold that it is paradigmatic, the only "true" 24-hour city in the United States, and that other cities qualify for this category to the degree that they resemble New York. That's a problem for quite a few reasons. It is a fairly Platonic way of looking at things: an "ideal form" approach that doesn't fit many of the facts about New York, which is far from ideal (take it from a life-long resident!). It is also problematic from a statistical standpoint. As several of the figures in this and earlier chapters have shown, New York's scale sometimes skews the numbers considerably, creating "fat tails" in the data distributions. And lastly, it is a dangerous perspective – verging on chauvinism – to set up New York as a model for what all cities should aspire to, rather than letting each develop its own identity and evolve according to its own dynamic.

The burden of this book's analysis and discussion up to this point – and the real estate data to be reviewed in Chapter 9 – makes clear the claim that New York has primacy in the group of 24-hour cities. It would be gilding the lily to press that point here. Rather, let me look at a few of the areas where even this remarkable place – my hometown – has room for improvement.

Although New York's many critics deride it as a playground for the rich, the fact is that income inequality is more pronounced in New York than in any other American city.[30] The impact of deep and pervasive poverty saps civic strength in many ways and is visible in housing, in education, in crime, and in social welfare costs. On the other hand, as I will argue in Chapter 10, the affluent end of the scale provides many resources of great benefit to New Yorkers as a whole. Some have attempted to pit the approaches of Mayor Mike Bloomberg (a self-made billionaire) against those of Mayor Bill DiBlasio (whose campaign theme was "a tale of two cities"), but there is no reason to assume that there is but a single right way to advance New York's future.

New York has massive infrastructure investment needs, some of which it shares with Miami as a coastal city. But mostly, the physical capital needs of New York stem from its age. Cities that have largely grown up since World War II simply don't have vital infrastructure that is more than a century old and stressed by a population of 8 million people. Water systems, roadways, mass transit, bridges, schools: all these are capital assets wearing down faster than they are being improved. That creates chokepoints, and those will continue to be complicated issues for government.

Lastly (for now), I would mention the political reality that New York cannot address its own needs independently of other levels of government. New York City is a net exporter of tax revenues to New York State and to the federal government, meaning that it receives less in returns of cash and services than it sends out. This hobbles the city's ability to improve education and to provide healthcare, for example. Regional transportation is another matter, with the Metropolitan Transportation Authority a creature of the State of New York, and the Port Authority reporting to the governors of both New York and

New Jersey. There's horse-trading between city and suburbs, as well as upstate–downstate within New York – and sharp (often petty) competition between the two states for Port Authority resources.

So, as much as my research has turned up reasons to celebrate New York, I'm far from persuaded it should rest on its laurels.[31]

Philadelphia

Philadelphia's boosters are frankly offended by suggestions that it is not a 24-hour city. They look at Center City and ask, "What could be missing?" Its downtown housing is on Birch's "fully developed" list. Wanamaker's may be gone, but Macy's inhabits its downtown emporium. There are world-class museums on Ben Franklin Parkway, and the Philadelphia Orchestra on Broad Street. Mass transit and close-in housing allow more than a third of all city residents to take the bus or subway or walk to work each day. The Universities of Pennsylvania, Drexel, and Temple are renowned higher-education institutions. And now crime has dropped to its lowest level since 1971.[32]

That's all true, and my overnight visits to Philadelpha are frequent enough for me to affirm that the streets are far from deserted as I walk back to my hotel from dinner. Nevertheless, the sidewalks are not exactly bustling. A Philadelphia-based investment manager whose offices are in Center City revealed that they prefer to own office parks in Philly's suburbs and remarked negatively about homelessness and panhandling ("a turn-off") in the core. Many interviewees remarked positively about the city's ties to history, but found it lacking in energy and contemporary feel. A hotelier commented, "It's a missed opportunity. It has all the elements but they are not capitalized on." A Chicago-based institutional investor mused, "Why doesn't it work? It has all the right ingredients, but the cake fell."

For many, Philadelphia seems like a weaker version of the robust cities in the Boston–Washington corridor. One observer went so far as to say, "Isn't it really just an Edge City for New York?"

I think that such judgments are unduly harsh, but do point out a very important feature. Creating a 24-hour city is not just about assembling a list of ingredients: it is about having a good recipe and executing it well. Philadelphia's relatively high nightlife scores may indicate that its reputation is getting stronger, or may actually reflect a shift in the way its urban elements are beginning to interact more vigorously.

Professor Emil Malizia of the University of North Carolina at Chapel Hill has been researching "vibrant places" and classifies Philadelphia's downtown in fourth place on his multifactor "vibrancy index," after New York, Chicago, and San Francisco, but ahead of Washington, DC, Boston, Seattle, and Los Angeles.[33] There's something happening here, and it appears that an analysis of factor interactions may be a fruitful path forward.

Phoenix

On a beautiful spring afternoon in 2013, I took the light rail from Tempe, Arizona, where I was attending a meeting at Arizona State University (ASU), into downtown Phoenix. It was an almost perfect 70°, clear and dry, on this particular Friday, and I timed my arrival for 4:30 p.m. My thought was to walk around the downtown, observe the beginning of the weekend, have a "people-watching dinner" somewhere, and take the tram back to Tempe around 10 p.m. As it happened, that would have been about four hours too long.

Now, what is in downtown Phoenix? Practically everything on an economic developer's shopping list. (Sort of like Philadelphia in that respect, only more so.) There is Chase Field, the home of the Arizona Diamondbacks. There's the US Airways arena, home to the Phoenix Suns NBA team. A couple of blocks north is the convention center, and just to the west of that is Symphony Hall. The Herberger Theater Center is just to the north of that. Phoenix's civic buildings are nearby, as are several lovely small parks, the city's historic Mercado, and the administrative offices of the Roman Catholic Diocese. The Phoenix campus of the University of Arizona School of Medicine is right there, too, as is the Prep School for ASU. There are plenty of hotels at all price points, and, of course, the quick and quiet light rail to facilitate movement. If you drew it up in city planning school, you couldn't ask for more.

I walked around for about 90 minutes, noticing that I didn't have very much company on the sidewalks. I watched cars streaming out of the parking garages on North Central Avenue, heading out on this Friday night to homes in Scottsdale, Glendale, Mesa, Carefree, Goodyear, and the other bedroom communities in the Valley of the Sun. I kept waiting for the 20- and 30-somethings to head for the bars and the restaurants to start the weekend off. I could be waiting still, but, when sheer quiet descended shortly after 6 p.m., I found a pizzeria, grabbed a couple of slices, and headed back to Tempe, where the ASU students *were* commencing a social evening in ways I'd recognize.

Eugenie Birch is, I think, generous in dubbing Phoenix a "slow growing residential downtown." My observation was that it fell far, far short of critical residential mass. And, hence, regardless of the actual utilization rates of the sports facilities, the arts and culture venues, the convention center, and the other chess pieces that had been positioned on Phoenix's street grid, they were not returning the key desirable outcome of the economic development investment – a lively, thriving, exciting downtown. Perhaps in none of the other cities I've focused on was it so clear how critical, how catalytic, the residential contribution is to a 24-hour downtown.

If downtown is not "home" to a significant number of citizens, those who work in the central core will exit at the end of the work day so that they *can* go home. Housing is key – housing for those who work in the immediate vicinity, housing at appropriate income levels, housing that is "a part of, not apart from" the other downtown functions. Unless there is enough housing to create truly urban densities – by which I mean approximately 10,000 residents per square mile, or 15 persons per acre, *in addition to the other land uses in the same area*, it is probable that a 24-hour ambience cannot be generated.

Phoenix makes this case wonderfully, precisely because it has brought together all the other elements – and still is sterile at its core.

San Francisco

Situated at the tip of a peninsula, San Francisco fits its population of 852,000 into just 47 square miles of land area, yielding a population density of 18,187 residents per square mile. Inward commutation accounts for a surplus of employment over residents, as the late 2014 jobs figure for San Francisco exceeded 1 million, indicating a daytime population that is 22 percent higher than the number of residents. San Francisco's compact physical footprint aids in creating the critical mass that releases its 24-hour-city energy.

BART, the Bay Area Rapid Transit System, has five lines that reach south on the pensinsula to San Francisco's international airport (and slightly beyond to Millbrae), as well as across the Bay to Fremont, Dublin/Pleasanton, Pittsburg/Bay Point, and Richmond. Critically, all five lines converge at San Francisco's downtown. Although it doesn't describe the dense capillary system of cities such as Boston, Chicago, and New York, BART is far more efficient at funnelling commuters into the core of the region than MARTA in Atlanta, or the limited light rail available in Dallas or Phoenix. Consequently, 14.6 percent of metro area workers use transit on their journey to work (versus the US average of 5.0 percent), and 31.8 percent of San Francisco city residents are transit commuters. Overcrowding on BART is emerging as an issue, in fact.[34]

Impressively, 10.3 percent of city residents walk to work (versus 2.9 percent nationally), the second highest proportion (after Boston's 14.1 percent) of on-foot commuting of the 14 cities at the heart of this study. I find the walk-to-work ratio particularly impressive once the city's daunting hills are taken into consideration. The city's overall Walk Score is a "very walkable" 86 (highest among the 14 cities), with the Financial District a nearly perfect 99, a "walkers' paradise." This no doubt helps San Franciscans take advantage of the temperate Northern California climate.[35]

San Francisco is consistently rated a 24-hour city in the *Emerging Trends* reports, and a top investment choice among US cities. One of the keys may be the barriers to entry that come from the limited amount of land and stringent zoning requirements imposed by the City of San Francisco. But San Francisco (and the Bay Area as a whole) has evolved a complex and interactive approach to managing growth – and decidedly does not encourage growth for growth's sake. Regulation is indeed employed, but so are economic incentives, design guidelines, citizen coordination, and sophisticated technological tools, including GIS and computer simulations of development proposals.[36]

Like New York, San Francisco has its jarring juxtapositions of affluence and poverty, with homeless encampments and street begging chockablock with the office towers of the Financial District, the upscale retailing along Market Street, and the public spaces around the Moscone Convention Center. This is just another reminder to look clearly at our cities, warts and all, as we consider what works and what doesn't for urban America.

Seattle

Seattle is another city for which there is divided opinion about its classification as 24-hour or 9-to-5. A prominent New York developer with worldwide interests called it, "a lovely city, but not 24-hour." The head of research for a European financial giant saw it as "up and coming; a notch below San Francisco with a high quality of life but not enough density." The chief investment officer of a large public pension fund characterized it as "the prettiest city in America on a clear day."[37] An Atlanta-based specialist in real estate debt placement compares Seattle to Philadelphia: "good bones, and doing the right things locally." And, most enthusiastically, a San Francisco-based urbanist affirmed, "Yes – a 24-hour place; walkable and tight downtown; diversity in population with an Asian and Canadian flavor."

Seattle is compact, with a land area of just 84 square miles supporting its population of 652,000 for a density of 7,762 per square mile. This is a lower density than is typical of the primary 24-hour cities. Seattle is no better than mid-rank in the Markusen–Schrock scale of regional distinctiveness. Transit commutation share for the city is 19.5 percent, and 8.7 percent at the metro level; these are strong transit shares for the 9-to-5 grouping of cities, but are below average for the set of seven 24-hour places being considered.

On the specifically diurnal measures presented earlier in this chapter, Seattle has little in the way of late-night automobile traffic and the lowest number of 24-hour drugstores of any of the 14 cities being reviewed. It has middling scores for art and culture, and below average scores – even for 9-to-5 cities – for food and restaurants. With the exception of live music, Seattle rates poorly for its nightlife in both the *Travel + Leisure* and city-data.com survey results. On balance, it seems safe to say that Seattle is "not there yet" in its claim for 24-hour status.

More positively, though, the technology industry – with behemoth Microsoft headquartered nearby – is driving tremendous economic activity in Seattle's downtown.[38] Birch solidly classifies the city as a strong "emerging downtown," and this seems to have accelerated since her 2005 study, with the increasing influence of the Millennial generation. Construction levels for multifamily units have been rising, and so have rents. These are positive signs for that key, catalytic factor of downtown housing that makes so many of the 24-hour features economically sustainable.

The vitality of Seattle's core rests upon its own strengths and weaknesses, of course. But it must be acknowledged that the core is located on a comparatively narrow piece of land between Puget Sound and Lake Washington. Substantial suburban competition exists in Bellevue, Everett and Renton (the home of Boeing), and Redmond (the location of Microsoft's main campus). On balance,

Seattle probably finds itself with a credible claim to 18-hour-city status, along with Los Angeles and Philadelphia (and other cities, such as Austin, Houston, Denver, and Nashville). But, for the purposes of this book – where a bright line between 24-hour and 9-to-5 cities is being tested – Seattle falls into the latter category.

Washington, DC

In 2005, Birch's "Who Lives Downtown?" classified Washington as a residential downtown "on the edge of takeoff." In the 10 years since then, the nation's capital has soared as a place to live well beyond the most optimistic of expectations. There have long been solid residential neighborhoods in reasonable proximity to the District's government and business offices. Georgetown, Dupont Circle, Adams Morgan, and the West End have attracted students and young professionals by their proximity to employment, the spin-off improvements of good universities, and a "hip" nightlife. Now Washington is finding that formerly desolated areas, including sections close to the US Capitol building and places such as Columbia Heights that were devastated in the 1968 riots, are being revitalized.

The 24-hour-city claim for Washington starts with the observation that the US government never shuts down. Congressional staffers, in particular, put in extremely long hours and have been known to flock to the bars and nightclubs once freed from their offices. Tourism to the many monuments and venues on the National Mall is a constant spur to activity, and night photography of many of the sites (such as the Lincoln and Jefferson Memorials) can be spectacular. Put the combination of young people and tourists together, and DC's high rating for ethnic restaurants and arts–culture venues comes into focus.

I recall having a late dinner in the Georgetown neighborhood, some years ago, and walking back to my hotel around 11 p.m. Realizing that I'd finished the book I had been reading, I was about to consider seeing if the TV selections included *The Colbert Report* when I walked by a Barnes & Noble bookstore that had open hours until midnight. In I went and found myself amid a throng of book browsers and buyers. Obviously, there was enough pedestrian traffic to justify such late hours.

The count of 24-hour drugstores confirms this. With 25 such all-night stores within 10 miles of the city center, Washington tallied more than any of the 9-to-5 cities, with the exception of Philadelphia.

Although Washington's metro subway does cease operation in the small hours of the morning (only New York City runs subways 24/7/365), its ridership is about 700,000 per day, and Washington ranks second (at 37.1 percent) only to New York (at 54.9 percent) in the proportion of city residents commuting to work by transit. Moreover, 11.1 percent of Washingtonians walk to work, keeping city-generated auto commutation to just 43.2 percent. (The traffic jams coming in from the suburbs, however, are legendary.)

The sprawling suburban ring in Northern Virginia and in Maryland does vitiate Washington's 24-hour energy somewhat. According to the US Census Bureau's 2005–2010 ACS of Daytime Population, no other city in America approaches the relative increase in workday population in Washington – at 79 percent above its resident-worker base. By comparison, Boston has 40 percent, Seattle has 27 percent, Dallas has 21 percent – and New York has 8 percent.

A veteran urbanist with a West Coast investment firm summarizes his take on Washington this way: "It is a top choice for institutional investors. It has great infrastructure in its parks, the Metro, and is a functional walkable city. Everyone in our shop loves D.C."

Notes

1 One intriguing effort to address this question can be found in Luis M.A. Bettencourt, José Lobo, Dirk Helbing, Christian Kuhnert, and Geoffrey B. West, "Growth, Innovation, Scaling, and the Pace of Life in Cities," *Proceedings of the National Academy of Science 104:17* (April 24, 2007), pp. 7301–7306. Accessible online at www.pnas.org_cgi_doi_10.1073_pnas.0610172104

2 I gratefully acknowledge the help of my daughter, Beth Kelly, in tackling the tedious job of tracking down this information and arraying it in a convenient spreadsheet for my analysis.

3 See Andres Duany, Elizabeth Plater-Zybeck, and Jeff Speck, *Suburban Nation: The Rise of Sprawl and the Decline of the American Dream*, North Point (New York, 2000); Peter J. Taylor, "Leading World Cities: Empirical Evaluations of Urban Nodes in Multiple Networks," *Urban Studies 42:9* (August 2005); Eugenie Birch, "Who Lives Downtown?" Brookings Institution (2005); Ann Markusen and Greg Schrock, "The Distinctive City: Divergent Patterns in Growth, Hierarchy, and Specialization," *Urban Studies 43:8* (July 2006); Jeff Brugmann, *Welcome to the Urban Revolution: How Cities Are Changing the World*, Bloomsbury Press (New York, 2009); Edward Glaeser, *The Triumph of the City*, Penguin Press (New York, 2011).

4 See Peter Hall, *Cities in Civilization*, Pantheon Publishing (New York, 1998).

5 John Montgomery, *The New Wealth of Cities: City Dynamics and the Fifth Wave*, Ashgate (Farnham, UK, 2007).

6 I have discussed the various levels of research, and the cognitive functions associated with them, in "Dimensions in Real Estate Research," *Real Estate Review 31:1* (Fall 2001).

7 Zagat also publishes *Nightlife* guides for Los Angeles and San Francisco in the US, and London in the UK.

8 Montgomery's discussion of the evening economy explicitly examines the phenomena of alcohol consumption and abuse as these affect entertainment districts and adjacent residential areas, with an eye to formulating planning and regulatory principles to maximize the benefits of a "cafe society," without the negative effects of concentrated public intoxication.

9 Quantcast data are as of May 2015. City-data.com was ranked 87th in traffic in 2010, indicating that it is nicely sustaining its place in the Internet hierarchy.

10 The results of the three polls should not be considered independent, as the same users may be voting in more than one survey. Likewise, this is not a "random sample," as participation is self-selected. Nevertheless, the city-data.com results seem very consistent with the *Travel + Leisure* evaluation of nightlife for the various cities.

11 Art. cit.

12 Of course, the fans of the Chicago Cubs may fairly note that they have been even more long-suffering, as they have been waiting since 1908 to celebrate a World Series victory.
13 Census population estimate as of 2013.
14 Visitor statistics for calendar year 2014, according to the Greater Boston Convention and Visitors Bureau.
15 Birch, art. cit.
16 See, among many examples, Salim Muwakkil, "Violent Crime in Black Neighborhoods Price of Debts Unpaid," *Chicago Tribune*, July 13, 2014 (op-ed); Monica Davie, "In a Soaring Homicide Rate, a Divide in Chicago," *New York Times*, January 2, 2013; Peter Slevin, "Chicago Grapples with Gun Violence; Murder Toll Soars," *Washington Post*, December 21, 2012.
17 For details of the Downtown Dallas 360 plan, visit the website online at www.downtowndallas360.com/
18 The City of Dallas itself has 1.28 million residents and covers 386 square miles, yielding a population density of 3,330 per square mile. It is intriguing to note that Dallas, frequently cited as one of America's "growth cities," has attained this population in more 160 years since the mid-19th century, whereas "slow growing" New York City has gained 1.4 million residents since the 1980 Census, for a population density *increase* of 4,255 per square mile.
19 Principal cities as defined in metropolitan area definitions from the US OMB.
20 Corporate locations are for 2013 Fortune 500 companies, as reported by the *Dallas Business Journal*, May 7, 2013.
21 Art. cit.
22 Richard Florida calls this subgroup the "experientials" in *Who's Your City?*
23 See Sara Corbett's article, "How Zappos' CEO Turned Las Vegas into a Start-up Fantasyland," *Wired*, January 21, 2014, online at www.wired.com/2014/01/zappos-tony-hsieh-las-vegas/
24 Joel Kotkin, "Why the Rush to Manhattanize L.A.?" *Los Angeles Times*, August 12, 2007 (op-ed).
25 See Brian Bandell, "International Banking Turned Brickell into the Wall Street of Miami," *South Florida Business Journal*, August 23, 2013; the website of the Florida International Banking Association (online at www.fiba.net/pages/Home) provides a comprehensive list.
26 "Ranking of the World's Cities Most Exposed to Coastal Flooding Today and in the Future," OECD Environment Working Paper 1 (2007).
27 There is even a restaurant guide to the Skyway using the online booking service Open Table, accessible online at Downtown Minneapolis Skyway Guide www.skywaymyway.com/
28 Birch, art. cit.
29 Kristen Leigh Painter, "Opus Updates Plans for 32-Story Residential Tower in Downtown Minneapolis," *Star Tribune*, February 27, 2015.
30 Joel Kotkin, "Where Inequality Is Worst in the United States," *Forbes*, March 20, 2014. But see also a Brookings Institution report, "Some Cities Are Still More Unequal than Others – an Update" (March 17, 2015), that employs a measure comparing the 95th percentile income to the 20th percentile income and finds NYC ranking sixth in inequality, after San Francisco, San Jose, Washington, DC, Atlanta, and Seattle.
31 *Premises*, the magazine that I edit for New York University's Schack Institute of Real Estate, has published many articles prodding New York toward improvement. These include Mary Ann Tighe, "New York: What We Have and What We Need" (Spring 2012); Corrine Packard, "The Bloomberg Way Out?" (Fall 2012); Richard Florida, "Ever Resilient New York" (Spring 2013); Frank Sciame, "New Ways of

Building a Greater New York" (Fall 2013); and John T. Farrell, "Undermining the Foundation: Dysfunctional Transit Funding Systems Compromise New York's Future" (Spring 2014).

32 See "Hard Graft Endangered: The Philadelphia Mayor's Race," *The Economist*, May 23–29, 2015, pp. 22–23.

33 Emil Malizia, "Vibrant Places: Clarifying the Terminology of Urbanism in the US Context," *BDC Journal 13* (2013), pp. 175–180. I should note that Dr. Malizia applied his concept of vibrant places to the 24-hour-city research conducted for my own doctoral dissertation (which is the starting point for this book) in "Office Property Performance in Live–Work–Play Places," *Journal of Real Estate Portfolio Management 20* (2014), pp. 79–84. Subsequently, we decided to undertake further research in this area as a joint project.

34 Mark Andrew Boyer, "BART Grapples with Crowding as Ridership Surges," *KQED News*, April 17, 2014, online at http://ww2.kqed.org/news/2014/04/16/132915/

35 Joel Garreau places San Francisco in his "Ecotopia" area in *The Nine Nations of North America*, Avon Books (New York, 1982).

36 For a detailed discussion of these issues, see Karina Pallagst, "Growth Management in the San Francisco Bay Area: Interdependence of Theory and Practice," Working Paper 2006–02, Institute of Urban and Regional Development, University of California, Berkeley, CA (2006).

37 However, downtown Seattle averages only 71 clear, sunny days a year, according to the National Climatic Data Center (report issued by the National Oceanic and Atmospheric Administration [NOAA], online at www.ncdc.noaa.gov/extremes/extreme-us-climates.php).

38 *Emerging Trends in Real Estate 2015*, p. 41.

9 Voting with the wallet

Paying a premium for the 24-hour city

We have many ways of framing a key concept in American idioms. "You get what you pay for," is oft cited and has been attributed to John Arbuckle.[1] A business school commonplace about "having a clear value proposition" can be used as a justification for expensive pricing based on the claim of superior quality.[2] The tattoo artist Norman "Sailor Jerry" Collins maintains, "Good work ain't cheap; cheap work ain't good."

Although value and price are clearly not the same thing, markets are supposed to be able to close any gaps between the two over time. Efficiency should drive overpricing or underpricing toward a more sustainable central value. If and when such gaps arise, market theory says that investors can and will exploit the difference to make money through a process called arbitrage, by betting that excessively high or low prices will correct themselves. Many observers anticipate that this will occur by a process called reversion to the mean, although, if there are real differences in quality, there is good reason for a related spectrum of prices.

There is little question that the 24-hour cities are expensive. The cost of living is elevated, especially by housing expenses, and wage rates tend to reflect the higher costs. This, in turn, means that employers find their payrolls more costly. Over the years, economic development and corporate relocation specialists have looked to such conditions as a vulnerability for the group of cities identified as 24-hour markets, arguing that a rational business executive would prefer lower-cost locations.[3]

Figure 9.1 shows that the 24-hour cities are uniformly above-average in cost of living, whereas the 9-to-5 cities are arrayed around the average, with mean and median costs higher in the 24-hour set than in the 9-to-5 set and higher than the remaining cities listed among the 25 largest in the United States.[4]

The comparison of seven 24-hour locales and seven 9-to-5 areas is silent about the more general question of whether either or both sets are distinctively different from US markets as a whole. To address this question, the balance of the top 25 metropolitan areas by population size were also studied and will be considered in this chapter.

The 24-hour and 9-to-5 sets of metropolitan areas featured rough geographic balance, east to west and north to south, and were very closely matched in

24-hour		9-to-5		Balance of top 25	
City	**Cost index**	**City**	**Cost index**	**City**	**Cost index**
Boston	140.1	Atlanta	93.5	Baltimore	111.3
Chicago	117.4	Dallas	95.7	Cincinnati	91.8
Las Vegas	102.7	Los Angeles	131.0	Cleveland	96.2
Miami	107.2	Minneapolis	110.3	Denver	104.0
New York	182.8	Philadelphia	121.2	Detroit	96.1
San Francisco	159.9	Phoenix	97.3	Houston	99.0
Washington, DC	141.6	Seattle	117.5	Portland, OR	119.1
				Riverside, CA	112.0
				St. Louis	94.4
				Sacramento	109.8
				San Diego	129.2
				Tampa	92.9
Mean	136.0	Mean	109.5	Mean	104.7
Median	140.1	Median	110.3	Median	101.5

Figure 9.1 Cost of living index: Major US cities (index: US average = 100)

Source: Council for Community and Economic Research, "Cost of Living Index – Third Quarter 2013," October 2013

aggregate population and employment at the start of the current century. The metro area was selected as the initial level of economic geography, as this level of urbanization captures commuter populations more completely and provides data relevant to CBD–suburb comparisons. The data are from the first quarter 2000, a snapshot from a time when the 24-hour city argument was being proposed, but had not yet been statistically analyzed.

The sets of 24-hour and 9-to-5 metros were within 1.8 percent of each other in aggregate population, and within 0.6 percent in total employment (Figure 9.2). By comparison with the balance of the 25 largest metro areas, the 24-hour and 9-to-5 sets had higher average populations and workforce. This reflects the inclusion of eight of the ten largest MSAs in the clusters of 24-hour and 9-to-5 markets to be studied. In 2000, the employment–population ratio for the United States was 46.6 percent. Both the 24-hour MSAs and the 9-to-5 MSAs surpassed this norm, at 47.5 percent and 48.1 percent, respectively. The balance of the 25 largest MSAs was slightly subpar on this measure, at 46.2 percent.

	Metro population	City population	Downtown population	City as percentage of MSA (%)	Downtown as percentage of city (%)	Downtown as percentage of metro (%)	City population density per square mile
24-hour							
New York	9,314,235	8,008,278	169,420	85.98	2.12	1.82	27,072
Chicago	8,272,768	2,896,016	72,843	35.01	2.52	0.88	12,544
Washington, DC	4,913,153	572,059	27,667	11.64	4.84	0.56	9,472
Boston	3,406,829	589,141	80,903	17.29	13.73	2.37	12,352
Miami	2,253,362	362,470	19,927	16.09	5.50	0.88	10,816
San Francisco	1,713,183	776,733	43,531	45.34	5.60	2.54	15,808
Las Vegas	1,563,282	478,434	31,834	30.60	6.65	2.04	4,800
Total	31,436,812	13,683,131	446,125				
Average	4,490,973	1,954,733	63,732	43.53	3.26	1.42	13,266
9-to-5							
Los Angeles	9,519,338	3,694,820	36,630	38.81	0.99	0.38	8,192
Philadelphia	5,100,931	1,517,550	78,349	29.75	5.16	1.54	10,816
Atlanta	4,112,198	416,474	24,931	10.13	5.99	0.61	3,648
Dallas	3,519,176	1,188,580	22,469	33.77	1.89	0.64	3,520
Phoenix	3,251,045	1,321,045	5,925	40.63	0.45	0.18	3,072
Minneapolis	2,968,806	382,618	36,334	12.89	9.50	1.22	6,784
Seattle	2,414,616	563,374	21,745	23.33	3.86	0.90	6,848
Total	30,886,110	9,084,461	226,383				
Average	4,412,301	1,297,780	32,340	29.41	2.49	0.73	6,126
Balance of large metros							
Houston	4,715,407	1,953,631	15,708	41.43	0.80	0.33	3,372
Riverside	3,254,821	255,166	14,139	7.84	5.54	0.43	3,267
San Diego	2,813,833	1,223,400	17,894	43.48	1.46	0.64	3,772
St. Louis	2,698,687	348,189	7,511	12.90	2.16	0.28	5,622
Baltimore	2,552,994	651,154	30,067	25.51	4.62	1.18	8,058
Tampa	2,395,997	303,447	7,845	12.66	2.59	0.33	2,708
Denver	2,176,240	554,636	4,230	25.49	0.76	0.19	3,617
Cleveland	2,148,143	478,403	9,599	22.27	2.01	0.45	6,166
Detroit	2,060,913	951,270	36,871	46.16	3.88	1.79	6,822
Cincinnati	1,979,202	331,285	3,189	16.74	0.96	0.16	4,247
Portland	1,927,888	529,121	12,902	27.45	2.44	0.67	3,940
Sacramento	1,796,715	407,018	27,021	22.65	6.64	1.50	4,189
Total	30,520,840	7,986,720	186,976				
Average	2,543,403	665,560	15,581	26.17	2.34	0.61	4,648

Figure 9.2 Population data at 2000

It is revealing to examine the degree to which populations are concentrated in the core cities of the metro areas and, within those cities, the degree of concentration of residents in the downtown districts themselves. City and metro area population data are both published by the Census Bureau, but it is important to note that, whereas metro areas are administratively defined according to criteria centrally determined by OMB, cities are political jurisdictions incorporated under the laws of the several states. Metro areas, for the most part, comprise several contiguous counties, although the Los Angeles metro area is coterminous with Los Angeles County. For the most part, US cities are subsets of counties; New York City, which was consolidated in 1898, is the exception, as this city encompasses five counties.

An important finding of this research is that urbanization seems considerably more intense in the 24-hour markets. As seen in Figure 9.2, population density for the 24-hour cities averages 13,266 persons per square mile, more than twice the density of the 6,126 persons per square mile that is the mean level of the 9-to-5 cities. The 12 cities rounding out the set of 25 largest MSAs have lower densities yet. At 4,648 persons per square mile, these cities display density averaging 24.1 percent lower than the 9-to-5 cities. The average city size for the 24-hour market grouping is 50.6 percent higher than the 9-to-5 market grouping. As the metro areas in the two groups have virtually identical population bases, city population as a percentage of metro total is also half again as great for the 24-hour cluster, 43.5 percent for the 24-hour places and 29.4 percent for the seven 9-to-5 cities. The mean population of the balance of large metros (BLM) is just 51.2 percent of the 9-to-5-city average, and core cities represent only 26.2 percent of the aggregate population of those 12 metros.

Downtown residential populations have been a key element of the descriptive definition of 24-hour cities since the term first surfaced in the industry literature. Eugenie Birch of the University of Pennsylvania, in her paper "Who Lives Downtown?" assembled data from the 2000 Census for residential populations within or adjacent to urban CBDs. Those data can be found in the column in Figure 9.2 labeled "Downtown population." As the industry participants in the annual *Emerging Trends* survey have correctly surmised, the mean downtown population of the cluster of 24-hour cities is much higher – double that of 9-to-5 cities. Similarly, the 24-hour downtowns have about twice the metro share of population (1.42 percent vs. 0.73 percent) of the 9-to-5 urban areas. As with the other variables, the balance of the top 25 MSAs have yet-lower downtown residential populations, averaging 15,581 persons, as well as a lower share of metro total at 0.61 percent.

Commercial office space: Metrics of the workplace

If you do a Google Image search using the simple keyword "city," what comes up? High-rise towers, principally office buildings. The skyline is the iconic image of the city. And even deep into the technological era that has bruited "the death of distance,"[5] the evidence on the ground still more strongly supports "the seduction of place."[6]

Trends in development and market size

24-hour cities have been characterized as live–work–play environments in the real estate industry literature, a terminology now increasingly being explored academically. Office buildings have become the workplace for more and more Americans in the past five or six decades. But office buildings are not distributed uniformly across the urban landscape of the United States.[7] Figure 9.2 shows that the populations of the three clusters of metro areas identified were approximately equal, ranging between 30.5 and 31.5 million as of the 2000 US Census. At the turn of this millennium, however, the seven 24-hour metros had nearly 1.2 billion square feet of commercial office space, whereas the 9-to-5 cluster had about 675,000 million square feet of offices in its seven metro markets, and the 12 metros in the BLM group had even less office space, 572 million square feet. Nearly half of all the office space tallied in the three groups resided in the 24-hour metros as of 2000.

In terms of downtown offices, the degree of concentration in the 24-hour cities is even more extreme. Some 696 million square feet of offices could be found in the 24-hour CBDs, compared with 194 million square feet in the 9-to-5 downtowns and 166 million square feet in the BLM downtowns. Looked at another way, the majority (59.1 percent) of office space in the 24-hour metros was located in the CBD as of 2000, whereas both other groupings had just a 29 percent concentration in their space.

Examining the trends over time (Figure 9.3), the disproportion increased through 2014. The 24-hour metros added 253 million square feet of offices since the turn of the millennium, whereas the 9-to-5 metros developed 133 million square feet and the BLM metros 143 million square feet. And the share of all office space located downtown moved up to 60.0 percent in the 24-hour markets, whereas it dropped to 27.0 percent in the 9-to-5 CBDs and even more steeply to 24.6 percent in the BLM downtowns. So, it is apparent that the hollowing-out process is still very much part of the urban story – except in places where the 24-hour city attributes exist.

Over a longer period of time, as the calculation of change 1987–2014 in Figure 9.3 shows, the 24-hour set has added more office space than the 9-to-5 and the BLM sets at the metro level, and also for the CBD and suburban subsets. More seriously, the latter two market sets have decelerated in their office workplace additions since 2000, and this is especially true in their CBDs. As places to work, the 9-to-5 and BLM markets are still growing in an absolute sense, true, but they are losing ground relative to the 24-hour markets. Further, the gravity of the 24-hour downtowns is doing better in maintaining the CBD–suburb equilibrium, whereas the entropic trend toward sprawl is accelerating in the other two sets of cities.

Year	Urban segment	24-hour group	9-to-5 group	BLM group
1987	Metro area	973,734	487,828	439,392
	CBD	624,489	152,009	144,115
	Suburbs	349,245	335,819	295,277
2000	Metro area	1,177,846	674,100	571,893
	CBD	695,885	194,322	166,770
	Suburbs	481,961	479,778	405,781
2014	Metro area	1,431,091	807,988	715,224
	CBD	800,887	218,187	175,896
	Suburbs	630,204	589,801	539,328
Change 1987–2014	Metro area	457,357	320,160	275,832
	CBD	176,398	66,178	31,781
	Suburbs	280,959	253,982	244,051
Change 2000–2014	Metro area	253,245	133,888	143,331
	CBD	105,002	23,865	9,126
	Suburbs	148,243	110,023	133,547

Figure 9.3 Office space inventory and change (thousands of square feet): 1987–2014

Source: CBRE database

Economic operations: Rent and vacancy comparisons

"Location, location, location" has long-since entered the lexicon of real estate clichés (as the answer to the question, "What are the three most important factors for real estate success?"). Superior locations are expected to command a price premium, and every residential real estate broker is fluent in the language explaining such superiority. The prestige of neighborhood or even of specific address is one selling point. The quality of schools or transportation access or amenities such as water or park views also enter into the discussion. Public safety, municipal services, and the educational attainment of local residents may also factor in.

All of these factors are what economists call "externalities," influences on the price of a property that are actually not features of the physical asset itself. For commercial properties such as office buildings, the urban attributes discussed in Chapters 7 and 8 constitute such externalities. Are businesses willing to pay higher real estate costs in the form of office rents to acquire those externalities? If so, what is the premium? What justifies the greater expenditure?

A review of the statistical data unambiguously answers the question about the willingness to pay. As Figures 9.4 and 9.5 show graphically, over time, the 24-hour downtowns have earned significantly higher rents – on average, 85 percent higher than the 9-to-5 CBDs, and 121 percent higher than the BLM downtowns. This is a huge premium that has been maintained in both the ebbs and flows of economic and market cycles. The likelihood that this is accidental is vanishingly small, from a statistical perspective. Businesses act as if there is a real value proposition here.

Suburbs in the 24-hour metros also command a premium when compared with the other clusters of locations. Although the premium has been maintained each and every year since 1987, it is, on average, less impressive than the

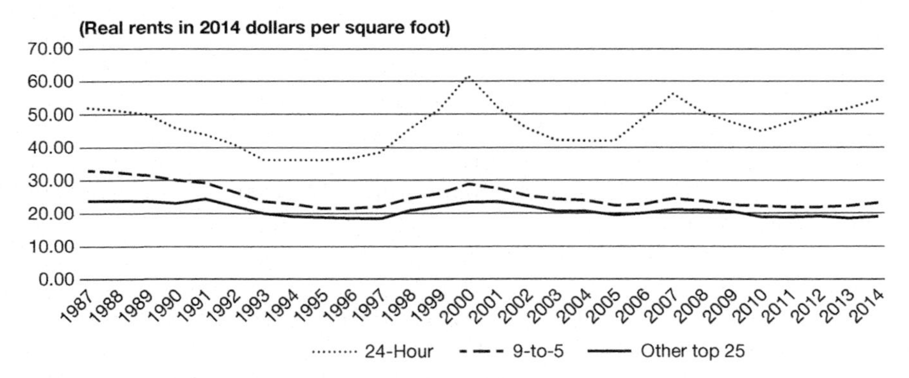

Figure 9.4 Downtown rents in 24-hour cities far surpass other CBD office markets

Source: Data from CBRE, Inc.; calculations by author

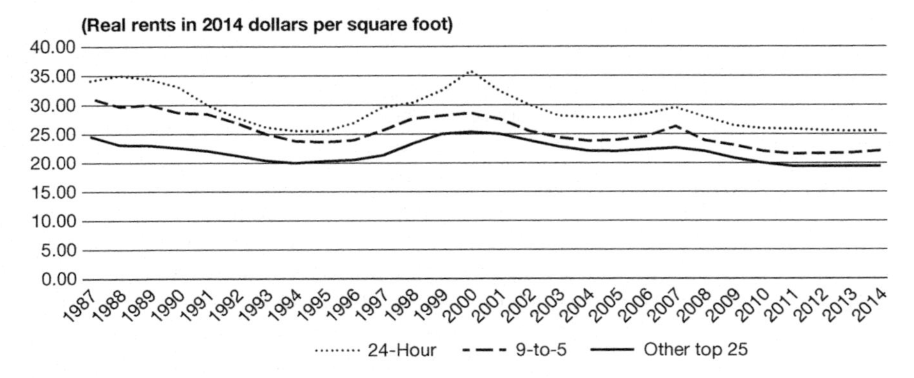

Figure 9.5 Suburban rents near 24-hour cities average more than 20 percent higher than other large suburban markets

Source: Data from CBRE, Inc.; calculations by author

downtown spread. Suburbs in the 24-hour cluster have averaged rents at 114 percent of those of 9-to-5 suburban markets, and 132 percent of BLM suburbs. In economic terms, the rent premium is strongest at the core and attenuates rapidly toward the periphery.[8]

Rosenthal and Strange, following standard economic theory since Alfred Marshall (1842–1924), consider agglomeration benefits to accrue to certain locations, producing business efficiencies that warrant a real estate premium where they are concentrated. Marshall specified the sources of such benefits: input sharing, knowledge spillovers, and labor pooling. It seems clear that the high-density CBDs of 24-hour cities would be ideal environments for cultivating such benefits.

Most of the research literature on agglomeration acknowledges that its benefits are not observed directly, and most frequently land prices are examined as proxies for those benefits. In the United States, land prices are themselves often unobserved. In built-up cities, land trades in a market that is relatively illiquid, and, in the more frequent sales of developed property, the land component is not typically isolated in the transaction price. My colleague Matthew Drennan, of UCLA, and I have argued that office rents provide a richer proxy for agglomeration benefits.[9] If this is the case, the spreads in CBD rents between the cluster of 24-hour markets and the other two groupings would indicate a powerful set of agglomeration factors at work. Dense input sharing, knowledge spillovers, and labor pooling would then be priced as agglomeration externalities by tenants accepting the manifestly expensive rents typical of the 24-hour downtowns.

One important note should be made about the office rents as I have analyzed them. The foregoing graphs express the rents in 2014 dollars, or "real rents," as economists term dollar values that have been adjusted for inflation.[10] Real estate has often been credited with having "inflation hedge" characteristics. This is because rents are set in the marketplace but are constrained at the lower level by property operation expenses. Those expenses must cover any inflation in material and labor costs, as well as taxes and financing costs. Rents are also inflation-sensitive at market peaks, as development only becomes feasible when rents are sufficient to provide a profit margin over and above increases in the price of building material, construction labor, and the aforesaid tax and financing factors.

A careful reader will note that, even given cyclical fluctuations, there has been a downward drift in real rents over the 1987–2014 period. CBD office rents in the 24-hour cities appear to be an exception to this general tendency. Density in the core areas is likely to be a key reason for the better capacity of downtown rents in the 24-hour cities to meet or exceed inflation over time. Construction of new office towers in the downtowns not only has to meet the test of supply and demand for office space itself, but also has to compete with housing, retail, hotel, amenity, and civic land uses as well. Under this pressure, coupled with land scarcity due to an already compact core and an attendant propensity toward careful land use regulation, many real estate professionals speak of "barriers to entry" in 24-hour downtowns that keep levels of new development constrained.[11]

The constraints on supply show up in rents, and also in occupancy. The factors interact, of course, but the classic supply/demand curves suggest that the quantity demanded (that is, space leased) should adjust downward as prices (rents) rise. Against this fundamental relationship, though, is the motivation of businesses to reap the agglomeration benefits available in the 24-hour downtowns. Thus far, the 24-hour downtowns have solved the difficult trade-off problem consistently, maintaining higher levels of occupancy (lower vacancy) in every year since 1987. On average, over the 28 years shown on the graph, vacancy in the 24-hour CBDs has averaged 10.3 percent, compared with 16.6 percent in the 9-to-5 downtowns and 15.4 percent in the BLM central areas.

However, as the agglomeration benefits dissipate away from the core, the suburbs in the 24-hour metros cannot rely upon this inelastic relationship between rents and vacancy to the same degree as the downtowns. As a matter of fact, since the disruption of the financial crisis late last decade, occupancy in the 24-hour metro suburbs has been flat, whereas the less expensive 9-to-5 and BLM suburbs have seen their vacancy rates decline. This might suggest that suburban office space is much more a "commodity" product, where the second type of value proposition (low cost) trumps the "higher quality is worth the higher price" dynamic that seems to apply in the 24-hour CBDs. Then, too, the exceptionally low level of downtown office supply increases since 2008 (a net increase of just 4.9 million square feet across the seven 9-to-5 markets; and a net *decrease* of 3.9 million square feet in the 12 BLM downtowns) may have triggered a diversion of demand from the core to the periphery in those markets. Such a move would be consistent with the forces of entropy in most metro areas and contrasts with the anentropic (or concentrating) forces that are supporting the economies of the 24-hour CBDs.

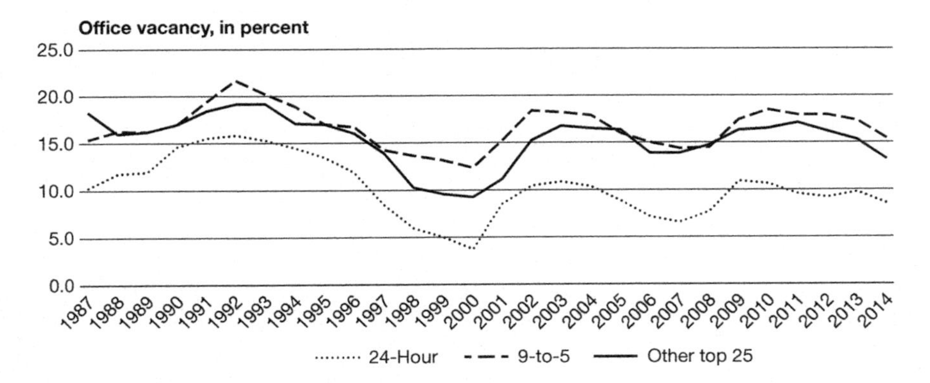

Figure 9.6 Downtown markets in 24-hour cities have sustained lower vacancies than other markets, regardless of cycles

Source: Data from CBRE, Inc.; calculations by author

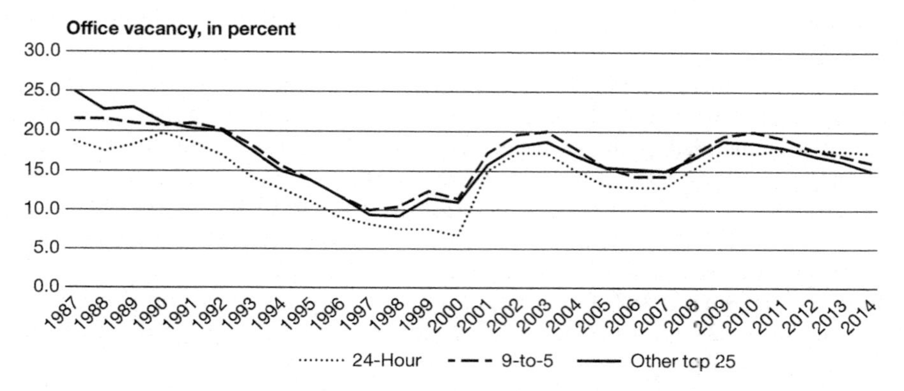

Figure 9.7 Financial crisis of past decade has attenuated recovery in suburbs of 24-hour cities, as other suburbs improve

Source: Data from CBRE, Inc.; calculations by author

Investment returns

Until the final quarter of the 20th century, commercial real estate investment was a largely local affair. Each city had a cadre of professional developers who knew their markets intimately. Commercial banks – then largely restricted to defined geographies by their charters – had relationships with those builders, who typically turned to the banks for construction loans while pre-arranging permanent financing ("take-out loans") with the nation's large insurance companies, such as Equitable, Prudential, Aetna, Met Life, New York Life, and Connecticut General. For many developers, commercial properties were held over the long term and became the foundation for family fortunes.

In the final 25 years of the century, an increasing wave of institutional ownership of commercial property redefined real estate investment across the US. A major spur to the growth in institutional ownership was the passage of the Employee Retirement Income Security Act of 1974 (ERISA). There had been some movement into commercial property investment prior to ERISA, most notably the establishment of the Prudential Realty Institutional Separate Account in 1970 and the First National Bank of Chicago Real Estate Fund in 1973. Under the leadership of Blake Eagle of the Frank Russell Company, data began to be aggregated to evaluate the suitability of commercial real estate as an asset class under ERISA's requirements that pension funds act as prudent fiduciaries for the retirement savings of their beneficiaries. The National Council of Real Estate Investment Fiduciaries (NCREIF) was established as the trade association for the industry, with the purposes (among others) of collecting reliable data in a secure environment and establishing performance measures and indexes useful for research and for investment decision-making.[12]

The NCREIF Property Index (NPI) had 233 properties valued at $580.9 million at its start date of December 31, 1977. During the turbulent decade of the 1980s, a boom-and-bust era for real estate, the NPI database expanded to 1,660 properties, worth $32.7 billion. As of year-end 2014, there were 7,062 properties included in the NPI database, with an aggregate value of $409.3 billion. As the database grew and broadened, its contents became more statistically reliable, and the volatility that characterized the NPI in its early years became more reflective of overall cycles, trends, and events in the economy and the wider real estate industry.

The larger database, in turn, became more and more the "go to" source for serious research in the commercial real estate field. A large academic literature was spawned. And, in response to the demands of sophisticated researchers, NCREIF honed its tools, providing the industry and institutional investments with a clearer window into the performance of the asset class.[13]

Figures 9.8 and 9.9 make use of NPI, using one of the tools supplied by NCREIF, its "custom query" facility. By using the custom query, a researcher can define specific portfolios and generate performance measures for those portfolios that are consistent with the index methodology of the NPI. I have drawn down data for the period from the start of 1987 to the end of 2014, for the CBD and the suburban office markets of each of the three sets of markets (24-hour, 9-to-5, and other top 25 metros), to calculate the cumulative total return for each of those six market segments.

Cumulative total return is a fairly simple concept that is very useful for performance measurement over time. The analysis sets a date for an initial investment (indicated by the index value of 100) and then tracks the return on that investment, assuming that it is held over time, with its performance tracking the market segment as a whole. For NCREIF, "total return" includes both an income component (the return attributable to the net operating income generated at the property level) and an appreciation component (the change in value, period to period, as measured by NCREIF's proprietary formula tracking market value).[14]

Shifting attention from real estate metrics such as rents and vacancy to investment returns indicates a transition from commercial property in the user-market perspective to a capital-market point of view. Real estate lives in both those worlds. Although the flavor of the city may seem to be carried more directly by the user market – where businesses choose to locate, how employees get to work, how the workplace is connected to its surroundings – such characteristics are closely bound to the capital market too.

Investors carefully scrutinize the operating statistics of office buildings (and all other forms of commercial property) in evaluating price. In doing so, potential purchasers and their lenders first consider what is explicit in the operating statement. Such revenue variables as the rent roll, the relation of contract rents to market rents, the level of occupancy, and the expiration schedule of leases are examined in detail. Then, the cost to run the operation must be matched to the revenue. Payroll, utility costs, insurance, property management fees, supplies, general overhead, real estate taxes, and needed capital improvements

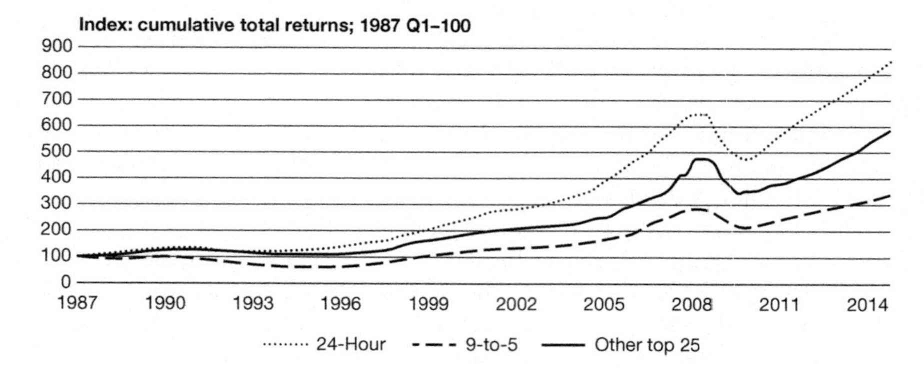

Figure 9.8 Investment returns in 24-hour downtown have far outpaced other large markets, with 9-to-5 CBDs lagging severely

Source: NCREIF (using database custom query facility)

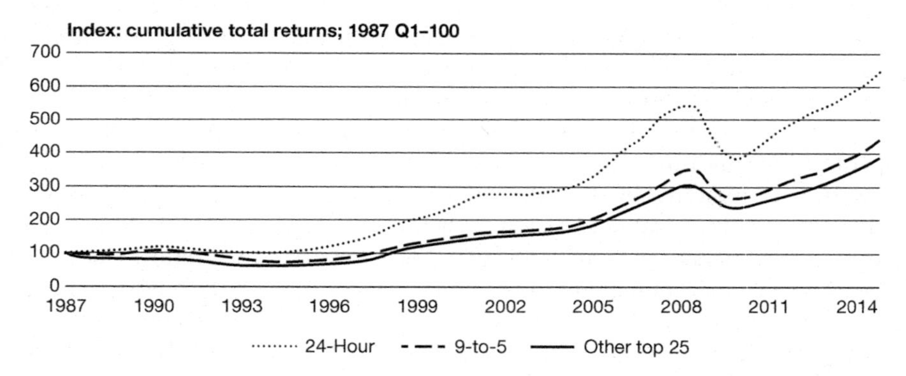

Figure 9.9 Suburbs of 24-hour metros share in CBD investment advantage, with 9-to-5 and other large markets producing similar low returns

Source: NCREIF (using database custom query facility)

have to be taken into consideration. It requires a lot of detailed work to accurately produce a revenue–expense statement and to estimate cash flow.

But that is just a step along the way. The operating property is also a capital asset that is priced and traded in the investment market. Investors require not only an internal operating analysis of the real estate, but also an understanding of the market conditions (including externalities) supporting, enhancing, or constraining operations and an evaluation of how property income is capitalized. This last element relates income and appreciation potential, factoring both into price.[15]

Although every investment prospectus is obligated to note that past performance does not guarantee future results, it should be the rare and foolish investor who ignores historical information. And yet, it happens. In a trenchant and valuable review of investment folly, John Kenneth Galbraith wrote, in *A Short History of Financial Euphoria*, that "the extreme brevity of financial memory" is a recurrent feature of investment market bubbles. He noted:

> There can be few fields of human endeavor in which history counts for so little as in the world of finance. Past experience, to the extent that it is part of memory at all, is dismissed as the primitive refuge of those who do not have the insight to appreciate the incredible wonders of the present.[16]

Though there is always the temptation to hubris that Galbraith alluded to, there is abundant information that can help guide commercial real estate investment choices, if attention is paid. Figures 9.8 and 9.9 present some of that information. Cumulative total returns on office investments in the 24-hour CBDs, shown in Figure 9.8, have exceeded returns in 9-to-5 downtowns by 152 percent since 1987, and have surpassed the total return in the remaining large markets in the US by 45 percent. That has certainly been a powerful factor in evaluating investment risk and opportunity in the office sector.

The long-term trend of superior performance is just one of the lessons to be drawn from Figure 9.8. The 9-to-5 downtowns were more seriously impacted by the real estate bubble of the late 1980s, which featured the unraveling of the real estate investment syndication industry and the related collapse of the savings and loan industry, with the takeover of the assets of the thrifts by the federal agency, the Resolution Trust Corporation. Where the 24-hour downtowns never saw their index of cumulative total returns fall below the 1987 Q1 value of 100, the 9-to-5 downtowns dropped to 90.9 by the second quarter of 1988, fell to 61.1 by the end of 1994, and didn't return to par until the final quarter of 1998.

And, though all three groups of CBDs suffered in the Great Recession that began in 2008, the 24-hour downtowns rebounded to their peak level of cumulative total return by the beginning of 2012 and have advanced 32 percent on that index since then. The 9-to-5 downtowns, on the other hand, required an additional year to regain their index peak and, at the end of 2014, were just 16 percent above that level, or half the overall gain of the 24-hour markets. As the slopes of the total return lines since 2010 show, the pace of recovery in the 9-to-5 CBDs has been slower than the balance of the large city office markets, as well as lagging the 24-hour cities.

This seems to illustrate the greater resiliency of 24-hour downtowns when faced with disruptive change. There is no suggestion that any market is "recession proof" or immune to cycles. But it does appear that 24-hour cities have accomplished something of a "change of state" that 9-to-5 markets have yet to attain. Some possible reasons for this will be offered in this book's final chapter.

That the change of state was anticipated in the 1995 edition of *Emerging Trends in Real Estate* seems all the more remarkable, considering that there was very little separation in the total return lines for 24-hour downtowns and the set of other top 25 CBDs. The cumulative returns started to diverge noticeably only in the late 1990s, but, since then, the gap depicting the superior investment performance of the 24-hour CBDs has been widening.

Part of the surprise has been that the key 24-hour cities – New York, Boston, San Francisco, Washington, DC – are among the oldest urban areas in America. They were all well-established cities by the mid-19th century. And, as Chapters 4 and 5 discussed, it was not at all clear that such "mature economies" had another growth spurt ahead. It did appear that the logistic (or sigmoid) curve had flattened out for these places, that other and younger cities were carrying the banner of the future, and that what growth these older places could sustain would continue to occur at their perimeter – in the so-called edge cities. It has not turned out that way, reminding us of a truth I try to convey to all my students and clients: "There is no point in doing research unless you are willing to be surprised."

Now, with one-seventh of the 21st century already in the books, it might be wise to consider what surprises may lie ahead in the economic geography of the next decade or so. Just as select urban cores were able to reverse the hollowing-out trends of decades of suburban dominance, we need to ask about comparative advantage in the suburban realm. There is not much we can predict confidently in the economic arena. But demography has a way of shaping economies in broad and powerful strokes. The Millennial Generation, Gen Y, is 83 million strong – larger than the Baby Boomer cohort – and its average age is 25 at the middle of the present decade. Gen Y preferences for urban living have helped drive the revival of the 24-hour city downtowns, and may continue to do so. But, assuredly, enough of the Millennials will be exploring suburban options as they pass through their 30s and 40s, seeking housing, jobs, and lifestyle amenities. This should create opportunities for communities beyond the metro core. Will Gen Y be satisfied with the conforming, commoditized suburbs of the early post-World War II era, the "boxes made of ticky-tacky," or will they be remolding suburbia in more heterogeneous ways?

If "follow the money" is a good guide to sustainable trends, it looks like the first suburbs to reinvent themselves may be those in the 24-hour metro areas. The cumulative total returns for that cluster of markets pencil out as a 46 percent premium over the 9-to-5 metro suburbs, and a 66 percent premium over the BLM set. This is evident in Figure 9.9.

Part of the advantage is the energy emanating from the thriving downtowns. But there are also significant synergies and externalities in such suburbs as Sunnyvale, Berkeley, and Walnut Creek near San Francisco; Bethesda, Old Town Alexandria, and Reston in the ring around Washington, DC; and Morristown, Tarrytown, and Purchase in the New York City area.

Infrastructure hubs, especially those supporting transit-oriented development, places with natural amenities such as rivers, lakes, or parkland, and suburbs fortunate enough to be home to high-quality colleges and universities are those most likely to step away from the pack. So, places such as Tempe, Arizona, the Denver Tech Center, or Beaverton, Oregon, also carry promise.

What the nearly overlapping lines for the 9-to-5 suburbs and the BLM suburbs seem to be indicating is that commodity real estate – the plain vanilla office park or freestanding glass box by the highway – produces commodity results. That is just not attractive to capital in a competitive world. Nothing, it seems, compromises future return on investment more than an abundance of cheap land at the perimeter of an urban area. Although this may attract cost-sensitive businesses and households looking for "more bang for the buck," the ability to create more and more real estate – as a commodity "substitute goods" – merely causes investors to worry about how seriously their returns can be eroded. And that directs their capital elsewhere.

Suburbs, and downtowns, that are starved of investment capital and that must compete as "low-cost providers" will find it hard to put together the ingredients for vibrancy, much less execute the recipe whereby the live–work–play formula can turn those ingredients into a superior community for the whole range of its citizens.

Investment transaction flows

The principle that capital seeks yield predicts that those urban areas that produce superior total returns should act as capital magnets over time. The first 14 years of the 21st century have been a crucible testing many basic assumptions in finance, and the evidence is strong that capital has indeed been seeking superior returns – more properly, risk-adjusted returns – in the 2001–2014 period.

Real Capital Analytics (RCA) has emerged as a dominant source of information about commercial real estate transaction markets, first in the US and increasingly around the world. Since 2001, RCA has tracked nearly 42,000 US office property sales, covering 365 million square feet of space, at an aggregate price of $1.153 trillion. The three market groups we have been monitoring have accounted for 60.5 percent of the deal count, 67.9 percent of the square footage, and a stunning 80.4 percent of the total price over this span of time.

Figure 9.10 incorporates a tremendous amount of information about the turbulent start to the 21st century for office properties. It is common to refer to the triggering events for the Great Recession as "a housing finance bubble" and a "Wall Street meltdown." Commercial real estate, regrettably, bears at least its fair share of the burden of blame. The Niagara of capital that washed over the nation brought a torrent of money, both for equity and debt investment, that compromised underwriting standards, inflated prices, and lured sophisticated and novice investors alike toward shipwreck.[17]

It is tempting to accept the indictment that real estate's troubles were simply the result of self-inflicted wounds. But it is more accurate to reflect on what brought the flood of capital into the commercial property sector in the first place, creating the situation with which investment managers needed to cope. At the beginning of the decade, the Standard & Poors 500 stock index stood at 1,498. By July 2002, the S&P 500 was down to 815, a decline of 46 percent. Investors were burned by the dot-com bubble and its collapse, a classic maturation failure in overenthusiasm for unproven investment vehicles, and shifted their attention to tangible assets that delivered contractually based cash flow. Real estate answered those needs. The S&P 500 would not regain its prior peak until April 2007.

The terrorism of September 11, 2001, also shook the investment world. Geopolitical tensions escalated with a highly publicized War on Terror that congealed into land wars in Iraq and Afghanistan. Corporate America was rocked by scandals, including the bankruptcy of Enron and the subsequent dissolution of its complicit accounting firm Arthur Andersen. A tsunami in Indonesia and Hurricane Katrina's devastation in New Orleans drew attention to environmental vulnerabilities with profound economic and social consequences. Such hazards were named "event risks" and came into focus for investors and the general public with the surprising publication success of Nassim Taleb's *The Black Swan*. The US real estate investment market was viewed as an arena that was large and diverse, linked to the broad-based, growing American economy and enjoying an era of asset appreciation. Momentum investors from around the globe flocked to the rally.

Their confidence – frankly, overconfidence – could not survive the cataclysm of 2007 and 2008. That disruption (no mere cycle) was presaged by an earlier pattern of crises, as described at the end of Chapter 6. But the investment community, by and large, no more saw the threat than did the Chairman of the Federal Reserve.[18] The commercial real estate professionals were swept away with the rest of the global financial system, having succumbed to the classic pattern of euphoria identified by Galbraith: over-leverage, enrapture by novelty, naïve extrapolation of short-term trends into the indefinite future, confusion of price with value, and the illusory "proof" that rapid wealth accumulation was a signal of superior intelligence. Humility would come hard and painfully, but, as wisdom has taught for thousands of years, humility is a virtue (Latin

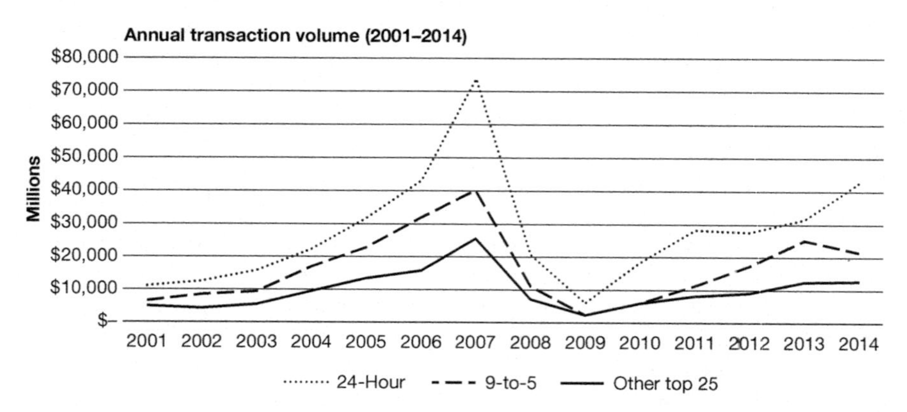

Figure 9.10 24-hour metros lead office investment throughout a volatile decade and a half

Source: Real Capital Analytics

virtus: strength) conducive to improved decision-making in a difficult and shifting world.

From an investment perspective, the Great Recession was a "risk-off liquidity crisis." Most commentary considered the illiquidity that drove down asset prices in the equities market, in commodities, and in real estate the emergence of a change of state that might be called "a Sahara Desert of capital" in contrast to the prior Niagara. I think it was more a "Hoover Dam of capital": investors sought capital preservation rather than capital growth – *return of* capital more than *return on* capital. In a risky world, capital flies to quality and to safety.

We still see this in the nearly insatiable demand for US Treasury notes and bonds and the related demand for top-rated sovereign debt, such as Switzerland's. Central banks in Denmark and Sweden have pushed their benchmark interest rates into negative territory, and the 1-month Euribor rate (the interbank lending rate for the Eurozone) went negative in early 2015, with the interesting effect of bringing mortgage rates in Portugal and Spain to levels where each month's mortgage payment meant a principal reduction greater than the amortization schedule for the loan.[19]

In the US, investment markets, one of the best signals of the "Hoover Dam" phenomenon has been the volume of capital residing in Institutional Money Market Mutual Funds. These data used to be counted in the M-3 money supply, until M-3 measurement was discontinued in March 2006. The Fed asserted that M-3's cost of collection did not justify the limited benefits of its increased information. But the Fed's money supply reports (H.6) continued to publish the Institutional Money Market data as an addendum item. This is capital directed to very safe, albeit low-yielding, assets such as US Treasury securities. It is where risk-averse capital goes in troubled times. In April 2006, there was about $1.25 trillion in such funds. By March 2009, that amount had doubled. Then, as it became apparent that pricing in the asset markets – including real estate – had plunged so far that they had "overcorrected" in all probability, the capital behind the Hoover Dam began slowly to be returned into the markets. Stock prices began to rise, as liquidity gradually returned. But, as of early summer 2015, there was still roughly $1.8 trillion of capital preserved in Institutional Money Market Mutual Funds.

Commercial real estate also benefited from a limited "risk-on" return of liquidity after 2009. Office investment, in particular, displayed patterns of selectivity and discipline as the century turned into its second decade. The "gateway cities" – a near-synonym for the 24-hour cities – were accorded preference for office investment, and the CBDs preference over the suburbs. That such performance was registered in the RCA database is significant. Where NCREIF monitors only the institutional investment community of its fiduciary members, RCA covers all buyers and sellers – private equity, hedge funds, international investors, local market players, developers, and REITS. So, in the first stage of recovery, the mutually reinforcing "flight to quality" and "flight to safety" directed a disproportionate share of office investment capital to the 24-hour downtowns. RCA data showed that, for 2012, for instance, 36.3 percent of all office (CBD and suburban) investment was concentrated in just five downtown markets: New York, Boston, Chicago, San Francisco, and Washington, DC – the prime 24-hour cities.

In the more recent past, lower yields in the 24-hour markets have motivated office investors to look for opportunities in so-called secondary and tertiary markets,[20] many of which are included in the 9-to-5 market set and BLM set in this chapter's analysis. This has raised questions from some observers about eroding investment discipline (nominal-yield-seeking behavior)[21] and the potential for another real estate asset bubble, as price becomes detached from underlying value. However, an array of 2014 investment volumes from RCA against the independently derived outlook for investment/development performance published by *Emerging Trends in Real Estate 2015*, based on its survey of 1,055 real estate experts, shows close alignment of capital deployment with fundamental market expectations.[22] This analysis (Figure 9.11) does not seem to be a case of indiscriminate spending. And the overall volume of office investment seen in Figure 9.10 remains substantially below the "Niagara of capital" peak level. Although prudent caution is ever advisable, it is hard to see evidence of euphoria or disequilibrium in office investment as this book is written.

In Figure 9.12, Sections A, B, and C summarize high-order investment measures for the 24-hour, 9-to-5, and BLM sets of markets. In addition to presenting the aggregated data from the RCA database for 2001–2014 for each of the market sets, the results for the individual metro areas are shown as well. From this welter of statistics, what important results emerge?

Consistent with the results on urban variables covered in Chapters 7 and 8, the real estate transaction data confirm that 24-hour cities have risen to the top as investment locations and are being priced as such. The seven 24-hour markets capture a 24.4 percent share of the number of office properties traded, and these are larger properties yielding a 31.1 percent share of total office workspace. But it is in value and in price per square foot that the 24-hour cluster shows its dominance, at $529.4 billion in investment volume over the period (49.5 percent of the national total) and a price per square foot of $292.32 that is 88.3 percent higher than all other US markets[23] taken together.

The 24-hour city markets have averaged deal prices above $51 million, twice the level of the 9-to-5 markets and 148 percent higher than the BLM market set. Virtually all the transaction metrics show the expected rank order of 24-hour markets first, the 9-to-5 markets next, and the BLM set third.

That is not to say that there is not some overlap in the data for individual cities. Total transaction volume for Los Angeles, for instance, ranks third among all markets, trailing only New York and Washington, DC. The price per transaction for Seattle ranks sixth (after New York, Boston, Chicago, San Francisco, and Washington), with Houston seventh and Los Angeles eighth. Seattle is in the top five in price per square foot (ahead of Chicago, Las Vegas, and Miami from the 24-hour cluster), and Los Angeles is sixth.

It would actually be fairly surprising to find any one set of cities "running the table" in all categories. But even the younger and smaller 24-hour markets, Miami and Las Vegas, take their place near the top in some key variables.

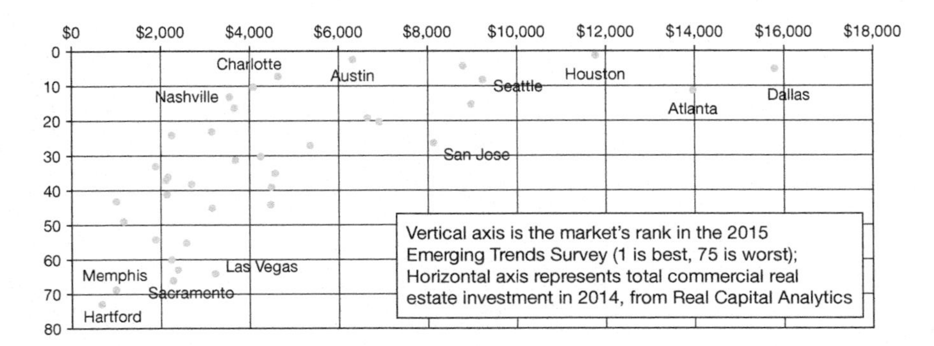

Figure 9.11 Secondary market investment: 2014

Source: Author's analysis based upon RCA and *Emerging Trends* data

New York, as in so many cases in the urban variables, also stands out as the dominant commercial office market. But, underneath its primacy, the other 24-hour cities justify their status both as a group and individually.

What the real estate data say

Critics of the 24-hour-city concept decry the high prices and presumed bias toward a demographic elite. And there is no hiding from the fact that, in cost of living, especially in housing, these cities are expensive places. That high-price attribute carries over into commercial office markets, as the data clearly show. This, however, begs the question of value. Price, it must be said again, is not the same as value, and low price is but one of four metrics for the value proposition of "You get what you pay for."

Businesses have voted with their wallets to make the high-rent 24-hour markets the best performing locations in terms of office building occupancy. The high prices paid by investors have sustained the greatest cumulative total returns on their acquisitions. Those superior returns have further motivated higher volumes of capital to move into the 24-hour markets for reasons of quality, conservation of capital, and greater expected liquidity over the long haul. Such factors appear to be more important than the lower initial yields (capitalization rates) that are typical of the core 24-hour downtowns.

Overall, the real estate data vindicate the original assertion in the 1995 edition of *Emerging Trends in Real Estate* that 24-hour locations would be the best-performing places for commercial property investment. Beyond that, the superior real estate performance looks to be well grounded in urban attributes that are not only observable but that are related to property performance in measurable ways. This leads to the expectation that investment and public policy conclusions can look to the combined evidence in Chapters 7–9 for decision-making guidance. That will be the task of Chapter 10.

A: 24-hour metros	Total investment activity	Averages	Shares of US total	
New York	# properties: 2,199 Volume: $208,721,931,066 Total units: 481,102,168 sf	Sf/deal: 218,782 Price/deal: $94,916,749 Price/sf: $433.84	5.2% transactions 18.1% volume 8.3% sf	
Boston	# properties: 1,402 Volume: Total units: 240,488,286 sf	Sf/deal: 171,532 $61,377,333,390 Price/sf: $255.22	3.3% transactions Price/deal: $43,778,412 4.1% sf	5.3% volume
Chicago	# properties: 1,572 Volume: Total units: 380,338,418 sf	Sf/deal: 241,946 $63,798,115,924 Price/sf: $167.74	3.7% transactions Price/deal: $40,584,043 6.5% sf	5.5% volume
Las Vegas	# properties: 421 Volume: Total units: 30,068,414 sf	Sf/deal: 71,421 $5,449,359,039 Price/sf: $181.23	1.0% transactions Price/deal: $12,943,846 0.5% sf	0.5% volume
Miami	# properties: 734 Volume: Total units: 78,743,629 sf	Sf/deal: 107,280 $14,391,665,211 Price/sf: $182.77	1.7% transactions Price/deal: $19,607,173 1.4% sf	1.2% volume
San Francisco	# properties: 1,588 Volume: Total units: 235,253,477 sf	Sf/deal: 148,145 $68,880,217,698 Price/sf: $292.79	3.8% transactions Price/deal: $43,375,452 4.0% sf	6.0% volume
Washington, DC	# properties: 2,342 Volume: Total units: 364,875,793 sf	Sf/deal: 155,797 $106,742,057,686 Price/sf: $292.54	5.6% transactions Price/deal: $45,577,309 6.3% sf	9.3% volume
All 24-hour markets	# properties: 10,258 Volume: Total units: 1,810,870,185 sf	Sf/deal: 176,532 $529,360,680,014 Price/sf: $292.32	24.4% transactions Price/deal: $51,604,668 31.1% sf	45.9% volume
B: 9-to-5 metros				
Atlanta	# properties: 1,277 Volume: Total units: 217,922,489 sf	Sf/deal: 170,652 $30,894,869,302 Price/sf: $141.77	3.0% transactions Price/deal: $24,193,320 3.7% sf	2.7% volume
Dallas	# properties: 1,523 Volume: Total units: 296,009,661 sf	Sf/deal: 194,360 $34,895,085,554 Price/sf: $117.88	3.6% transactions Price/deal: $22,912,072 5.1% sf	3.0% volume
Los Angeles	# properties: 2,751 Volume: Total units: 349,196,166 sf	Sf/deal: 126,934 $80,723,249,703 Price/sf: $231.17	6.6% transactions Price/deal: $29,343,239 6.0% sf	7.0% volume
Minneapolis	# properties: 527 Volume: Total units: 102,151,212 sf	Sf/deal: 193,835 $12,281,576,592 Price/sf: $120.23	1.3% transactions Price/deal: $23,304,699 1.8% sf	1.1% volume
Philadelphia	# properties: 684 Volume: Total units: 125,439,834 sf	Sf/deal: 183,392 $16,248,229,923 Price/sf: $129.53	1.6% transactions Price/deal: $23,754,722 2.2% sf	1.4% volume
Phoenix	# properties: 1,258 Volume: Total units: 132,807,927 sf	Sf/deal: 105,571 $20,769,052,974 Price/sf: $156.38	3.0% transactions Price/deal: $16,509,581 2.3% sf	1.8% volume
Seattle	# properties: 1,077 Volume:	Sf/deal: 124,738 $34,883,711,391	2.6% transactions Price/deal: $32,389,704	3.0% volume

	Total units: 134,342,601 sf	Price/sf: $259.66	2.3% sf	
All 9–5 markets	# properties: 9,097	Sf/deal: 149,266	21.7% transactions	
	Volume:	$230,695,775,438	Price/deal: 25,359,544	20.0% volume
	Total units: 1,357,869,890 sf	Price/sf: $169.90	23.3% sf	
C: Other top 25 metros				
Baltimore	# properties: 462	Sf/deal: 106,339	1.1% transactions	
	Volume:	$7,072,880,969	Price/deal: $15,309,266	0.6% volume
	Total units: 49,128,824 sf	Price/sf: $143.97	0.8% sf	
Cincinnati	# properties: 213	Sf/deal: 141,486	0.5% transactions	
	Volume:	$2,820,255,295	Price/deal: $13,240,635	0.2% volume
	Total units: 30,136,492 sf	Price/sf: $93.58	0.5% sf	
Cleveland	# properties: 214	Sf/deal: 156,132	0.5% transactions	
	Volume:	$2,913,085,895	Price/deal: $13,612,551	0.3% volume
	Total units: 33,412,218 sf	Price/sf: $87.19	0.6% sf	
Denver	# properties: 1,025	Sf/deal: 147,754	2.4% transactions	
	Volume:	$22,345,990,740	Price/deal: $21,800,967	1.9% volume
	Total units: 151,447,784 sf	Price/sf: $147.55	2.6% sf	
Detroit	# properties: 418	Sf/deal: 180,093	1.0% transactions	
	Volume:	$6,089,234,879	Price/deal: $14,567,548	0.5% volume
	Total units: 75,278,814 sf	Price/sf: $80.89	1.3% sf	
Houston	# properties: 1,193	Sf/deal: 217,524	2.8% transactions	
	Volume:	$35,973,737,746	Price/deal: $30,154,013	3.1% volume
	Total units: 259,505,743 sf	Price/sf: $138.62	4.5% sf	
Portland	# properties: 416	Sf/deal: 111,294	1.0% transactions	
	Volume:	$8,576,366,270	Price/deal: $20,616,265	0.7% volume
	Total units: 46,298,474 sf	Price/sf: $185.24	0.8% sf	
Riverside	# properties: 371	Sf/deal: 57,857	0.9% transactions	
	Volume:	$3,216,881,593	Price/deal: $8,670,840	0.3% volume
	Total units: 21,464,901 sf	Price/sf: $149.87	0.4% sf	
St. Louis	# properties: 383	Sf/deal: 152,861	0.9% transactions	
	Volume:	$6,439,909,780	Price/deal: $16,814,386	0.6% volume
	Total units: 58,545,655 sf	Price/sf: $110.00	1.0% sf	
Sacramento	# properties: 530	Sf/deal: 97,914	1.3% transactions	
	Volume:	$9,344,866,524	Price/deal: $17,631,824	0.8% volume
	Total units: 51,894,527 sf	Price/sf: $180.07	0.9% sf	
San Diego	# properties: 1,101	Sf/deal: 92,114	2.6% transactions	
	Volume:	$25,214,768,099	Price/deal: $22,901,697	2.2% volume
	Total units: 101,417,171 sf	Price/sf: $248.62	1.7% sf	
Tampa	# properties: 532	Sf/deal: 110,624	1.3% transactions	
	Volume:	$7,655,270,299	Price/deal: $14,389,606	0.7% volume
	Total units: 58,852,134 sf	Price/sf: $130.08	1.0% sf	
All balance of large markets	# properties: 6,022	Sf/deal: 130,658	14.4% transactions	
	Volume:	$125,484,778,331	Price/deal: $20,837,725	10.9% volume
	Total units: 786,825,109 sf	Price/sf: $159.48	13.5% sf	

Figure 9.12 Office transactions: 2001–2014

Source: Real Capital Analytics

Notes

1. Arbuckle was a coffee magnate in the late 19th and early 20th centuries whose business locations included the Empire Stores just under the Brooklyn Bridge, in what is now known as the DUMBO neighborhood on the Brooklyn waterfront. This has become one of the nation's most notable urban regeneration areas and – in tribute to Arbuckle's principle – one of its most expensive.
2. This is only one of four approaches to the value proposition, according to Michael Treacy and Fred Wierseman, *The Discipline of Market Leaders*, Addison-Wesley (New York, 1995). The others are: being the low-cost provider, providing the greatest convenience, and taking full responsibility for producing results for the customer. There has been a frequently repeated exhortation to "better, faster, cheaper." Most often, however, there must be a choice that acknowledges, "You may pick two out of three."
3. Edward Glaeser offers a more nuanced analysis in *The Triumph of the City*, when he compares prices, incomes, and local amenities and suggests that individuals and businesses have a complex set of trade-offs to consider (pp. 116–133). Even Glaeser acknowledges, though, that the trade-offs are different for higher-, middle-, and lower-income households and suggests that it is the middle class – and the worker groups that constitute the middle-income job market – that feel the cost-of-living pressure most severely (pp. 180–183).
4. Those cities, listed on the right in Figure 9.1, will be used in this chapter as a benchmark to compare the 14 cities discussed in Chapter 8. As is the case in cost of living, the 24-hour cities emerge as distinct from the 9-to-5 set, and the seven 9-to-5 cities examined show much resemblance to the balance of the top 25 cities. This, it will be argued, further justifies speaking of 24-hour cities as a differentiated group, an urban category with special characteristics.
5. Cairncross, op. cit.
6. Rykwert, op.cit.
7. Data on office-space volume, rents, and vacancy rates are drawn from the CBRE proprietary database for 52 metro markets for the period 1987–2014.
8. Stuart Rosenthal and William Strange, "The Micro-Empirics of Agglomeration Economies," in the *Blackwell Companion to Urban Economics*, Wiley-Blackwell (Hoboken, NJ, 2005), and, with particular reference to New York, "The Geography of Entrepreneurship in New York," *Economic Policy Review 11* (2005), pp. 29–53, Federal Reserve Bank of New York.
9. Drennan and Kelly, "Measuring Urban Agglomeration Economies with Office Rents," *Journal of Economic Geography 11: 3* (2011), pp. 481–507.
10. For this analysis, rents were adjusted using the US GDP deflator, a broader inflation measure than the CPI. CPI is based upon a fixed "basket" of goods and services. The GDP deflator reflects changes in prices while also accounting for changes in patterns of consumption and investment in the economy.
11. David Lynn, Bohdy Hedgecock, and Jeff Organisciak, "Supply Constrained Markets," *Real Estate Issues 35:2* (2010), pp. 20–27. The authors calculate a price elasticity of supply index and find that New York and San Francisco are less elastic in supply relative to price change, implying a greater level of constraint in introducing new development into these markets.
12. See NCREIF website, online at http://ncreif.org/about.aspx
13. A retrospective look at the evolution of NCREIF and its data can be found online at www.ncreif.org/. . ./NCREIF_Academy/History-of-Institutional-Real
14. Information about the NPI and its calculation can be accessed online at http://ncreif.org/faqsproperty.aspx

15 This classic relationship between the user and investor market is well discussed in John B. Bailey, Peter F. Spies, and Marilyn Kramer Weitzman, "Market Study + Financial Analysis = Feasibility Report," *Appraisal Journal* (October 1977), pp. 550–577. The relationship between income and value is known as the "capitalization rate."
16 Galbraith, op. cit., p. 13. On December 28, 2011, the *Financial Times* reported that the Chartered Financial Analyst Society of the UK sought to address the deficiency of memory by recommending that the study of financial history be required for the certification of its members.
17 See Anthony Downs, *Niagara of Capital: How Global Capital Has Transformed Housing and Real Estate Markets*, Urban Land Institute (Washington, DC, 2007). In the 2006 edition of *Emerging Trends in Real Estate*, Chapter 1 was headlined, "As Long as Capital Keeps Flowing, Everything Will Be Alright." Few grasped in advance just how "not alright" things were quickly going to become.
18 Alan Greenspan, "Never Saw It Coming," *Foreign Affairs* (November–December 2013), pp. 88–96.
19 "Why Negative Interest Rates Have Arrived – And Why They Won't Save the Global Economy," *The Economist*, February 8, 2015, accessed online at www.economist.com/blogs/economist-explains/2015/02/economist-explains-15. See also, "Tumbling Interest Rates in Europe Leave Some Banks Owing Money on Loans to Borrowers," *Wall Street Journal*, April 13, 2015, accessed online at www.wsj.com/articles/as-interest-benchmarks-go-negative-banks-may-have-to-pay-borrowers-1428939338
20 RCA classifies the Los Angeles metro area within its "six major markets" and includes 24-hour cities Miami and Las Vegas in its grouping of "secondary markets."
21 Nominal yield is contrasted with the concept of "risk-adjusted returns."
22 The Pearson rank-order correlation of the two variables is 0.6989. This analysis first appeared in the February 2015 newsletter I prepared for Real Estate Capital Partners, a firm that intermediates European private capital into US real estate markets.
23 Calculated by the author from data including secondary and tertiary markets in the RCA database, including but not restricted to the 9-to-5 and BLM market sets.

10 Gleanings from the past and present; glimmers for the future

On July 13, 1977, at about 8:30 in the evening, lightning strikes began to hit electrical-generation facilities and power lines north of New York City. Power company staff began to juggle the load on the system, which was already stressed by the demands of a hot and humid night. Despite their efforts, the entire Con Edison system shut down about an hour after the first bolt hit the Buchanan, New York, substation. With the exception of a few neighborhoods in southern Queens, New York City was entirely blacked out.

In contrast to other, even more widespread, blackouts in 1965 and 2003, the blackout of 1977 was an urban catastrophe. The economic and demographic strains of that difficult era – discussed in Chapter 5 – exploded into several nights of looting and arson that devastated whole communities, poor communities especially. I was commissioned to do a photo study of an arc of the most impacted impoverished areas: Williamsburg, Bushwick, Brownsville, and East New York. It was a stark experience, fascinating if not entirely frightening, and certainly eye-opening. At the time, I could have no inkling that, by the early 21st century, Williamsburg and Bushwick would be discussed as two of the hottest upcoming neighborhoods in New York City, with their bars, clubs, galleries, and music performance venues vigorously participating in New York's 24-hour character.

Even if Brownsville and East New York are not in the same category of "newly chic," they are nonetheless vastly improved. One of the linchpins of progress was the organization of East Brooklyn Congregations (EBC), under the professional guidance of the Industrial Areas Foundation. During my photographic assignment, I was able to take pictures of some neighborhood buildings – principally the churches – from six or seven blocks away. This was possible because all that intervened were vacant lots strewn with rubble, land that had previously been crowded with four-to-six-story walkup apartment houses. As in the South Bronx, whole swaths of land were cleared of buildings and of people, and the culprit was frequently arson for profit, the profit coming in the form of property and casualty insurance proceeds.[1]

One of EBC's great contributions was to see that vacant land for what it was – a precious urban resource – instead of the occasion for hand wringing that most others did. The churches asked me to do a market study, which my firm,

Landauer Associates, permitted me to undertake *pro bono publico*. EBC then partnered with builder I.D. Robbins in developing low-cost, single-family-ownership housing, along the lines of traditional Brooklyn row houses. So great was the pent-up demand from households doubled up in public projects that a sell-out was virtually guaranteed, even though cash equity had to be produced and conventional financing secured in the form of bank mortgages. The current phase of the program,[2] named Nehemiah after the biblical prophet who rebuilt the walls of Jerusalem, will bring the number of homes built in East Brooklyn by EBC to 4,525, creating more than $1 billion of wealth in this once-blighted slum. Remarkably, fewer than a dozen Nehemiah homeowners suffered foreclosure in the housing crisis of the past decade.[3] Nehemiah's success has stimulated public and private investment in East Brooklyn, and, between 2013 and 2014, land prices per buildable square foot more than tripled.[4]

The initial Nehemiah site is just 5 miles – 40 minutes by local bus – from the Kensington neighborhood where I now live, described in the Introduction to this book.

Such a revival is not exclusively a New York story, to be sure. Downtown Houston, in the mid-1980s, was struggling in one of the periodic busts of the energy industry. Business activity was so low that, on one consulting trip there, I stayed at a Four Seasons Hotel for $85 a night. Thirty-five percent of the downtown's land area was devoted to surface parking lots, and there were fewer than a thousand housing units there. As commercial development expanded in the Houston CBD, the City Council sought to promote multifamily residential construction with a tax incentive of $15,000 per housing unit. The residential population downtown has grown to 4,000, and the incentive program recently expanded its target to 5,000 new units, amounting to a subsidy of $75 million to create the housing necessary to promote a 24-hour community.

Manifestly, cities are not willing to accept the hollowing-out phenomenon as an inevitability. The successes of the 24-hour cities in turning their fortunes around has emboldened economic development officials and planners. For many urban areas, this will be a daunting task.[5] Long-stagnant cities such as Hartford, cities with persistent population loss such as Buffalo, and those that have essentially run out of money such as Detroit will find the road ahead a tough one. But one key lesson to be learned from the history of the 24-hour cities – including New York, Miami, San Francisco, and Washington, DC – is that it is very hard to kill a big city. And it is demonstrably possible to revitalize one. That is a signal hope based upon the history of the late 20th and early 21st centuries.

Change (before you have to beg for change)

Success in this domain is predicated upon a willingness to change, which will begin for many cities with confronting a need to redefine themselves. If a city has not only hit the leveling out phase of maturation but is already steeply in the decline of urban old age, such change is not only hard but wrenching in terms of identity. Moreover, it is likely to face resistance from local interests clinging nostalgically to the past. Further obstacles may come at higher levels of government, where competing jurisdictions – especially suburbs – see economic development as a zero-sum game and seek to deny or restrict funding for urban programs ranging from infrastructure to education. The politics of urban revitalization are not purely local: they extend upward to state legislatures and to federal programs.

Willingness is not enough, either. Change needs to be strategically conceived. In his classic military work *Strategy*,[6] B.H. Liddell Hart proposes a brief series of historically proven principles of strategic success. Several of them can be illustrated by effective actions taken by cities over the last quarter-century.[7]

Keep your object always in mind while adapting your plan to circumstances. Los Angeles' ambition to develop a modern downtown core was so little achieved in 1990 that Joel Garreau made LA the avatar of the edge city metro, saying, "Every single American city that is growing, is growing in the fashion of Los Angeles, with multiple urban cores."[8] Yet Los Angeles has stuck to a consistent set of design guidelines for its downtown, through several economic cycles and political administrations. Those guidelines emphasize "the object in mind":

- Employment opportunities: Maintain and enhance the concentration of jobs, in both the public and private sectors, that provides the foundation of a sustainable downtown.
- Housing choices: Provide a range of housing types and price levels that offer a full range of choices, including home ownership, and bring people of diverse ages, ethnicities, household sizes, and incomes into daily interaction.
- Transportation choices: Enable people to move around easily, on foot or by bicycle, transit, and auto. Accommodate cars, but fewer than in the suburbs, and allow people to live easily without one.
- Shops and services within walking distance: Provide shops and services for everyday needs, including groceries, day care, cafes and restaurants, banks and drugstores, within an easy walk from home.
- Safe, shared streets: Design streets not just for vehicles, but as usable, outdoor spaces for walking, bicycling, and visual enjoyment.
- Gathering places: Provide places for people to socialize, including parks, sidewalks, courtyards, and plazas, that are combined with shops and services. Program places for events and gatherings.
- Active recreation areas: Provide adequate public recreational open space, including joint-use open space, within walking distance of residents.
- A rich cultural environment: Integrate public art and contribute to the civic and cultural life of the City.[9]

Although Los Angeles' downtown may not yet have passed the threshold as a 24-hour city, there is no question about its objective, the long-run commitment to its strategic goal, and its progress to date. No wonder, then, that Los Angeles is the 9-to-5 city that most frequently meets some of the socioeconomic and real estate investment characteristics of a 24-hour locale.

Adjust your ends to your means. Investors have greatly benefited from San Francisco's restriction on high-rise development, passed in 1986 as a citizen referendum known as Proposition M, or Prop M. Prior to its passage, though, the real estate community had strongly opposed the measure and several earlier initiatives to control the pace of construction in the space-constrained core of the city. Prop M stressed affordable housing, neighborhood preservation, and small business growth over a "rush to the sky."

Earlier initiatives had sought to enforce height limits at 80 feet (Proposition T, 1971) and 160 feet (Proposition P, 1972). After these defeats, the proponents of limited development regrouped and restrategized, eventually putting forward, in 1979, a ballot initiative dubbed Proposition O that limited height in some areas to 260 feet and in others to 150 feet. As in the previous propositions, Prop O was defeated, as a coalition of large downtown businesses, real estate developers, and labor effectively argued that the growth limitations would cost the city jobs, induce sprawl, and increase the tax burden on homeowners.[10] Clearly, the focus on height limitation itself was not an effective "end," as the proponents were not able to summon the means to achieve it.

Recognizing the wisdom of Liddell Hart's principle, *Do not renew an attack along the same line (or in the same form) once it has failed*, the limited-growth advocates devised Prop M, with provisions to cap high-rise development at a fixed volume per year (475,000 square feet), and linked this limitation to master planning elements targeting neighborhood and homeowner benefits. The timing was fortuitous, as a nationwide construction boom was nearing an end, and the development limitation helped San Francisco maintain relatively strong office occupancies (about 88 percent) at a time when most other downtowns saw occupancies drop to 75–80 percent.

From the 24-hour-city perspective, the impact was to halt the monotonic development of high-rise office towers, opening an opportunity for San Francisco to create a richer mix of uses in its downtown and in the adjacent South of Market district. Both the supply-side constraint and the development of complementary land uses downtown have helped support strong user-market and investment-market performance for San Francisco's office space.

Ensure that both plans and dispositions are flexible – adaptable to circumstance. Choose the line of least expectations. Chicago provides an excellent instance of a downtown revival that was aided by some pioneering efforts only loosely linked to commerce per se. The South Loop, in particular, was in decline, as the post-industrial shift reduced the demand for the older loft buildings in that area. Local colleges and universities, however, had need of dormitory space. South of the Chicago River, such institutions included DePaul University, Roosevelt University, as well as the Art Institute of Chicago. Adaptive reuse of the loft buildings included student housing, but also expanded classroom space. The net effect was keeping students proximate to downtown, which has proven a positive force for 24-hour cities elsewhere (including New York, Boston, and Miami).[11]

One key element to keep in mind in developing strategies for change is that it takes money and time to accomplish the desired positive results. Cities must be willing to make investments – looking without fear to tapping tax revenues today to create a more sustainable economic and fiscal base for the future. And they must do the even more difficult job of looking beyond an election cycle, because a successful economic development strategy takes time well beyond the timeframe needed to set up photo ops or ribbon-cutting ceremonies.

A list of ingredients does not constitute a recipe

In an era of communication by sound-bite, it is imperative to educate the citizenry and the local political structure that an effective strategy has to be based on a number of interactive elements. A simple recipe for bread requires yeast, flour, oil, sugar, salt, and water. Leave any one out as a cost-cutting measure, and you don't get bread – just goop. Then, too, there is a sequence to the recipe and an absolute need for patience. The bread dough needs time to rise, not just once, but twice. It is no accident that the cities that have successfully achieved 24-hour status (or are manifestly advancing in that direction, such as Houston and Los Angeles), in a kind of change of state, have taken decades to follow their recipes effectively.

Let's take a look at what the recipes might entail.

Diversity is a key ingredient, diversity in population, in economic base, and in planning/zoning approach. Limited diversity means limited interaction, with the result being a flat urban experience. Diversity in ***population*** reflects the multiethnic character of the United States. Brown University has tabulated 2010 Census data in its American Communities Project, and those data have been analyzed by Nate Silver's team at fivethirtyeight.com. The data place New York, Chicago, and Boston in the top five cities for population diversity.[12] Looking at the three sets of cities detailed in Chapter 9, we find the average diversity scores in the descending rank order we have come to expect: the 24-hour cities are the most diverse, with an average of 64.5; the 9-to-5 cities come next, with a 61.2 average; and then there is the balance of the large markets averaging 57.2.

With the exception of Miami (a low diversity score of 46.8), all the 24-hour cities score above 60 (moderately high diversity), with New York and Chicago having very high diversity scores, above 70. None of the 9-to-5 cities attains a score of 70; four are in the 60s (Dallas, Los Angeles, Philadelphia, and Phoenix), but three have moderately low diversity scores, in the 50s (Atlanta, Minneapolis, and Seattle). The benchmark set of the remaining cities in the top 25 population metros shows Detroit with an extremely low score of 29.0 and Portland with 45.5, just below Miami's score. Four cities are in the 50s: Baltimore, Cincinnati, Cleveland, and St. Louis (low–moderate diversity). Five cities are in the high–moderate diversity 60s: Denver, Houston, Riverside, San Diego, and Tampa. Sacramento edges out New York for the highest diversity score, with 73.8.

Diversity in a city's economic base shields an urban area from exposure to cycles proper to a dominant industry – whether that be banking, autos, energy, or gambling. However, Kris Hartley argues that economic development strategies targeting the "diversification of industries" is only a partial approach, with attention to specialization also important. He recommends "structural flexibility" in identifying a city's comparative advantage, claiming such an approach enables a city:

> to reorient towards emerging opportunities and maintain development potential across economic cycles. Furthermore, flexibility gives cities of any size hope for transformative growth. Not every city has the native advantages to meaningfully diversify, but flexibility can be their wild-card strategy.[13]

The design firm Gensler notes that population and economic diversity require, in turn, an approach to *place-making that maximizes choice and interaction* in urban locations. Gensler's Shawn Gehle terms this a perspectival shift, wherein, "work is integrated into the city rather than located within it." He cites work in the firm's Raleigh office that:

> identified 100 acres of residual downtown space, left-over during this past century of sprawl, which they propose filling-in with flexible mixed-use developments. In their proposal, previously unused and neglected spaces will be refit to spawn a dense and vibrant environment.

Some of the proposed space is semi-permanent, whereas other uses are "pop-ups" that may shift use seasonally or even by time of day. In Baltimore, Gehle sees this concept moving a step further toward "an innovation landscape where government, industry and higher education create partnerships with and become anchors for the larger city, a city where entrepreneurs and students of the future will want to live and work."[14] Gensler's planning is an instance of what Bruce Katz and Julie Wagner of the Brookings Institution term innovation districts:

> geographic areas where leading-edge anchor institutions and companies cluster and connect with start-ups, business incubators, and accelerators. They are . . . physically compact, transit-accessible, and technically-wired and offer mixed-use housing, office, and retail . . . altering the location preferences of people and firms and, in the process, re-conceiving the very link between economy shaping, place making and social networking.[15]

This sounds very much like a young person's game. And it is. There may be no better change agent than the young. Chapter 5 discussed how powerfully the Baby Boom generation impacted cities, beginning in the 1980s and 1990s. The even larger Gen Y is shaping the urban landscape with equal force today. Like the yeast in the bread recipe, the young-worker cohort is the leavening agent of the economy of cities.

Much has been written about the sociological characteristics of Gen Y, and this is no place to repeat common observations. One of the consequences of the Millennials' slower entry into family life and a long-term career path is that they are in school longer. In the decade from 2003 to 2013, undergraduate enrollment in US colleges and universities grew by 3 million students, or 1.9 percent per year.[16] The total number of graduate students in US institutions grew 1.5 percent per year from 2003 to 2013.[17] The US Department of Education projects that, by 2023, an additional 2.5 million students will be in college or graduate school, a 12.3 percent increase to 22.9 million students.

Chapter 7 discussed the impact of ***higher education*** on the development of "skilled cities"[18] and the historical evidence that the presence of universities is a strong predictor of future urban growth. My investigations have shown that the number of students enrolled in colleges and universities in an urban area is highly correlated with other key socioeconomic variables. The correlations of total post-secondary students to such variables are as follows: total downtown population (0.863); index crime rate (–0.681); city population density (0.738); and employment in producer services industries (0.891).

Katz and Wagner underscore the importance of the linkage of research universities as anchor institutions in effective economic development programming:

> Innovation districts help their city and metropolis move up the value chain of global competitiveness by growing the firms, networks, and traded sectors that drive broad-based prosperity . . . creating a dynamic physical realm that strengthens proximity and knowledge spillovers. . . . Innovation districts represent an intentional effort to create new products, technologies and market solutions through the convergence of disparate sectors and specializations (e.g., information technology and bioscience, energy, or education).[19]

For the long haul, however, the ability to retain the advantages of a young, educated population is going to require a focus on education for the children of the Millennials. As my survey results discussed in Chapter 7 (Figure 7.11) show, it is ***K–12 school quality*** that real estate professionals considered the most vital educational attribute for sustaining 24-hour cities. So often, debates about school funding simplistically draw a bright line between urban and suburban schools, or public and private education. A more fundamentally sound approach is taken by David Card and Alan Krueger, who seek to measure the rate of return on education using empirical metrics such as teacher–pupil ratios, average term length, and teacher pay. They find positive effects (controlling for other variables, including geographic location, race, family income, and education levels) for all these variables, as well as a positive correlation of teachers' educational attainment to students' future incomes. Moreover, they find that the differential of private schooling to public schooling is insignificant when the statistics are rigorously analyzed.[20]

As in so many cases, the political discourse often mistakenly categorizes educational expenditure as an "operating expense," when it should more properly be considered a capital investment – specifically an *investment in human capital*. Looked at in such a light, short-run metrics are eclipsed by longer-run returns on the money committed. Suburbs truly understood this in their heyday. Cities now have the opportunity to take that lesson to heart.

Thinking about live–work–play cities

The 24-hour cities, their close cousins the 18-hour cities, and Malizia's "vibrant places" [21]all can be encompassed under the rubric "live–work–play." If it is not too obvious, let's note that the triad begins with "live." Vitality, quality of life, starts at home.

A city population density of 9,000 residents per square mile, coupled with a downtown population base of more than 25,000, is, generally speaking, a basic ingredient for a place seeking the 24-hour advantage. Both volume and concentration are needed to support other critical elements, such as retailing, restaurants, walk-to-work employment opportunities, sufficient mass transit ridership, and urban recreational and cultural amenities. Also, as Jane Jacobs pointed out and those real estate experts surveyed confirmed, it is the residential base that provides reliable "eyes on the street" that enhance public safety. Density is negatively correlated with crime (-0.616), as well as being positively correlated with office prices per square foot (0.792).

The failure to develop sufficient critical mass in downtown housing has inhibited Phoenix, Atlanta, and Dallas from energizing their CBDs and, until recently, retarded the development of a 24-hour downtown in Miami. The unique case of Las Vegas is a statistical outlier, as its diurnal character is shaped by the all-night casino industry. But, even here, we find a downtown population of 31,000 – mostly in moderately priced rental housing.

As many have pointed out, the high densities and competing land uses in 24-hour locations elevate the cost of housing and place stress on affordability.[22] But, for this issue, as for others, perceptions about the cost and income disparity are shaped by the way statistics are reported. Most calculations of affordability compare median costs with median incomes, but pay little if any attention to the *range* of rents and incomes.

The range of options at any given time can be quite large and quite diverse. For example, apartment asking-rent information accessed in November 2014 via the website Zillow.com showed 32,771 NYC apartments seeking new tenants (Figure 10.1). The median rent was $2,214 per month, meaning that one-half of the units could be leased for less; 9,175 apartments were available for $1,750 per month, or less. Although pricey by national standards, those 9,000+ apartments were "affordable" to those earning the area median income, if 30 percent of income was dedicated to rent. This analysis is not intended in any way to diminish the severity of the housing burden facing many New Yorkers, but it is intended to indicate that options do exist, to a degree not

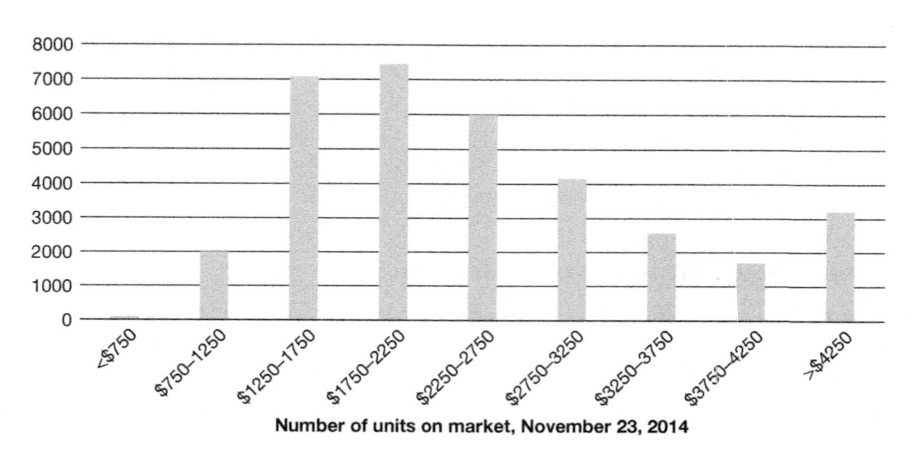

Figure 10.1 Array of NYC apartment rents available
Source: Zillow.com

generally appreciated by public advocates or even by some housing policy specialists.

Within a 6-mile arc around City Hall in Manhattan, practically the whole span of housing choices, by rent and price range, can be found, from super-luxury condominiums, to working-class neighborhoods still resembling "Archie Bunker" housing, to Section 8-supported subsidized apartments and New York City Housing Authority projects. In the public theatre that plays out the New York housing policy debate, the emphasis is far more melodrama than it is analysis, allowing the range of choices to be effectively ignored.[23]

Again, it may seem obvious, but discussing the "work" component of the live–work–play triad needs to begin with the observation that work is not normally a solitary activity. There is a persistent romantic narrative that technology is "freeing" workers from the workplace. Perhaps this is consonant with the myth of "rugged individualism" that runs from the frontiersman to Ayn Rand's character Howard Roark. Perhaps it is the kind of wishful thinking that produced television commercials featuring day-traders putatively making millions while sitting on decks in Malibu or Aspen. Maybe it stems from the 19th-century celebration of the singular genius, whether poet, composer, or inventor. Or it might be a side effect of the same impulse that causes people to stop enjoying life at the moment to set up a "selfie" photograph and Instagram it out to the world.

For most people, though, work means working with others. Figure 7.6, earlier in this book, indicates that only 1 out of 20 American workers is a "telecommuter" based in the home. Collaboration gets the most out of human-capital resources, and, despite more than two decades of attempts to radically decentralize operations, most businesses still convene people in specific workplaces. Belying its title, "From Freelancers to Telecommuters: Succeeding in the New World of Solitary Work," a blogpost from the Wharton School of Business takes most of its time explaining the difficulty of working alone:

- Lone workers must take responsibility for their own professional image, networking, training, and motivation.
- Despite apparent cost savings, employers relying on remote workers have increased risk of miscommunication and higher threats to productivity.
- The necessary balance between work and home life becomes more tenuous.
- The need for internal discipline is a *sine qua non* for solitary-worker success.
- In most organizations, there is a premium on face time. The more management sees you, the more value you are perceived to add.
- Skills can easily diminish, and off-site workers can become obsolete more quickly.
- Non-task communication goes missing with distance work – and that communication is a key foundation for trust.[24]

The so-called "death of distance" presumption was discussed in Chapter 7, and I noted that author Frances Cairncross (who must be credited with popularizing, if not coining, the phrase) acknowledged that downtowns with attributes closely aligned with the 24-hour-city array of characteristics retain great attractiveness for high-value-adding workers. Kevin Kelly, a founder of *Wired* magazine speaks of "conviviality" as one of the hallmarks of technological tools that have a multiplier effect on individual effort.[25] Office space design firms are now vying to find the most exciting ways to create "collaborative work environments." The 24-hour downtown is already that concept writ large.

Chapter 9 presents the evidence from 1987 to 2014 that employers and investors both vote with their wallets that office buildings in highly concentrated live–work–play downtowns pay dividends in profitable operations and in return on real estate investment, and do so to a greater degree than 9-to-5 locations, a sample of 12 other major US cities, and even the suburbs proximate to the 24-hour CBDs. If housing is the ingredient that is the foundation of the 24-hour-city recipe, it is the workplace that is the robust, taste-defining main ingredient.

Mayor John V. Lindsay remains a polarizing figure, even though he left office as New York City's mayor at the end of 1973. To his critics, there will always be a vein of irony in this patrician figure's connection with the black and poor, dating even from his congressional term as a Republican representative of the "Silk Stocking District." But perhaps there has been no remark of Lindsay's more gleefully seized upon for derision than his remark that New York was "a fun city," even in the midst of a paralyzing transit workers' strike in January 1966.[26]

Irony, however, has a way of turning back on itself over time. The "fun city" comment finds its echo in the "play" element of the live–work–play triad. It turns out that a city that is not fun in some basic way is missing a critical element of zest in its recipe. Recreation, entertainment, parks, sports, art, and culture are the seasonings that bring out the best in cities when they interact with home and work. In my interviewing, the theme of fun came up spontaneously when senior real estate executives were queried about 24-hour cities, particularly New York. A Chicago-based investment manager said, "It is very vibrant, exciting and fun. It's noisy. People aren't as bad as their reputation." And an association manager in Washington, DC, echoed those themes: "Crowded, noisy, exciting, and fun."

It is not just a New York story, of course. Chicago has Rush Street's bars and clubs, as well as the Art Institute and the theatres in the Loop. Boston has the Commons and the Public Gardens and Fenway Park and the TD Bank Garden for the basketball Celtics and the hockey Bruins. Washington's attractions are legion: the monuments and museums on the National Mall, the Kennedy Center, the nightlife of Dupont Circle. In other cities, Denver has gelled nicely in its LoDo district, and San Diego in its Gaslamp Quarter. I have already given a nod to the excellent developments in Los Angeles and Houston, with "fun" playing an integral role.

On the other hand, we can see examples of downtowns that have gone for the icing without having the cake completed first. Phoenix is a prime example. Or there are cities such as Atlanta and Dallas that have separated the main ingredients (housing and workplace) with insufficient concentration and then placed the "play" elements at a distance from the downtown. It is hard to have the ingredients interact if they are kept in separate bowls.

How cities learn to compete intelligently – and how they forget

Scholars and professionals in the field of economic development speak of three waves of policy approaches: business attraction, business retention, and community-based economic strategies.[27]

The first wave characterized most activity in the field into the early 21st century and principally consisted of incentives such as subsidized loans, tax exemptions, and outright grants targeted directly to favored firms sought by economic developers.

The second wave broadened attention to firms already in the local economy and adopted a defensive strategy, recognizing that it could be more cost effective to retain existing firms in the area than to tempt firms to migrate into an area. In the second wave, economic development strategies included indirect industry assistance such as marketing, revolving loan funds, and technical innovation support. The second wave stressed entrepreneurship, industry clusters, public–private partnerships, and agglomeration economies.

The third wave highlights public investments to improve quality of life, social justice, and community empowerment. Greater attention is paid to small businesses and micro-enterprise, especially in neighborhoods with high poverty and poorer economic prospects.

Between 1994 and 2004, local economic development officials were generally reducing their use of first-wave tools, substantially increasing the use of second-wave tools, and marginally adding third-wave tools to the mix. However, the intense alarm generated by the financial collapse of 2007–2008 led those officials to revert to first-wave approaches by 2009, with the use of such firm-oriented incentives increasing from 72 percent of respondents (2004) to 95 percent of respondents (2009). Warner and Zheng caution that "interlocal competition" causes local governments to bid against their neighbors (near neighbors in 78 percent of the cases), prompting a destructive "race to the bottom." This has emerged as a key concern of economic developers, but, in stressful times, those concerns are trumped by the short-run objective of shoring up the local tax base. The impact of economic cycles is manifest, as firm-oriented incentives are strongly correlated with reported fear of imminent economic contraction.

On balance, though, third-wave strategies appear to be well synchronized with 24-hour-city characteristics. Warner and Zheng observe that governments with higher property tax bases use fewer attraction incentives, whereas lower local tax revenues spur heavier reliance on incentives, triggering a beggar-thy-neighbor approach and downward spiral of destructive competition. The demonstrably higher commercial property values generated by 24-hour cities mitigate the need for such an approach. Moreover, there is a strong positive correlation (0.740) between office investment prices per square foot in the CBD and the number of edge cities in the region. This was one of the more surprising results of my research into 24-hour cities and commercial investment performance. It suggests that strong downtowns support economic health in their regions. By sustaining robust market conditions in their downtowns, the attributes of 24-hour cities contribute to overall economic growth regionally, making economic development something other than a zero-sum game.

To summarize, Warner and Zheng say, "heavier incentive users (first wave strategies) face more competition, slower economic growth, and lower property tax revenue . . . they need help in identifying strategies that can address their more challenging economic circumstances." Such strategies center on investment in fundamentals such as schools, physical infrastructure, and social services that make communities more vital and support their well-being over the long term.[28]

The lesson is "Dare to be Different." The 24-hour cities are living proof that dramatic turnarounds are possible, and other cities, including Los Angeles, Denver, San Diego, Houston, and (possibly) Philadelphia, are making progress. It is not only the largest cities, either. Austin has attracted widespread attention and respect, even with a city slogan that reads, "Keep Austin Weird." Nashville has bet its chips on an approach that markets its reputation as "Music City USA," coupled with a mayor determined to make education his economic development priority. The small city of Greenville, South Carolina, has succeeded in drawing businesses with 20,000 employees back downtown by its attention to residential development and quality of life, supporting 100 retailers and 120 restaurants and registering a Walk Score of 82 ("very walkable").

The evidence recommends an approach that emphasizes regional distinctiveness and third-wave strategies, over homogeneity and race-to-the-bottom incentives. Sometimes, being the low-cost alternative means you are being paid what you are worth.

A final observation on cities and their problems[29]

So, what does it all mean, as we try to imagine our cities in the years ahead? We do well to keep in mind the sage advice of Yogi Berra, the former player and manager of the New York Yankees, who said, "Forecasting is hard, especially when it is about the future." Here are a few observations, which may or may not rise to the level of prediction.

In the broadest context, the world will have more cities, and many cities will be much bigger. We are at about the 50 percent level of urbanization globally, at a population base of 7 billion. By 2050, the United Nations projects 70 percent urbanization and a population of 9 billion. That means the world's urban population grows from 3.5 billion to 6.3 billion in 35 years.[30] 2030 is a milestone on that journey. That observation is neither new, nor is its forecast improbable. The amount of developed real estate on Earth will certainly increase immensely in the next decade and a half. So will the strain on global resources and the likely level of global population migration.

Within the US, already heavily urbanized, cities will capture most of the projected population growth of 42 million by 2030, a combination of natural increase and openness to immigration. By 2050, the US will be approaching a population threshold of 400 million. According to Census Bureau projections, the age cohort between 18 and 44 years will increase by nearly 17 million between 2014 and 2050, with most of that increase coming by 2030. The cohort of 45–64 years of age will have expanded by 14.5 million at the half-century mark, though all that increase will come after 2030, as the Millennials age.[31] The demographic changes will generate substantial real estate demand domestically, in both the residential and commercial markets. With a 13.5 percent population increase by 2030, followed by an additional 10.8 percent growth between 2030 and 2050, the urban, suburban, and exurban land assets of the nation will have even greater need of wise resource management.

Against the background of such growth, some financial realities need to be confronted. Capital will be distributed unevenly. It will flow to cities disproportionately. And, among cities, it will flow to those places that use the capital most efficiently. That refers to all forms of capital: physical, financial, and human. Here's the rub: even with the globalization of capital, it is not arbitraged toward the mean or distributed equitably. Capital gravitates toward the optimum return. That's not likely to shift radically in so short a period of time as a decade and a half.

Those three forms of capital are not of equal weight, as far as cities are concerned. Evidence is accumulating that now, and into the future, it will be human-capital attributes that will drive the allocation of physical and financial capital. That is a major shift in emphasis for development, requiring the rethinking of incentives that heretofore have focused on subsidies for hard assets and for the owners of financial capital – large employers. Incentives need to be targeted to populations, encouraging the development of talent and of innovation, supporting human capital development while taking care not to compromise risk-taking and creativity, and the discipline of failure. That entails a third-wave approach to economic development, and it favors 24-hour cities, which should increase their attractiveness as magnets for capital in all dimensions.

As a final point, let's acknowledge that the process will be messy. Change is frequently surprising and often disruptive. In the 1970s, when Walt Disney established Disney World in Orlando, he set aside an area for EPCOT – the Experimental Prototype City of Tomorrow. No real city has evolved that bears any resemblance to EPCOT. Why? Because Disney had – and his company still has – no tolerance for messiness.

Inarguably, the future for cities will be messy. The years 2014 and 2015 have seen racial tensions flare in incidents in the Staten Island borough of New York City, in Cleveland, in the St. Louis suburb of Ferguson, in Baltimore, and in Charleston, South Carolina. Dogmatic aversion to taxation has created a condition of revenue starvation, exacerbating cities' budgetary dilemma as they struggle with expense pressures in items such as pension liabilities, bond indebtedness, education, and public safety. Similarly, capital budgeting for important but unglamorous improvements to infrastructure is hindered by a combination of short-term thinking by political leaders and competing priorities for funds, priorities often aligned with the interests of campaign donors.

Basic changes in socioeconomic conditions – in all five forms of change: cycles, trends, maturation, change of state, and disruption – will challenge cities to think freshly about the future. Fresh thinking is typically in short supply, as the default stance is to stay comfortably with the familiar until an emergency arises. Climate change, in particular, poses threats to cities requiring current actions whose benefits may not be visible for decades. The mayors of the C40 cities – 12 of which are in the United States – have taken remarkable leadership steps in this realm, but execution is made more difficult by constrained budgets and foot-dragging at the national level. Cities will be the places where, for good or ill, the consequences will be played out in a nation and now a world that is majority urban.

Let me commit heresy here: messiness is a good thing. Central control, rigid rules, and micromanaged processes are favored in all sorts of organizational hierarchies, cities and their subsidiary elements included. Sanitized, ownership association-controlled, gated communities and their new urbanism analogs eschew messiness. So do monotonic suburbs and relatively homogeneous cities. But it is messiness – emergent problems – that generates new solutions. The urban laboratory is not, and should not be, a sterile environment. The most successful cities will not be problem-free cities, but problem-solving cities.

So, that's our urban challenge: To invent – to produce by creative thought and action – and to invest – to convert money into assets in hope of profit. Not a simple task. But 3,000 years of history, from Babylon and Athens to New York, London, and Hong Kong, suggest we'd be foolish to bet against our cities: Especially cities whose attributes of density and diversity, talent and tolerance, innovation and opportunity set them ahead of the game, even today.

Notes

1 The newspaper columnist Jimmy Breslin, who wrote for the *New York Herald-Tribune*, the *Daily News*, and *Newsday* for 40 years, had a recurrent character named "Marvin the Torch" in his pieces. Here's a brief sketch: "Marvin the Torch never could keep his hands off somebody else's business, particularly if the business was losing money. Now this is accepted behavior in Marvin's profession, which is arson." Cited in the *New York Times*, November 14, 2004, accessed online at www.nytimes.com/2004/11/14/nyregion/thecity/14bres.html?_r=0
2 Recent phases of the Nehemiah program have used housing modules prefabricated at the Brooklyn Navy Yard, keeping down costs, speeding production, and enhancing local employment.
3 Information and links about the Nehemiah program can be found online at http://ebc-iaf.org/content/affordable-housing
4 Andrew Rice, "The Red Hot Rubble of East New York," *New York Magazine*, January 26, 2015.
5 For example, see the discussion of the Downtown Dallas 360 plan in Chapter 8.
6 B.H. Lidell Hart, *Strategy* (2nd ed.), Meridian (New York, 1991). See especially Chapter XX, "The Concentrated Essence of Strategy and Tactics," pp. 334–337.
7 Liddell Hart's principles are noted in italics.
8 Garreau, op. cit., p. 3. Garreau's chapter on Southern California (pp. 261–301), incredibly, fails to mention downtown Los Angeles.
9 From Downtown Design Guidelines, City of Los Angeles, p. 16, June 15, 2009, online at http://planning.lacity.org/urbanization/dwntwndesign/TableC.pdf
10 See online at www.spur.org/publications/article/1999-07-01/proposition-m-and-downtown-growth-battle
11 See the discussions in Chapters 5 and 8 about attractiveness to the college and young-worker cohorts and the positive influences of strong universities in cities.
12 The lowest possible citywide diversity index is 0 percent, which is what you get if everyone is the same race. The highest possible one is 80 percent. Why not 100 percent? Because the Brown data only include five racial groups (white, black, Hispanic, Asian, and other, principally Native American). Even if the population is divided exactly evenly between these groups, you'll still have 20 percent of the people belong to the same race as you. Silver is the author of *The Signal and the Noise: Why Most Predictions Fail but Some Don't*, Penguin Press (New York, 2012).
13 Kris Hartley, online at www.newgeography.com/content/004934-flexible-economic-opportunism-beyond-diversification-urban-revival, June 1, 2015
14 These concepts are taken from Gensler's web posting, online at www.gensleron.com/work/2014/9/12/work-in-the-city-urban-diversification.html
15 Bruce Katz and Julie Wagner, "The Rise of Innovation Districts: A New Geography of Innovation in America," Metropolitan Policy Program at Brookings (Washington, DC, May 2014).
16 National Center for Education Statistics, *Digest of Education Statistics*, Table 303.70, online at http://nces.ed.gov/programs/digest/d14/tables/dt14_303.70.asp

17 Jeff Allum, *Graduate Enrollment and Degrees: 2003 to 2013*, Council of Graduate Schools (Washington, DC, 2014).
18 Glaeser and Saiz, art. cit.
19 Katz and Wagner, op. cit.
20 David Card and Allan B. Krueger, "Does School Quality Matter? Returns to Education and the Characteristics of Public Schools in the United States," *Journal of Political Economy 100:1* (February 1992), pp. 1–40. Lest the Card and Krueger study be suspected of being out of date, more recent studies suggest that educational outcomes are more similar than different between private and public schools. A 2006 report from the US Department of Education reported that public schools registered better performance on 4th-grade math scores, whereas private schools scored better in 8th-grade reading. There was no statistically significant difference in 4th-grade reading or 8th-grade math between private and public schools (Henry Braun, Frank Jenkins, and Wendy Grigg, *Comparing Private Schools and Public Schools Using Hierarchical Linear Modeling*, (NCES 2006–461), US Department of Education (2006)). What of performance at urban high schools? A Center on Education Policy (CEP) study, which focused on inner-city high-school students from low-income households and looked both at prior educational performance and after-graduation results, led to the conclusion that, "students who attend private high schools receive neither immediate academic advantages nor longer-term advantages in attending college, finding satisfaction in the job market, or participating in civic life" (Harold Wenglinski, *Are Private High Schools Better Academically Than Public High Schools?* CEP, October 2007).
21 Malizia, art. cit. (2014). Also see Emil Malizia and David Stebbins, "Making Downtowns What They Used to Be," *Urban Land* (June 26, 2015), online at http://urbanland.uli.org/development-business/making-downtowns-used/
22 Glaeser, in *Triumph of the City*, adds New York's planning and zoning constraints as factors in creating the housing cost burden on the city's households.
23 See Michael Zisser, "Affordable Housing's Abiding Mythology," *Crain's New York Business* (June 29–July 12, 2015), p. 9., especially comments on middle-class communities such as Hudson Heights in Manhattan and Sunnyside in Queens.
24 Accessible online at http://knowledge.wharton.upenn.edu/article/from-freelancers-to-telecommuters-succeeding-in-the-new-world-of-solitary-work/
25 Kevin Kelly, op. cit., pp. 239–265. Kelly credits Ivan Illich's book *Tools for Conviviality* (1973), available online as a pdf at http://monoskop.org/File:Illich_Ivan_Tools_for_Conviviality.pdf
26 A laudatory examination of the Lindsay administration can be found in Sam Roberts (Ed.), *America's Mayor: John V. Lindsay and the Reinvention of New York*, Museum of the City of New York/Columbia University Press (New York, 2010).
27 In this section, I rely principally upon two studies by Cornell University researchers Lingwen Zheng and Mildred Warner: "Business Incentive Use Among US Local Governments: A Story of Accountability and Policy Learning," *Economic Development Quarterly 24:4* (August 2010), pp. 325–336, and "Economic Development Strategies for Recessionary Times: Survey Results from 2009," in *ICMA Municipal Year Book 2011*, International City/County Management Association, pp. 33–42. The 2014 survey results, in raw form, are available online at http://icma.org/en/icma/priorities/surveying/aggregate_survey_results
28 Laura Reese, "Creative Class or Procreative Class: Achieving Sustainable Cities," paper presented at International Making Cities Livable Conference, Charleston, SC, (October 17–21, 2010). Cited in Zheng and Warner (2011).
29 These final remarks were presented in somewhat greater detail at the RE+D Conference in Athens, October 13, 2014. That speech has been published: "Building

the 24-hour City: Just Imagine Real Estate in 2030," *Real Estate Issues 40:1* (2015), online at www.cre.org/publications/rei.cfm

30 Population projections from the United Nations are online at http://esa.un.org/wpp/unpp/panel_population.htm

31 US Census Bureau data are projections published March 2015, accessible online at www.census.gov/content/dam/Census/library/publications/2015/demo/p25-1143.pdf

Index